Subhajit Paul

Metric Spaces

Subhajit Paul
Department of Mathematics
Salesian College (Autonomous), Siliguri
Campus
Siliguri, West Bengal, India

ISSN 2731-9318 ISSN 2731-9326 (electronic)
University Texts in the Mathematical Sciences
ISBN 978-981-96-9258-3 ISBN 978-981-96-9259-0 (eBook)
https://doi.org/10.1007/978-981-96-9259-0

Mathematics Subject Classification: 46A45, 46A50, 54C05, 54C30, 54C35, 54C40, 54E45, 54E52, 26A03, 54E35, 54E50, 26A06

This Springer imprint is published by the registered company Springer Nature Singapore Pte Ltd.
The registered company address is: 152 Beach Road, #21-01/04 Gateway East, Singapore 189721, Singapore

University Texts in the Mathematical Sciences

Textbooks in this series cover a wide variety of courses in mathematics, statistics and computational methods. Ranging across undergraduate and graduate levels, books may focus on theoretical or applied aspects. All texts include frequent examples and exercises of varying complexity. Illustrations, projects, historical remarks, program code and real-world examples may offer additional opportunities for engagement. Texts may be used as a primary or supplemental resource for coursework and are often suitable for independent study.

To my love,
Mon.

Foreword

This is a textbook. This is intended for students and their instructors. It is positioned as a bridge between real analysis and general topology. It has seven chapters, including an introductory chapter, bringing the students up to date with what has preceded their study and bringing them to the point where the real content of the book begins. A good teacher's task is to revise the previous day's class—in this case, probably the entire background to calculus covered before entering the graduation programme. The author, a meticulous teacher, in introducing the basics of metric spaces gaining momentum after a gap of 250 years makes an inspirational reference to the work of the mathematicians Niels Henrik Abel (1802–1829) and Galois Evariste (1811–1832) his contemporary in France.

He brings their cases in the Introduction by way of motivation in grasping the significance of the generalisation process in mathematics. The theory of metric spaces is obtained by looking at the existing subject of real analysis from a higher perspective. The generic term used in this book to indicate this shift in perspective in mathematics is "generalisation process." Abel and Galois had generalised a different field (algebra) of mathematics to solve a problem that stayed at the lower level. He exploits this fruit of the generalisation process in a different field to motivate the reader.

On my reading up on these two, as a student of the history of mathematics, I came across the class notes of the former in his early student days. It was posthumously recovered after he became famous. He died in penury relying on charity despite his mother having been from a wealthy background. His appointment as Professor at the University of Berlin arrived two days after he died back in his hometown, the present-day capital of Norway, Oslo. It came after a long hopeful tour of the centres of learning in Germany and France. Of course, they predated the King of Mathematics Leonard Euler (1707–1783), the Swiss child prodigy who entered Basel University at the age of 13 for a master's completing it in three years with a dissertation comparing Rene Descartes and Isaac Newton and were followed by Henri Poincare (1854–1912) regarded the father of metric spaces. What might probably intrigue a mathematician caught my attention as one looking at reality from a philosophical perspective when I came across Abel's note: $1 + 0 = 0$. If the entire edifice of measurements in the digital space and the computation and programming of AI revolves around 0 and 1, $(0 + 1)$ and $(1/0)$, of the two—"0" zero representing "nothing" or "emptiness" and "1" one representing "fullness" or "everything" which gets priority is a battle in thought on which the verdict is not yet out, at least not among the philosophers.

Abel and Galois, whom the author rightfully recalls, have a pedigree not only in algebra and mathematics, but I came to know from my attempt to self-learn that they had religion as a family background with father and grandfather being pastors (for the former) and politics as a passion of a student activist of the post-Napoleonic era (for the latter). If changes have occurred anywhere—be it in mathematics, religion or politics—it is because of individuals daring to think through as of now unthought solutions and possibilities. The emergence of metric spaces and its ramifications for the measurements of space are indicative of the solutions that are yet to be sought for the warring spaces of human affinities encountering limits as beings circumscribed by finitude on the one end and exploring through thought and imagination the endless infinities on the other. The finite, as Georg Cantor (1845–1918) another mathematical giant reminds us, is something consciously created and metaphorised because the infinite is too much (too vast or too great an expanse) to grapple with. Of course, he commented thus after a lifetime of grappling with Infinite which baffles mathematicians and philosophers alike to this day.

Reading through the bits and pieces of the book that would make sense to a non-mathematician, I was reminded of Rudy Rucker and went back to the preface to the 2019 edition of his *Infinity and the Mind: The Science and Philosophy of the Infinite*, a book I had read in its first edition of 1984 as a master's student of philosophy. There is something that mesmerises a practitioner of philosophy in mathematics, and my occasional welcome dialogues (on the side of administrative and academic business) with the author of this book have been precious moments of realisation of the reverse (the mathematician's admiration for philosophy) also as fact.

Just like $1 = 0 = 0 = 1$ is a thought pattern that infinitises the finite, so too, it maximises the one into many and minimises the many into one. In thinking it through, something is going on in the reflective mind of the mathematician as well as the philosopher. Their cognitive processes are purportedly akin to a reality outside their minds which the physicist—be he of astrophysics or nanophysics variety—will strive to empirically verify. Taking on the clue from Rudy Rucker this is my way of placing Cantor's definition of "[a] set is a many which allows itself to be thought of as a one". The slant in the choice between the "rational many" and the affective or emotive feeling of the "mystical one" will depend on how relaxed and contented one is just to be enjoying the moment versus how disturbed and stressed one is to make a mountain of a molehill.

The book is a baby step in the journey of making a mathematician. But where it not for our moms and aunts, in some, as in the case of the students reading this, the modern dads, changing our nappies and feeding us when we began to wail our wits out—at least for peace's sake—we wouldn't have reached this far. Therefore, the author, it will be evident to the reader, has made painstaking efforts to fine-tune the material for the benefit of the beginner. He is one among those dedicated sets of maths teachers who have made a difference.

The book in your hand, dear reader, is the "proof of that pudding" of an assertion. It is true from the early days of Euler, Abel, Galois and Poincare down to present-day campuses and school classrooms across the world. An author of a textbook is a teacher par excellence, doubly an expert for her students who had the opportunity to hear and test her mettle and equally for the co-instructors who will use it alongside their students. Every reader as a student/beneficiary of a textbook in a way collaborates with the author/teacher in her venture to generate understanding.

The author deserves much appreciation along with all who assisted him in the process of getting the book written, edited, proofread and published. May it unveil ever new hopes for a better world, a more wholesome space of infinite peace beginning from the classrooms of Salesian College.

General Coordinator, Salesian Institutions of Higher Education, Rome, Italy
George Thadathil
30 March 2025

Preface

One of the main tasks, mathematicians set out to execute, is the study of continuous functions. The primary part of this study stems into the subject of classical real analysis, which eventually flourishes as the subject of general topology. The intermediate stage of this endeavour is metric spaces, which successfully amalgamates the abstraction of topology with the intuitions developed in the classical real analysis. This book provides a systematic development of metric spaces to bridge the existing gaps between classical real analysis and general topology.

The book caters to both students and instructors, providing formal definitions, theorems, proofs, examples, remarks, exercises and explanatory anchors for students. Instructors benefit from abundant examples and miscellaneous results to chart their teaching approach effortlessly.

This book comprises seven chapters, with the first devoted to listing the primary results (without proofs) from its prerequisite undergraduate course in real analysis. Each subsequent chapter builds upon insights gained from the previous level, appropriating them to fit in the framework of metric spaces. The second chapter introduces the reader to the definition and some classical examples of metric spaces. It also presents some related terminologies such as pseudometric, diameter and distance between subsets of a metric space. The third chapter covers the topology of metric spaces, including discussions on open sets, closed sets, dense sets, subspaces and sequences. It delves into equivalent metrics and provides an overview of separable, first and second countable spaces. The fourth chapter addresses the quintessential property of real numbers: the completeness axiom, and its adaptation to metric spaces, with a thorough rationalisation of the process. It includes a significant discussion on the powerful Baire category theorem and several applications, concluding with establishing the completion theorem for metric spaces. The fifth chapter delves deeply into the concept of continuity. Beginning with a comprehensive justification of the definition of continuous functions in real analysis, it then adapts this concept to metric spaces. The chapter also extensively covers uniform continuity, homeomorphism and contraction mappings. The sixth chapter focuses on compactness. It begins by appropriating the intuitive necessary and sufficient condition for compact sets in $\mathbb{R}$ provided by the Heine-Borel theorem into metric spaces. The second section explores some notable results, motivated by progressively simpler proofs of the Heine-Borel theorem in $\mathbb{R}^n$. The chapter concludes with discussions on continuous functions on compact metric spaces and examples of compact subsets in specific spaces, including the Arzelà-Ascoli theorem. The last chapter talks about connectedness. As a relatively new topic compared to its predecessors in relevance to classical analysis, it carefully nurtures the intuition to frame its definition. The subsequent sections talk about disconnected spaces and path-connected spaces in detail.

In addition to the main chapters, the book features four appendix chapters: Hölder's and Minkowski's inequalities, Stereographic projection, Weierstrass approximation theorem and the Cantor set. The inclusion of the first three is essential to maintain the flow of discussion in the formal chapters, separating the less relevant proofs into the appendices. Conversely, the chapter on the Cantor set is included to satisfy the reader's curiosity about this fascinating mathematical object.

I want to take this opportunity to thank all my teachers and the authors of some great texts in mathematics, a few of which have been listed in the bibliography. I also want to thank the former principal of my college, Fr. (Prof.) George Thadathil SDB, for constantly encouraging me to finish the project. Even after being a person from a non-mathematical background, he gracefully accepted my request to write a foreword and, in the process, beautifully amalgamated his philosophical thoughts with the context of this book. My colleagues, friends and students have also played significant parts in this endeavour. Special grateful mentions to Gangotryi Sorcar, Saikat Das, Susobhan Mazumdar, Raisa Dsouza, Tulsi Srinivasan and Swarup Paul.

The reader is welcome to send in suggestions or responses and report typographical errors to subhajitp08@salesiancollege.net.

Siliguri, West Bengal, India Subhajit Paul
19 May 2025

Declarations

Competing Interests The author has no competing interests to declare that are relevant to the content of this manuscript.

Contents

Chapter 1
Recollections

This book aims to bridge the gap between the introductory courses of real analysis and general topology. In this chapter on prerequisites, we intend to provide a comprehensive outline of the fundamental concepts and principles covered in the first course of real analysis without halting much for their proofs. The chapter is divided into two sections: The first focuses on the basics of set theory, and the second encompasses important results from single-variable Calculus. Additionally, various exercises are included to house more facts from the respective domains.

1.1 Basics of Set Theory

Modern Mathematics is written in the language of sets. It provides the subject with the abstraction required for its universal applications. In this opening section of this book, we will explore the foundations of set theory. Key concepts will be stated, laying the groundwork for future topics in analysis and topology.

1.1.1 Sets

Definition 1.1 A *set* is a well-defined collection of distinct objects.

The term "well-defined" in the definition above means that, for any given object, determining its membership in the set yields a clear, definitive answer: either "Yes" or "No". The term "distinct" indicates that each element appears only once in the set, without duplication.

S. Paul, *Metric Spaces*, University Texts in the Mathematical Sciences,
https://doi.org/10.1007/978-981-96-9259-0_1

Remarks

(a) *Notations*: We usually use a lowercase letter (for instance, "a") to denote an element in a set; an uppercase letter in normal font to denote a set (for instance, "A"); and an uppercase letter in a scripted font to denote a collection of sets (for instance, "$\mathcal{A}$").
(b) When the elements of a set A are chosen based on satisfying a property P, we write

$$A = \{x : x \text{ satisfies } P\}.$$

(c) *Terminologies*: An element *belongs to* the set. We write $a \in A$. A smaller set (subset) is *contained in* a bigger set. See Definition 1.3. A bigger set (superset) *contains* a smaller set. When an element a does not belong to a set A, we write $a \notin A$.

Definition 1.2 Two *sets are called equal* if both of them contain the exact same elements.

Definition 1.3 A set A is called a *subset* of another set B if every element x of A is also an element of B. We write $A \subset B$. In this case, we also call B as a *superset* of A and write $B \supset A$. In notations,

$$A \subset B \text{ (or } B \supset A\text{) if for every element } x \in A \implies x \in B.$$

The set of all subsets of a set A is called the *power set* of A and is denoted by $\mathcal{P}(A)$.

Remark A is called a *proper* subset of B if B has all elements of A and some more. Unfortunately, unlike the standard strict and non-strict inequality symbols " $<$" and " $\leq$", the corresponding subset symbols " $\subset$" and " $\subseteq$" do not carry fixed senses. In this book, we will use the symbol " $\subset$" to denote both proper and improper subsets. To specifically indicate a proper subset, we will use the symbol " $\subsetneq$".

Theorem 1.4 *Two sets A and B are equal if and only if $A \subset B$ and $B \subset A$.*

Definition 1.5 The set containing all objects in a given framework is called the *universal set*. The set containing nothing is called the *empty set* and is denoted by $\varnothing$.

The term "all objects" in the definition above requires some clarification. It refers to everything relevant and involved in the discussion within a fixed context. For instance, if we are discussing the planets in our solar system, the set

{Mercury, Venus, Earth, Mars, Jupiter, Saturn, Uranus, Neptune}

serves as the universal set. Neither you nor I *need* to belong to this set. Conversely, if the discussion pertains to a study of the human population on Earth, then you, me, and every other human on the planet would form the universal set. However, the planet Earth itself would not qualify as an element of this set. When we discuss sets, we always consider them as subsets of a predefined universal set.

Lemma 1.6 *The empty set $\varnothing$ satisfies the following properties:*

i) *$\varnothing$ is a subset of every other set.*
ii) *$\varnothing$ is unique.*

We create new sets from the existing ones in the following ways.

Definition 1.7 Let A and B be two subsets of the universal set X. The unary/binary operations on the sets are defined as follows:

(a) *Union*: $A \cup B = \{x : x \in A \text{ or } x \in B\}$.
(b) *Intersection*: $A \cap B = \{x : x \in A \text{ and } x \in B\}$.
(c) *Difference*: $A \setminus B = \{x : x \in A \text{ and } x \notin B\}$.
(d) *Complement*: $A^c = X \setminus A = \{x : x \notin A\}$.
(e) *Symmetric difference*: $A \triangle B = (A \setminus B) \cup (B \setminus A)$.

Definition 1.8 A set whose elements are the indices (labels) corresponding to the members of another set is called an *index set*.

For example, let $X = \{x_n : n \in \mathbb{N}\}$ denote the collection of all terms of the sequence $\{x_n\}$. Here, $\mathbb{N}$ is the index set. In practice, we generally use the index set when we work with a family of sets.

Theorem 1.9 (De Morgan's Laws) *Let X be the universal set and $\{A_\alpha : \alpha \in \Lambda\}$ be a family of subsets of X (here Λ is the index set). Then:*

i) $\left(\bigcup_{\alpha \in \Lambda} A_\alpha\right)^c = \bigcap_{\alpha \in \Lambda} A_\alpha^c$.
ii) $\left(\bigcap_{\alpha \in \Lambda} A_\alpha\right)^c = \bigcup_{\alpha \in \Lambda} A_\alpha^c$.

1.1.2 Relations

Definition 1.10 Let A and B be two non-empty sets. The *Cartesian product* of A and B, written as $A \times B$, is defined as the set of all *ordered pairs* (a, b), where the first component a is an element of A and the second component b is an element of B.

$$A \times B = \{(a, b) : a \in A, b \in B\}.$$

If $A_1, A_2, \ldots, A_n$ is a finite collection of non-empty sets, the *Cartesian product* of the collection, denoted by $A_1 \times \cdots \times A_n$, is the set defined by

$$A_1 \times \cdots \times A_n = \{(a_1, \ldots, a_n) : a_i \in A_i,\ i = 1, \ldots, n\}.$$

In particular, if $A_1 = \cdots = A_n = A$, the Cartesian product, denoted by A^n, is the set of all *ordered n-tuples*

$$A^n = \{(a_1, \dots, a_n) : a_i \in A,\ i = 1, \dots, n\}.$$

Definition 1.11 Let A and B be two non-empty sets. A *relation* ρ between the sets A and B is a subset of $A \times B$.

If $A = B$, we say ρ is a *binary relation on* A.

Definition 1.12 Let A be a non-empty set. A binary relation ρ on A is called an *equivalence relation* if it satisfies the following properties:

(a) ρ is *reflexive*: For every $a \in A$, $(a, a) \in \rho$.
(b) ρ is *symmetric*: Whenever $(a, b) \in \rho$, (b, a) also belongs to ρ.
(c) ρ is *transitive*: Whenever $(a, b) \in \rho$ and $(b, c) \in \rho$, they together imply $(a, c) \in \rho$.

Definition 1.13 Let ρ be an equivalence relation on a non-empty set A. Let $a \in A$. The ***ρ**-equivalence class* containing a, denoted by $[a]$ (or $\mathrm{cl}(a)$, or $\langle a\rangle$), is the set of all elements from A related to a, i.e.

$$[a] = \{x \in A : (x, a) \in \rho\}.$$

Definition 1.14 Let A be a non-empty set. A *partition* of A is a collection of pairwise disjoint subsets of A, whose union is A.

Theorem 1.15 *Let A be a non-empty set. An equivalence relation on A produces a partition of A. Conversely, every partition of A corresponds to an equivalence relation on A.*

1.1.3 Mappings

Definition 1.16 Let A and B be two non-empty sets. A relation f with the property that each element $a \in A$ is related to exactly one element $b \in B$ is called a *mapping* or a *function* from A to B. We write $f: A \to B$. If $(a, b) \in f$, we almost always write $f(a) = b$.

If $f(a) = b$, then a is called *a pre-image* of b, whereas b is called *the image* of a. $f^{-1}(b)$ denotes the collection of all pre-images of b, i.e.

$$f^{-1}(b) = \{x \in A : f(x) = b\}.$$

The set A is called the *domain* of f, and the set B is called the *co-domain*. The collection of all image points is called the *range* of f.

$$\text{Range of } f = f(A) = \{f(x) : x \in A\}.$$

Remarks

(a) For different elements $b \in B$, the set $f^{-1}(b)$ can be empty, singleton, or containing multiple points.
(b) *Notations:*

- For $P \subset A$, $f(P) = \{f(x) : x \in P\}$.
- For $Q \subset B$, $f^{-1}(Q) = \{x \in A : f(x) \in Q\}$.

Definition 1.17 Let $f: A \to B$.

(a) f is said to be *injective* or *one to one* if for any two elements $a_1, a_2 \in A$, $a_1 \neq a_2$ implies $f(a_1) \neq f(a_2)$.
(b) f is said to be *surjective* or *onto* if for every element $b \in B$, there is some element $a \in A$ so that $f(a) = b$, i.e. $f(A) = B$.
(c) f is said to be *bijective* or an *one-to-one correspondence* if f is injective and surjective.

Remark If f is injective, then each element of B has at most one pre-image. When f is surjective, every element of B has at least one pre-image. Thus, when f is bijective, each element of B has exactly one pre-image. Consequently, for each element $b \in B$, $f^{-1}(b)$ is a definite element in A. Therefore, f^{-1} acts as a map from B to A. We develop the formality in the following way.

Definition 1.18 Two functions f and g are said to be *equal* if:

(a) f and g have the same sets A and B as their domains and co-domains, respectively.
(b) For each point $a \in A$, $f(a) = g(a)$.

Definition 1.19 Let A, B and C be non-empty sets and $f: A \to B$ and $g: B \to C$. The *composition of the maps* f and g is defined as the map $g \circ f: A \to C$ given by

$$(g \circ f)(a) = g(f(a)), \quad a \in A.$$

Definition 1.20 Let A be a non-empty set. The function $i_A: A \to A$ is called the *identity map* on A if it fixes every point of A, i.e.

$$\text{for every element } a \in A, \quad i_A(a) = a.$$

Definition 1.21 Let A and B be non-empty sets and $f: A \to B$. f is said to be *invertible* if there exists a map $g: B \to A$ so that

$$g \circ f = i_A \quad \text{and} \quad f \circ g = i_B,$$

where i_A and i_B are identity maps on A and B, respectively. The map g is usually denoted by f^{-1}.

Theorem 1.22 *A map is invertible if and only if it is bijective.*

Remark When f is invertible, there are two different meanings of the notation "$f^{-1}(Q)$", where $Q \subset B$, viz.,

$$f^{-1}(Q) = \begin{cases} \{x \in A : f(x) \in Q\}, & \text{as the collection of pre-images of} \\ & \text{the elements from } Q \text{ by the function } f, \\ \{f^{-1}(y) : y \in Q\}, & \text{as the range of the function } f^{-1} \\ \quad \text{of the subset } Q \end{cases} .$$

However, in this case, the last theorem guarantees that f establishes a one-to-one correspondence between the elements of A and B. Thus, the two meanings agree.

Theorem 1.23 *Let A and B be non-empty sets and $f: A \to B$ be a bijection. Then the inverse map $f^{-1}: B \to A$ is also a bijection and $[f^{-1}]^{-1} = f$.*

1.1.4 Cardinality of Sets

Cardinality of a set is a term to express the size of a set. For the empty set, the cardinality is 0; for non-empty finite sets, it is a natural number. For infinite sets, the cardinality is different sizes of infinity. We make the concept precise in the following way.

Definition 1.24 Two sets A and B are called *equipotent* if there is a bijection $f: A \to B$.

A and B are said to have the same *cardinality* if they are equipotent. Cardinality of A is denoted by $|A|$.

Axiom 1.25 The cardinality of the set $\{0, 1, 2, \ldots, n\}$ is $n + 1$.

Definition 1.26 A set A is called a *finite set* if A is equipotent to the set $\{0, 1, 2, \ldots, n\}$ for some non-negative integer n.

A is called an *infinite set* if it is not finite.

Definition 1.27 For any set A, the set of all subsets of A is called the *power set* of A and is denoted by $\mathcal{P}(A)$.

Theorem 1.28 (Cantor's Theorem) *There cannot exist a surjective map from a set to its power set.*

Definition 1.29 For two sets A and B, the cardinality of A is defined to be *smaller or equal* to that of B, ($|A| \leq |B|$) if there exists an one-to-one map from A to B. It is *strictly smaller*, ($|A| < |B|$), if the one-to-one map can never be surjective.

Theorem 1.30 (Schroöeder-Bernstein Theorem) *If for two sets A and B, $|A| \leq |B|$ and $|B| \leq |A|$, then $|A| = |B|$, i.e. if there is a one-to-one (injective) map from A to B and another one-to-one map from B to A, then there exists a one-to-one (bijection) correspondence between the elements of A and B.*

Definition 1.31 A set A is called *enumerable* if A is equipotent to the set $\mathbb{N}$ of all natural numbers.

A is called *countable* if A is either finite or enumerable.

A is called *uncountable* if it is not countable.

Theorem 1.32 *An infinite subset of an enumerable set is enumerable.*

This theorem guarantees that countable infinity is the "smallest" infinity.

Corollary 1.32.1 *A subset of a countable set is countable.*

Theorem 1.33 *(i) Countable union of countable sets is countable.*
(ii) Finite (Cartesian) product of countable sets is countable.

Corollary 1.33.1 *The sets $\mathbb{Z}$ of all integers and the set $\mathbb{Q}$ of all rationals are enumerable.*

Theorem 1.34 *The closed interval* $[0, 1]$ *is uncountable.*

Corollary 1.34.1 *The set $\mathbb{R}$ of all real numbers is uncountable.*

Corollary 1.34.2 *The cardinality of the set $\mathbb{N}$ is denoted by $\aleph_0$ (read "$\aleph$" as "aleph"), and that of the set $\mathbb{R}$ is denoted by $\aleph_1$. Then $\aleph_0 < \aleph_1$.*

Exercises: 1.1

1.1. For three sets A, B and C, prove that:

(a) $A \cap (B \cup C) = (A \cap B) \cup (A \cap C)$.
(b) $A \cup (B \cap C) = (A \cup B) \cap (A \cup C)$.
(c) $A \triangle B = (A \cup B) \setminus (A \cap B)$.
(d) $A \setminus (B \cup C) = (A \setminus B) \cap (A \setminus C)$.
(e) $A \setminus (B \cap C) = (A \setminus B) \cup (A \setminus C)$.

1.2. Let $f: A \to B$ be a mapping and P and Q be subsets of A. Prove the followings:

(a) If $P \subset Q$, then $f(P) \subset f(Q)$.
(b) $f(P \cup Q) = f(P) \cup f(Q)$.
(c) $f(P \cap Q) \subset f(P) \cap f(Q)$. Find an example where $f(P \cap Q) \subsetneq f(P) \cap f(Q)$ and hence deduce a condition on f so that $f(P \cap Q) = f(P) \cap f(Q)$.
(e) $B \setminus f(P) \subset f(A \setminus P)$, i.e. $[f(P)]^c \subset f(P^c)$, when f is surjective.

(*i*) Find an example of a non-surjective function f where $B \setminus f(P) \not\subset f(A \setminus P)$.
(*ii*) Find an example of a surjective function f where $B \setminus f(P) \subsetneq f(A \setminus P)$. Hence deduce a condition on f so that $[f(P)]^c = f(P^c)$.

1.3. Let $f: A \to B$ be a mapping and S and T be subsets of B. Prove the followings:

(a) If $S \subset T$, then $f^{-1}(S) \subset f^{-1}(T)$.
(b) $f^{-1}(S \cup T) = f^{-1}(A) \cup f^{-1}(T)$.
(c) $f^{-1}(S \cap T) = f^{-1}(S) \cap f^{-1}(T)$.
(e) $f^{-1}(B \setminus S) = A \setminus f^{-1}(S)$, i.e. $f^{-1}(S^c) = [f^{-1}(S)]^c$.

1.4. Let $f: A \to B$ be a mapping and P and Q be subsets of A and B, respectively. Prove the followings:

(a) $f[f^{-1}(Q)] \subset Q$. Find an example where $f[f^{-1}(Q)] \subsetneq Q$ and hence deduce a condition on f so that $f[f^{-1}(Q)] = Q$.
(b) $P \subset f^{-1}[f(P)]$. Find an example where $P \subsetneq f^{-1}[f(P)]$ and hence deduce a condition on f so that $P = f^{-1}[f(P)]$.

1.5. Let $f: A \to B$ and $g: B \to C$. Prove the followings:

(a) Let $P \subset C$. Then $[g \circ f]^{-1}(P) = f^{-1}[g^{-1}(P)]$.
(b) If both f and g are bijective maps, then $g \circ f$ is also bijective. Also, $(g \circ f)^{-1} = f^{-1} \circ g^{-1}$.
(c) If $g \circ f$ is bijective, show that f is injective and g is surjective.

1.6. A real number is called *algebraic* if it is a root of a polynomial equation with integer coefficients. Show that the set of all algebraic numbers is enumerable.
Remark: The above result, together with Theorem 1.33(i) and Corollary 1.34.1, shows that the set of all *transcendental* numbers (the real numbers that are not algebraic numbers) is uncountable.

1.2 Calculus of Single Variable

This section serves as a review of essential topics from single-variable calculus. Building on the principles of set theory, we will revisit key concepts typically introduced in an introductory real analysis course. The aim is to provide a comprehensive overview of the calculus material required for the chapters ahead.

1.2.1 Axiomatic Construction of Real Numbers

Real numbers are elements of a set that adhere to specific rules, referred to as *axioms*. To fully characterise the set of real numbers, we consider three types of axioms: algebraic axioms, order axioms, and the completeness axiom. Any set that satisfies all three will be recognised as the set $\mathbb{R}$ of real numbers. Each element of this set is termed a "real number". Let us now outline these axioms.

The set $\mathbb{R}$ of real numbers is equipped with two binary operations "+" and "·" and satisfies the following algebraic axioms:

Axiom 1.35 (Algebraic Axioms)

1. (a) For any two elements $x, y \in \mathbb{R}$, $x + y$ also belongs to $\mathbb{R}$.
   (b) For any three elements $x, y, z \in \mathbb{R}$, $x + (y + z) = (x + y) + z$.
   (c) There is an element 0 in $\mathbb{R}$ such that for any element $x \in \mathbb{R}$, $0 + x = x = x + 0$.
   (d) For every element $x \in \mathbb{R}$, there exists an element $-x \in \mathbb{R}$ such that $x + (-x) = 0 = (-x) + x$.
   (e) For any two elements $x, y \in \mathbb{R}$, $x + y = y + x$.
2. (a) For any two elements $x, y \in \mathbb{R}$, $x \cdot y$ also belongs to $\mathbb{R}$.
   (b) For any three elements $x, y, z \in \mathbb{R}$, $x \cdot (y \cdot z) = (x \cdot y) \cdot z$.
   (c) There is an element $1 \in \mathbb{R}$, *distinct from* 0, such that for any element $x \in \mathbb{R}$, $1 \cdot x = x = x \cdot 1$.
   (d) For every non-zero element $x \in \mathbb{R}$, there is an element $x^{-1} \in \mathbb{R}$ such that $x \cdot x^{-1} = 1 = x^{-1} \cdot x$.
   (e) For any two elements $x, y \in \mathbb{R}$, $x \cdot y = y \cdot x$.
3. For any three elements $x, y, z \in \mathbb{R}$:

   (a) $(x + y) \cdot z = x \cdot z + y \cdot z$.
   (b) $x \cdot (y + z) = x \cdot y + x \cdot z$.

The set $\mathbb{R}$ of real numbers satisfies the order axioms listed as:

Axiom 1.36 (Order Axioms) There is a non-empty subset $\mathbb{P}$ of $\mathbb{R}$, called the set of *positive real numbers*, satisfying the following properties:

(a) If $x, y \in \mathbb{P}$, then $x + y \in \mathbb{P}$.
(b) If $x, y \in \mathbb{P}$, then $x \cdot y \in \mathbb{P}$.
(c) *Trichotomy property*: For any $x \in \mathbb{R}$, *exactly one* of the followings holds:

$$x \in \mathbb{P}, \quad x = 0, \quad -x \in \mathbb{P}.$$

Definition 1.37 For $x, y \in \mathbb{R}$:

(a) If $x - y \in \mathbb{P}$, we say x is *greater than* y, or y is *smaller than* x, and write $\boldsymbol{x > y}$ or $\boldsymbol{y < x}$.
(b) If $x - y \in \mathbb{P} \cup \{0\}$, we say x is *greater than or equal to* y, or y is *smaller than or equal to* x, and write $\boldsymbol{x \geq y}$ or $\boldsymbol{y \leq x}$.

Definition 1.38 Let A be a non-empty subset of $\mathbb{R}$. A is called *bounded above* if there is some real number M so that for every element a from A, $a \leq M$. The number M is called *an upper bound* for A.

A is called *bounded below* if there is some real number m so that for every element a from A, $m \leq a$. The number m is called *a lower bound* for A.

A is called *bounded* if A is both bounded above and bounded below.

A is called *unbounded* if A is not bounded, i.e. either not bounded above or not bounded below.

Definition 1.39 Let A be a non-empty bounded above subset of $\mathbb{R}$. A real number μ^* is called the *least upper bound* (*lub*) or *supremum* of A, written as $\sup A$, if:

(a) For each element $a \in A, a \leq \mu^*$.
(b) For every $\epsilon > 0$, there exists some element $a_\epsilon \in A$ so that $\mu^* - \epsilon < a_\epsilon \leq \mu^*$.

Let A be a non-empty bounded below subset of $\mathbb{R}$. A real number μ_* is called the *greatest lower bound* (*glb*) or *infimum* of A, written as $\inf A$, if:

(a) For each element $a \in A, \mu_* \leq a$.
(b) For every $\epsilon > 0$, there exists some element $a_\epsilon \in A$ so that $\mu_* \leq a_\epsilon < \mu_* + \epsilon$.

The set $\mathbb{R}$ of all real numbers satisfies the completeness axiom listed as:

Axiom 1.40 (Completeness Axiom) Every non-empty bounded above subset of $\mathbb{R}$ has a least upper bound in $\mathbb{R}$.

Remarks

(a) The completeness axiom is a fundamental principle of the real numbers, underpinning many of the intuitive properties we naturally associate with them. However, a significant limitation of the classical approach to defining real numbers through an empirical set of axioms is that this crucial axiom must be accepted without proof. Modern approaches address this issue by constructing the real numbers in a way that establishes the completeness axiom as a provable property of the set.

   It is also worth noting that in the axiomatic development of the natural numbers, the set $\mathbb{N}$ begins from 1. In other frameworks, however, $\mathbb{N}$ may start from 0. In this book, we adopt the former convention, defining $\mathbb{N} = \{1, 2, 3, \ldots\}$.
(b) In the relevance of the above remark, we define the set $\mathbb{Z}$ of all integers as

$$\mathbb{Z} = \{\ldots, -3, -2, -1, 0, 1, 2, 3, \ldots\},$$

whereas the set $\mathbb{Q}$ of all rational numbers as the set

$$\mathbb{Q} = \left\{\frac{p}{q} : p \in \mathbb{Z},\ q \in \mathbb{N},\ p \text{ and } q \text{ are in their lowest terms}\right\}.$$

Results and Consequences

Proposition 1.41

(i) Well ordering principle of $\mathbb{N}$: *Any non-empty subset of* $\mathbb{N}$ *has a least element*[1].
(ii) Principle of mathematical induction: *If A is a non-empty subset of* $\mathbb{N}$ *so that:*

(a) $1 \in A$.
(b) *If $a \in A$, then $a + 1 \in A$.*

Then $A = \mathbb{N}$.
(iii) *The real number $\sqrt{2}$ is not rational.*
(iv) *For any real number x, $x^2 \geq 0$.*
(v) Triangle inequality: *For any two real numbers x, y, $|x + y| \leq |x| + |y|$*[2].
(vi) Archimedean property: *For any two real numbers x and y with $x > 0$, there exists a natural number n so that $nx > y$.*
(vii) *Let $n \geq 2$ be a positive integer and $a > 0$ be a real number. Then there exists a unique positive real number x so that $x^n = a$. x is called the nth* root *of a.*
(viii) Density properties: *For any two real numbers x, y with $x < y$, there exist a rational number r and an irrational number q so that*

$$x < r < y \qquad \textit{and} \qquad x < q < y.$$

1.2.2 Topology of Real Numbers

Definition 1.42 The set $(a, b) = \{x \in \mathbb{R} : a < x < b\}$ is called an *open interval*.
The set $[a, b] = \{x \in \mathbb{R} : a \leq x \leq b\}$ is called a *closed interval*.

Definition 1.43 For $x \in \mathbb{R}$ and for any positive δ, the open interval

$$N(x; \delta) = (x - \delta, x + \delta)$$

is called the ***δ-neighbourhood*** of x. A set U is called a *neighbourhood* of x if U contains a δ-neighbourhood around x for some $\delta > 0$.

The set $\hat{N}(x; \delta) = N(x; \delta) \setminus \{x\}$ is called the *deleted **δ**-neighbourhood* of x. A set $\hat{U}$ is called a *deleted neighbourhood* of x if $\hat{U} = U \setminus \{x\}$, where U is a neighbourhood of x.

[1] The infimum of a set may or may not belong to the set. When it lies in the set, it is called the least or smallest or minimum element of the set. Similarly, when the supremum of a set belongs to the set, it is called the greatest or maximum element of the set. If for a set A, $\inf A$ or $\sup A$ does not lie in A, we say the $\min A$ or $\max A$ does not exist.

[2] The *absolute value function* $|x|$ is defined as $|x| = \begin{cases} x, & \text{when } x \geq 0 \\ -x, & \text{when } x < 0 \end{cases}$.

Definition 1.44 Let $A \subset \mathbb{R}$. A point $x \in A$ is called an *interior point* if A contains some neighbourhood of x. The set of all interior points of A is called the *interior* of A and is denoted by $\operatorname{int} A$ or A°.

Definition 1.45 A subset $A \subset \mathbb{R}$ is called an *open set* if every point of A is an interior point, i.e. $\operatorname{int} A = A$.

Remark The empty set $\varnothing$ is considered to be an open set. The entire set $\mathbb{R}$ of all real numbers is also open.

Theorem 1.46 *i) Arbitrary union of open sets is open.*
ii) Finite intersection of open sets is open.

Definition 1.47 Let $A \subset \mathbb{R}$. A is called a *closed set* if its complement $A^c = \mathbb{R} \setminus A$ is an open set.

Remark The empty sets $\varnothing$ and $\mathbb{R}$ are closed sets as well.

Theorem 1.48

(i) Finite union of closed sets is closed.
(ii) Arbitrary intersection of closed sets is closed.

Definition 1.49 Let $A \subset \mathbb{R}$. A point $x \in \mathbb{R}$ is called a *limit point* of A if for every $\epsilon > 0$, the deleted ϵ-neighbourhood $\hat{N}(x; \epsilon)$ of x intersects A, i.e. $\hat{N}(x; \epsilon) \cap A \neq \varnothing$. The set A' of all limit points of A is called the *derived set* of A.

Example 1.49.1 0 is the only limit point of the set $\{\frac{1}{n} : n \in \mathbb{N}\}$.

Theorem 1.50 (Bolzano-Weierstrass Theorem) *Every bounded infinite subset of $\mathbb{R}$ has a limit point in $\mathbb{R}$.*

Theorem 1.51 *A subset $A \subset \mathbb{R}$ is closed if and only if A contains all its limit points, i.e. $A' \subset A$.*

Definition 1.52 Let $A \subset \mathbb{R}$. The smallest closed set containing A is called the *closure* of A and is denoted by $\bar{A}$. Thus, A is closed if and only if $\bar{A} = A$.

Theorem 1.53 *Let $A \subset \mathbb{R}$. Then $\bar{A} = A \cup A'$.*

Definition 1.54 Let $A \subset \mathbb{R}$. A point $x \in \mathbb{R}$ is called an *adherent point* of A if for every $\epsilon > 0$, the ϵ-neighbourhood $N(x; \epsilon)$ of x intersects A, i.e. $N(x; \epsilon) \cap A \neq \varnothing$.

The last theorem guarantees that the set of all adherent points of A is precisely the closure $\bar{A}$ of A.

Definition 1.55 Let $A \subset \mathbb{R}$. A collection $\mathcal{G}$ of open sets in $\mathbb{R}$ is called an *open cover* of A if

$$A \subset \bigcup_{G \in \mathcal{G}} G.$$

A finite subcollection $\mathcal{G}'$ of $\mathcal{G}$ is called a *finite subcover* of A by $\mathcal{G}$ if

$$A \subset \bigcup_{G \in \mathcal{G}'} G.$$

Definition 1.56 Let $A \subset \mathbb{R}$. A is called a *compact set* if every open cover of A admits a finite subcover.

Theorem 1.57 *Let $A \subset \mathbb{R}$. The followings are equivalent:*

(i) *A is compact.*
(ii) Heine-Borel theorem: *A is closed and bounded.*
(iii) Bolzano-Weierstrass compactness: *Every infinite subset of A has at least one limit point in A.*
(iv) Sequential compactness: *Every sequence in A has a subsequence converging to a point in A.*

1.2.3 Sequences of Real Numbers

Definition 1.58 A function $f : \mathbb{N} \to \mathbb{R}$ is called a *sequence* of real numbers.

Let $\mathbb{N}_1$ be an infinite subset of $\mathbb{N}$ with the natural order, i.e. $\mathbb{N}_1 = \{n_1, n_2, \dots\}$, where $n_1 < n_2 < \dots$. Then the restricted function $f\big|_{\mathbb{N}_1} : \mathbb{N}_1 \to \mathbb{R}$ is called a *subsequence* of the sequence f.

Notation If $f : \mathbb{N} \to \mathbb{R}$ is a sequence, writing the nth term $f(n)$ as x_n, we generally use the notation $\{x_n\}_{n=1}^{\infty}$ to denote the sequence. However, often this notation reduces to $\{x_n\}_{n\in\mathbb{N}}$ or $\{x_n\}$ for brevity. If n_k is the kth term of an infinite subset $\mathbb{N}_1$ of $\mathbb{N}$ with the natural order, then the subsequence $f\big|_{\mathbb{N}_1} : \mathbb{N}_1 \to \mathbb{R}$ is denoted as $\{x_{n_k}\}_{k=1}^{\infty}$, which often reduces to $\{x_{n_k}\}$.

Also, the phrase "$\{x_n\}$ is a sequence of real numbers" is often abbreviated to "$\{x_n\} \subset \mathbb{R}$ is a sequence".

Definition 1.59 Let $\{x_n\} \subset \mathbb{R}$ be a sequence. $\{x_n\}$ is said to be *convergent* if there is some real number x, so that for every $\epsilon > 0$, there exists a positive integer k, so that

$$\text{for every } n \geq k, \quad |x_n - x| < \epsilon.$$

In that case, the sequence $\{x_n\}$ is said to *converge* to x and is denoted as $\lim_{n\to\infty} x_n = x$, which often reduces to $\lim x_n = x$ or $x_n \to x$.

The sequence $\{x_n\}$ is said to be *divergent* if it is not convergent. $\{x_n\}$ is said to *diverge to* ∞ if for any $G > 0$, there exists $k \in \mathbb{N}$ so that for any $n \geq k$, $x_n > G$. It is said to *diverge to* $-\infty$ if for any $G > 0$, there exists $k \in \mathbb{N}$ so that for any $n \geq k$, $x_n < -G$. We write $x_n \to \infty$ or $x_n \to -\infty$ in respective cases.

Theorem 1.60 *Let $\{x_n\}, \{y_n\} \subset \mathbb{R}$ be two convergent sequences. Let $c \in \mathbb{R}$ be a constant. Then:*

i) $\lim(c \cdot x_n) = c \cdot \lim x_n$.
ii) $\lim(x_n \pm y_n) = \lim x_n \pm \lim y_n$.
iii) $\lim(x_n y_n) = (\lim x_n)(\lim y_n)$.
iv) $\lim(\frac{x_n}{y_n}) = \frac{\lim x_n}{\lim y_n}$, *provided, for each* $n \in \mathbb{N}$, $y_n \neq 0$, *and* $\lim y_n \neq 0$.

Definition 1.61 Let $\{x_n\} \subset \mathbb{R}$ be a sequence. $\{x_n\}$ is said to be *bounded below* if there exists some number A so that for each n, $A \leq x_n$. It is called *bounded above* if there exists some number B so that for each n, $x_n \leq B$.

Finally, $\{x_n\}$ is called a *bounded sequence* if $\{x_n\}$ is both bounded below and bounded above.

Definition 1.62 Let $\{x_n\} \subset \mathbb{R}$ be a sequence. $\{x_n\}$ is called a *monotone increasing* sequence if for every n, $x_n \leq x_{n+1}$. It is called *monotone decreasing* if for every n, $x_n \geq x_{n+1}$. In either cases, it is called a *monotone* sequence.

Theorem 1.63 (Monotone Convergence Theorem) *A monotone increasing and bounded above sequence of real numbers is convergent.*

Similarly, a monotone decreasing and bounded below sequence of real numbers is convergent.

Theorem 1.64 (Sandwich Theorem) *Let $\{x_n\}$, $\{y_n\}$, and $\{z_n\}$ be three sequences of real numbers so that for each $n \in \mathbb{N}$, $x_n \leq y_n \leq z_n$. If the sequences $\{x_n\}$ and $\{z_n\}$ converge to the same limit l, then the sequence $\{y_n\}$ is also convergent and converges to the same limit l.*

Theorem 1.65 (Cantor's Nested Interval Theorem) *Let $\{I_n\}$ be a nested sequence of closed and bounded intervals, i.e. $I_n = [a_n, b_n]$, and $I_{n+1} \subset I_n$. Let $|I_n| = b_n - a_n$ denote the length of I_n. If $\inf I_n = 0$, then the intersection $\bigcap_{n=1}^{\infty} I_n$ contains precisely one point.*

Definition 1.66 Let $\{x_n\} \subset \mathbb{R}$ be a sequence. A number l is called a *subsequential limit* of $\{x_n\}$ if there is a subsequence of $\{x_n\}$ converging to l.

Theorem 1.67 (Bolzano-Weierstrass Theorem) *Every bounded sequence of real numbers has a convergent subsequence.*

Definition 1.68 Let $\{x_n\} \subset \mathbb{R}$ be a bounded sequence. Let S be the set of subsequential limits of $\{x_n\}$. Then S is a non-empty and bounded subset of $\mathbb{R}$. The supremum of S is called the *limit superior* of $\{x_n\}$ and is denoted by $\overline{\lim}\, x_n$. The infimum of S is called the *limit inferior* of $\{x_n\}$ and is denoted by $\underline{\lim}\, x_n$.

If $\{x_n\}$ is unbounded above, then $\overline{\lim}\, x_n := \infty$.
If $\{x_n\}$ is unbounded below, then $\underline{\lim}\, x_n := -\infty$.
If $x_n \to \infty$, then $\underline{\lim}\, x_n = \infty$.
If $x_n \to -\infty$, then $\overline{\lim}\, x_n = -\infty$.

Theorem 1.69 *Let $\{x_n\} \subset \mathbb{R}$ be a sequence. There are subsequences of $\{x_n\}$ approaching to both $\overline{\lim}\, x_n$ and $\underline{\lim}\, x_n$.*

Theorem 1.70 *Let $\{x_n\} \subset \mathbb{R}$ be a bounded sequence. Then $\{x_n\}$ is convergent if and only if $\overline{\lim}\, x_n = \underline{\lim}\, x_n$.*

Definition 1.71 Let $\{x_n\} \subset \mathbb{R}$ be a sequence. $\{x_n\}$ is said to be a *Cauchy sequence* if for every $\epsilon > 0$, there exists some $k \in \mathbb{N}$ so that for any $m, n \geq k$, $|x_m - x_n| < \epsilon$.

Theorem 1.72 *A Cauchy sequence of real numbers is convergent and vice versa.*

1.2.4 Series of Real Numbers

Definition 1.73 Let $\{x_n\} \subset \mathbb{R}$ be a sequence. The sum $\sum_{n=1}^{\infty} x_n$ of all terms of the sequence is called a *series* of real numbers, usually denoted as $\sum x_n$. Let

$$s_n = x_1 + x_2 + \cdots + x_n$$

denote the sum of first n terms of the sequence. Then the sequence $\{s_n\}$ is called the *sequence of partial sums*. The nature of convergence of the series $\sum x_n$ is defined as the nature of convergence of the sequence $\{s_n\}$. If $\{s_n\}$ converges to a real number s, we call the *sum* of the series $\sum x_n$ is s and write $\sum x_n = s$.

Proposition 1.74 *If the series $\sum x_n$ is convergent, then* $\lim x_n = 0$.

Theorem 1.75 (Cauchy's Principle for Convergence of a Series) *Let $\sum x_n$ be a series of real numbers. Then $\sum x_n$ is convergent if and only if for any $\epsilon > 0$, there exists some $k \in \mathbb{N}$ so that for each $n \geq k$, and any $p \in \mathbb{N}$,*

$$|x_{n+1} + \cdots + x_{n+p}| < \epsilon.$$

Theorem 1.76 (Tests for Convergence of a Series of Positive Numbers) *Let $\sum x_n$ be a series of positive real numbers:*

(i) Comparison test: *Let $\sum y_n$ be another series of positive real numbers:*

  (a) First type: *Let there exist $k \in \mathbb{N}$ and $\alpha > 0$ so that for each $n \geq k$, $x_n \leq \alpha y_n$. Then:*

  - *$\sum x_n$ is convergent if $\sum y_n$ is convergent.*
  - *$\sum y_n$ is divergent if $\sum x_n$ is divergent.*

  (b) Second type: *Let there exist $k \in \mathbb{N}$ and $\alpha > 0$ so that for each $n \geq k$, $\frac{x_{n+1}}{x_n} \leq \alpha \frac{y_{n+1}}{y_n}$. Then:*

  - *$\sum x_n$ is convergent if $\sum y_n$ is convergent.*
  - *$\sum y_n$ is divergent if $\sum x_n$ is divergent.*

  (c) Limit form: *Let $\lim \frac{x_n}{y_n} = l$, where l is a non-zero real number. Then the series $\sum x_n$ and $\sum y_n$ converge or diverge together.*

(ii) D'Alembert's ratio test: *Let* $\lim \frac{x_{n+1}}{x_n} = l$*:*

- *If* $0 < l < 1$*, then* $\sum x_n$ *is convergent.*
- *If* $l > 1$*, then* $\sum x_n$ *is divergent.*
- *If* $l = 1$*, no definite conclusion can be made.*

(iii) Cauchy's root test: *Let* $\lim \sqrt[n]{x_n} = l$*.*

- *If* $0 < l < 1$*, then* $\sum x_n$ *is convergent.*
- *If* $l > 1$*, then* $\sum x_n$ *is divergent.*
- *If* $l = 1$*, no definite conclusion can be made.*

(iv) Cauchy's condensation test: *Write* $x_n = f(n)$*, where* $f\colon \mathbb{N} \to \mathbb{R}$ *is a monotone decreasing sequence. Let* $\alpha > 1$*. Then the series* $\sum f(n)$ *and* $\sum \alpha^n f(\alpha^n)$ *converge or diverge together.*

v) Raabe's test: *Let* $\lim n(\frac{x_n}{x_{n+1}} - 1) = l$*.*

- *If* $l > 1$*, then* $\sum x_n$ *is convergent.*
- *If* $0 < l < 1$*, then* $\sum x_n$ *is divergent.*
- *If* $l = 1$*, no definite conclusion can be made.*

(vi) Logarithmic test: *Let* $\lim n \log \frac{x_n}{x_{n+1}} = l$*.*

- *If* $l > 1$*, then* $\sum x_n$ *is convergent.*
- *If* $0 < l < 1$*, then* $\sum x_n$ *is divergent.*
- *If* $l = 1$*, no definite conclusion can be made.*

(vii) Gauss's test: *For each n, let* $\frac{x_n}{x_{n+1}} = 1 + \frac{a}{n} + \frac{b_n}{n^p}$*, where* $p > 1$ *and the sequence* $\{b_n\}$ *is bounded.*

- *If* $a > 1$*, then* $\sum x_n$ *is convergent.*
- *If* $a \leq 1$*, then* $\sum x_n$ *is divergent.*

Theorem 1.77 (Convergence of p-Series) *The series* $\sum \frac{1}{n^p}$ *is convergent for* $p > 1$ *and divergent for* $p \leq 1$*.*

Definition 1.78 Let $\{x_n\} \subset \mathbb{R}$ be a sequence of positive terms. Then the series

$$\sum_{n=1}^{\infty} (-1)^{n+1} x_n = x_1 - x_2 + x_3 - x_4 + \cdots$$

is called an *alternating series*.

Theorem 1.79 (Leibnitz's Test) *Let* $\{x_n\}$ *be a monotone decreasing sequence of positive real numbers and* $\lim x_n = 0$*. Then the alternating series* $\sum (-1)^{n+1} x_n$ *is convergent.*

Definition 1.80 Let $\sum x_n$ be a series of arbitrary reals. It is called *absolutely convergent* if the series $\sum |x_n|$ is convergent.

$\sum x_n$ is called a *conditionally convergent* series if it is convergent but not absolutely convergent.

Theorem 1.81 (Riemann's Theorem) *Let $\sum x_n$ be a conditionally convergent series. By appropriate rearrangement of terms, the series can be made to:*

(i) Converge to any real number l.
(ii) Diverge to ∞.
(iii) Diverge to $-\infty$.
(iv) Oscillate finitely.
(v) Oscillate infinitely.

1.2.5 Limit of a Function

Definition 1.82 Let D be a non-empty subset of $\mathbb{R}$, and $f: D \to \mathbb{R}$ be a map. Let c be a limit point of D. A real number l is said to be a *limit* of f *at* c if for every $\epsilon > 0$, there exists $\delta > 0$, such that

$$\text{for every point } x \in \hat{N}(c;\delta) \cap D, \qquad |f(x) - l| < \epsilon,$$

where $\hat{N}(c;\delta)$ is the deleted δ-neighbourhood of c. We write $\lim_{x\to a} f(x) = l$.

Theorem 1.83 (Sequential Criterion) *Let $(\varnothing \neq)\ D \subset \mathbb{R}$ and $f: D \to \mathbb{R}$. Let $c \in D'$. Then for some $l \in \mathbb{R}$, $\lim_{x\to c} f(x) = l$ holds if and only if for every sequence $\{x_n\} \subset D \setminus \{c\}$ converging to c, the sequence $\{f(x_n)\}$ converges to l.*

Theorem 1.84 (Cauchy Criterion) *Let $(\varnothing \neq)\ D \subset \mathbb{R}$ and $f: D \to \mathbb{R}$. Let $c \in D'$. Then $\lim_{x\to c} f(x)$ exists if and only if for every $\epsilon > 0$, there exists a $\delta > 0$, such that for every pair of points $x, x' \in \hat{N}(c;\delta) \cap D$, $|f(x) - f(x')| < \epsilon$.*

Definition 1.85 Let $(\varnothing \neq)\ D \subset \mathbb{R}$ and $f: D \to \mathbb{R}$. Let $c \in D_1'$, where $D_1 := D \cap (-\infty, c)$. Then f is said to have the *left-hand limit* l_1 at c if for every $\epsilon > 0$, there exists $\delta_1 > 0$, such that for every point $x \in \hat{N}(c;\delta_1) \cap D_1$, $|f(x) - l_1| < \epsilon$. We write $\lim_{x\to c-} f(x) = l_1$ or $f(c-) = l_1$.

Let $(\varnothing \neq)\ D \subset \mathbb{R}$ and $f: D \to \mathbb{R}$. Let $c \in D_2'$, where $D_2 := D \cap (c, \infty)$. Then f is said to have the *right-hand limit* l_2 at c if for every $\epsilon > 0$, there exists $\delta_2 > 0$, such that for every point $x \in \hat{N}(c;\delta_2) \cap D_2$, $|f(x) - l_2| < \epsilon$. We write $\lim_{x\to c+} f(x) = l_2$, or $f(c+) = l_2$.

Theorem 1.86 *Let $(\varnothing \neq)\ D \subset \mathbb{R}$ and $f: D \to \mathbb{R}$. Let $c \in D_1' \cap D_2'$, where $D_1 := D \cap (-\infty, c)$ and $D_2 := D \cap (c, \infty)$. Then for a real number l, $\lim_{x\to c} f(x) = l$ if and only if*

$$\lim_{x\to c-} f(x) = l = \lim_{x\to c+} f(x).$$

Definition 1.87 Let $(\varnothing \neq)\ D \subset \mathbb{R}$ and $f: D \to \mathbb{R}$. Let $c \in D'$. Then $\lim\limits_{x\to c} f(x) = \infty$ (or $-\infty$, respectively) if for every $G > 0$, there exists $\delta > 0$, such that for every point $x \in \hat{N}(c;\delta) \cap D$, $f(x) > G$ (or, $f(x) < -G$, respectively).

Let $D \subset \mathbb{R}$ be a domain so that $(a, \infty) \subset D$ for some $a \in \mathbb{R}$ and $f: D \to \mathbb{R}$. For a real number l, $\lim\limits_{x\to\infty} f(x) = l$ if for each $\epsilon > 0$, there exists $B > 0$ such that for every point $x \in D$,

$$x \geq B \implies |f(x) - l| < \epsilon.$$

Let $D \subset \mathbb{R}$ be a domain so that $(-\infty, a) \subset D$ for some $a \in \mathbb{R}$ and $f: D \to \mathbb{R}$. For a real number l, $\lim\limits_{x\to-\infty} f(x) = l$ if for each $\epsilon > 0$, there exists $B > 0$ such that for every point $x \in D$,

$$x \leq -B \implies |f(x) - l| < \epsilon.$$

Let $D \subset \mathbb{R}$ be a domain so that $(a, \infty) \subset D$ for some $a \in \mathbb{R}$ and $f: D \to \mathbb{R}$. Then $\lim\limits_{x\to\infty} f(x) = \infty$ if for every $G > 0$, there exists $B > 0$ such that for each point $x \in D$, $x \geq B$ implies $f(x) > G$.

Theorem 1.88 *Let $D \subset \mathbb{R}$ be a domain unbounded above such that for some real number $a > 0$, the ray $(a, \infty) \subset D$. Let $f: D \to \mathbb{R}$. Then for some real number l, $\lim\limits_{x\to\infty} f(x) = l$ if and only if $\lim\limits_{x\to 0+} f\left(\frac{1}{x}\right) = l$. Also, $\lim\limits_{x\to\infty} f(x) = \pm\infty$ if and only if $\lim\limits_{x\to 0+} f\left(\frac{1}{x}\right) = \pm\infty$.*

1.2.6 Continuity

Definition 1.89 Let D be a non-empty subset of $\mathbb{R}$. A function $f: D \to \mathbb{R}$ is said to be *continuous at the point* $c \in D$ if for every $\epsilon > 0$, there exists $\delta > 0$, such that for each point $x \in N(c;\delta) \cap D$, $|f(x) - f(c)| < \epsilon$.

f is said to be *continuous over* D if it is continuous at every point $c \in D$.

Remark If $c \in D$ is also a limit point of D, then f is continuous at c iff $\lim\limits_{x\to c} f(x) = f(c)$. f is called *discontinuous* at c if f is not continuous at c. If a function is discontinuous at a point c in its domain, then c is necessarily a limit point of the domain.

Theorem 1.90 (Sequential Criterion) *Let $(\varnothing \neq)\ D \subset \mathbb{R}$ and $f: D \to \mathbb{R}$. Then f will be continuous at a point $c \in D$ if and only if for every sequence $\{x_n\} \subset D$ converging to c, the sequence $\{f(x_n)\}$ converges to $f(c)$.*

Definition 1.91 Let $f: D \to \mathbb{R}$ be discontinuous at a point $c \in D$. f is said to have a *Type I discontinuity* if both (or the one which is well-defined) the one-sided limits

$f(c+)$ and $f(c-)$ exist finitely. f is said to have a *Type II discontinuity* if at least one of the one-sided limits $f(c+)$ and $f(c-)$ does not exist finitely.

Definition 1.92 A function $f: [a, b] \to \mathbb{R}$ is said to satisfy the *intermediate value property (IVP)* on $[a, b]$ if for every x_1, x_2 satisfying $a \le x_1 < x_2 \le b$ and for every μ between $f(x_1)$ and $f(x_2)$, there exists a point $\xi \in (x_1, x_2)$, such that $f(\xi) = \mu$.

Definition 1.93 Let $(\varnothing \neq)\ D \subset \mathbb{R}$ and $f: D \to \mathbb{R}$ be a map. f is said to be *uniformly continuous* on D if for every $\epsilon > 0$, there exists a $\delta > 0$ such that for any pair of points $x, x' \in D$ satisfying $|x - x'| < \delta$, we have $|f(x) - f(x')| < \epsilon$.

Definition 1.94 Let f be a continuous function over a domain $D \subset \mathbb{R}$. A continuous function $g: \mathbb{R} \to \mathbb{R}$ is said to be a *continuous extension* of f to $\mathbb{R}$ if g is continuous over $\mathbb{R}$ and $g = f$ on D.

Theorem 1.95 (Properties of Continuous Functions)

(i) *Let $(\varnothing \neq)\ A \subset \mathbb{R}$ and $f: A \to \mathbb{R}$. Let $g: D \to \mathbb{R}$, where $f(A) \subset D$. Let $c \in A'$ and $\lim_{x \to c} f(x) = l$. If $l \in D$ and g is continuous at l, then*

$$\lim_{x \to c} g\big(f(x)\big) = g(l) = g\left(\lim_{x \to c} f(x)\right).$$

(ii) *Let $f: D \to \mathbb{R}$ be continuous at $c \in D$. If $f(c) \neq 0$, there exists a neighbourhood of c in D, where f retains its sign same as $f(c)$.*

(iii) Bolzano's theorem: *Let $[a, b]$ be a closed and bounded interval and $f: [a, b] \to \mathbb{R}$ be continuous on $[a, b]$. If $f(a)f(b) < 0$, then there exists a point $\xi \in (a, b)$ such that $f(\xi) = 0$.*

(iv) Intermediate value theorem: *Let $[a, b]$ be a closed and bounded interval and $f: [a, b] \to \mathbb{R}$ be continuous on $[a, b]$. If $f(a) \neq f(b)$, then f attains every value between $f(a)$ and $f(b)$.*

(v) *Let $f: [a, b] \to \mathbb{R}$ be continuous. Then f satisfies the intermediate value property.*

(vi) *A continuous function defined over a compact domain is bounded.*

(vii) *A continuous function defined over a compact domain attains its bounds.*

(viii) *Continuous image of a compact set is compact.*

(ix) *A continuous function defined over a compact domain is uniformly continuous.*

(x) *A uniformly continuous function maps Cauchy sequences to Cauchy sequences.*

(xi) *Let $f: (a, b) \to \mathbb{R}$ be continuous. Then f admits a continuous extension to $\mathbb{R}$ if and only if f is uniformly continuous.*

1.2.7 Differentiation

Definition and Properties

Definition 1.96 Let $I \subset \mathbb{R}$ be an interval and $f: I \to \mathbb{R}$ be a map. Let $c \in I$. Then f is called *differentiable at* $\boldsymbol{c}$ if the limit

$$\lim_{h\to 0} \frac{f(c+h) - f(c)}{h}$$

exists *finitely*. This limit is called the *derivative* of f at c and is denoted by $f'(c)$.

f is said to be *differentiable on* $\boldsymbol{I}$ if it is differentiable at every point $c \in I$.

f is said to be *left differentiable* at c if the limit

$$\lim_{h\to 0-} \frac{f(c+h) - f(c)}{h}$$

exists *finitely*. The value of this limit is called the *left derivative* of f at c and is denoted by $Lf'(c)$.

Similarly, f is said to be *right differentiable* at c if the limit

$$\lim_{h\to 0+} \frac{f(c+h) - f(c)}{h}$$

exists *finitely*. The value of this limit is called the *right derivative* of f at c and is denoted by $Rf'(c)$.

Remarks

(a) If $I = [a, b]$, then differentiability of f at a will be guaranteed by its right differentiability at a. Similarly, differentiability at b will be implied by the left differentiability at b.

(b) If f is differentiable on I, and the derivative function f' is differentiable at $c \in I$, then we say f is *twice differentiable* at c. We denote the derivative of f' at c by $f''(c)$ and call it the *second-order derivative* of f at c. Inductively, we can differentiate a function $(k+1)$ times if the kth order derivative function is differentiable. We usually denote the kth order derivate of f at c by $f^{(k)}(c)$, where k is a positive integer.

Theorem 1.97 *Let $f: [a, b] \to \mathbb{R}$ be differentiable at a point $c \in [a, b]$. Then f is continuous at c.*

Theorem 1.98 *Let $f, g: [a, b] \to \mathbb{R}$ be differentiable at a point $c \in [a, b]$. Then:*

i) The function $f \pm g$ is differentiable at c and $(f \pm g)'(c) = f'(c) \pm g'(c)$.
ii) The function fg is differentiable at c and $(fg)'(c) = f'(c)g(c) + f(c)g'(c)$.

iii) The function $\frac{f}{g}$ is also differentiable at c and

$$\left(\frac{f}{g}\right)'(c) = \frac{f'(c)g(c) - f(c)g'(c)}{[g(c)]^2},$$

provided $g(c) \neq 0$.

Theorem 1.99 (Chain Rule) *Let f and g be two functions such that $f \circ g$ is well-defined over a neighbourhood of some point c. If g is differentiable at c and f is differentiable at $g(c)$, then $f \circ g$ is differentiable at c and*

$$(f \circ g)'(c) = f'(g(c))g'(c).$$

Corollary 1.99.1 *Let f be an invertible function and differentiable in some neighbourhood of c. If $f'(c) \neq 0$, then f^{-1} is differentiable at $f(c) = d$, say, and $(f^{-1})'(d) = \frac{1}{f'(c)}$.*

Definition 1.100 Let $D = [a, b]$ and $f: D \to \mathbb{R}$. Let $c \in D$. f is called *(locally) increasing at* **c** if there is some $\delta > 0$ such that:

(a) For every $x \in (c - \delta, c) \cap D$, $f(x) < f(c)$.
(b) For every $x \in (c, c + \delta) \cap D$, $f(x) > f(c)$.

f is called *(locally) decreasing at* **c** if there is some $\delta > 0$ such that:

(a) For every $x \in (c - \delta, c) \cap D$, $f(x) > f(c)$.
(b) For every $x \in (c, c + \delta) \cap D$, $f(x) < f(c)$.

Proposition 1.101 *Let $f: [a, b] \to \mathbb{R}$ and $c \in [a, b]$. Let f be differentiable at c. In that case:*

(i) If $f'(c) > 0$, then f is increasing at c.
(ii) If $f'(c) < 0$, then f is decreasing at c.

Theorem 1.102 (Darboux's Theorem) *Let $f: [a, b] \to \mathbb{R}$ be a differentiable function. Let $f'(a) \neq f'(b)$. If μ is a real number lying between $f'(a)$ and $f'(b)$, then there is a point $\xi \in (a, b)$ such that $f'(\xi) = \mu$.*

Mean Value Theorems (MVT)

Imagine the following scenario (Fig. 1.1 might help):

> An ant is moving along a straight line. At some point, it takes a detour along a smooth, attached string and eventually returns to the original line. Then, while on the string, the ant moves parallel to the initial line at a certain point.

The intuitive scenario described above serves as a thought experiment for any mean value theorem (MVT). Variations of the theorem arise depending on the orientation of the initial straight line and the nature of the attached string.

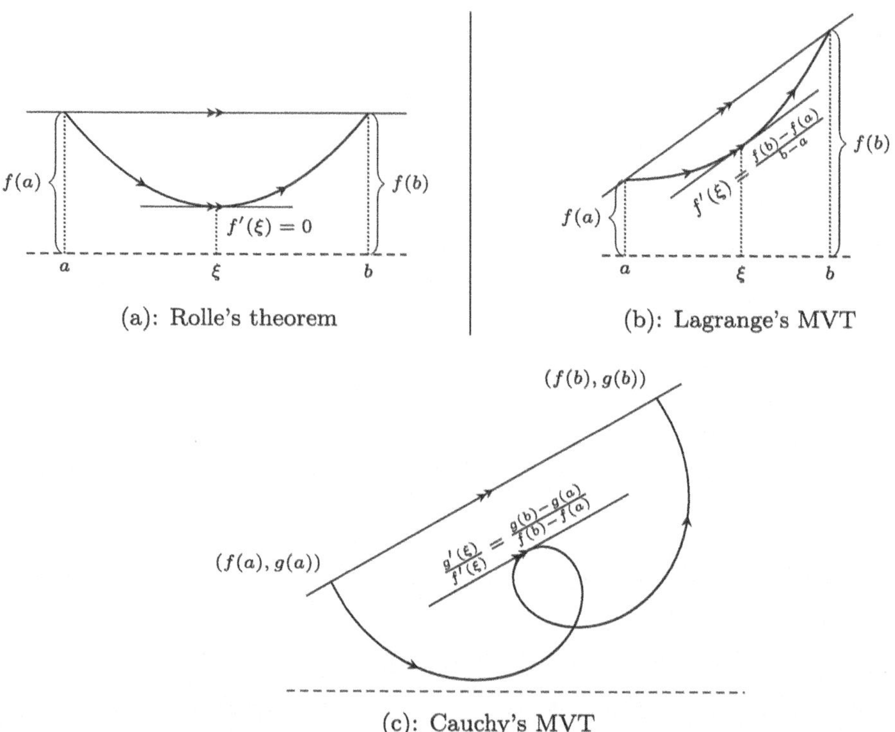

Fig. 1.1 Geometric interpretation of the mean value theorems

To obtain Rolle's theorem, assume the initial straight line to be parallel to the x-axis, with the string representing the graph of a function f. Then at some point $(\xi, f(\xi))$ on the string, the ant moves parallel to the x-axis, i.e. $f'(\xi) = 0$. Under this arrangement, if $A \equiv (a, f(a))$ and $B \equiv (b, f(b))$ are the starting and ending points of the detour, then $f(a) = f(b)$.

Theorem 1.103 (Rolle's Theorem) *Let* $f : [a, b] \to \mathbb{R}$ *be such that:*

(a) f *is continuous on* $[a, b]$.
(b) f *is differentiable on* (a, b).
(c) $f(a) = f(b)$.

Then there is a point $\xi \in (a, b)$ *such that* $f'(\xi) = 0$.

To obtain Lagrange's MVT from the earlier visualisation, we relax the constraint in Rolle's theorem that the straight line is parallel to the x-axis. If A and B represent the starting and ending points of the detour as before, the slope of the straight line connecting these points is $\frac{f(b)-f(a)}{b-a}$. Consequently, there will be a point $(\xi, f(\xi))$ on the string where $f'(\xi) = \frac{f(b)-f(a)}{b-a}$.

Theorem 1.104 (Lagrange's MVT) *Let $f\colon [a,b] \to \mathbb{R}$ be such that:*

(a) f is continuous on $[a,b]$.
(b) f is differentiable on (a,b).

Then there is a point $\xi \in (a,b)$ such that

$$f'(\xi) = \frac{f(b) - f(a)}{b - a}.$$

The most general form of MVT is given by Cauchy. To extend Lagrange's MVT, we remove the constraint that the attached string must be the graph of a single function. Instead, let the string be parametrised by two functions, $f(x)$ and $g(x)$. The slope of the straight line connecting $A \equiv (f(a), g(a))$ and $B \equiv (f(b), g(b))$ is given by $\frac{g(b)-g(a)}{f(b)-f(a)}$, whereas that of the tangent to the string at any point $(f(x), g(x))$ is $\frac{g'(x)}{f'(x)}$. Consequently, there exists a point ξ between a and b where $\frac{g'(\xi)}{f'(\xi)} = \frac{g(b)-g(a)}{f(b)-f(a)}$. To include the special cases when the denominators may be zero, we rewrite the last relation as $g'(\xi)(f(b) - f(a)) = f'(\xi)(g(b) - g(a))$.

Theorem 1.105 (Cauchy's MVT) *Let $f, g\colon [a,b] \to \mathbb{R}$ be such that:*

(a) Both f and g are continuous on $[a,b]$.
(b) Both f and g are differentiable on (a,b).

Then there is a point $\xi \in (a,b)$ such that

$$g'(\xi)[f(b) - f(a)] = f'(\xi)[g(b) - g(a)].$$

Taylor's Theorem and Taylor Series

Taylor's theorem assumes all the information about a function at a particular point and predicts its behaviour at neighbouring points.

Theorem 1.106 (Taylor's Theorem) *Let a function $f\colon [a, a+h] \to \mathbb{R}$ be such that:*

(a) $f^{(n-1)}$ is continuous on $[a, a+h]$.
(b) $f^{(n-1)}$ is differentiable on $(a, a+h)$.

Then there exists a real number $\theta \in (0,1)$ such that

$$f(a+h) = f(a) + hf'(a) + \frac{h^2}{2!} f''(a) + \cdots + \frac{h^{n-1}}{(n-1)!} f^{(n-1)}(a) + \frac{h^n(1-\theta)^{n-p}}{p(n-1)!} f^{(n)}(a + \theta h),$$

where $p \leq n$ is a positive integer.

Remark The last term $\frac{h^n(1-\theta)^{n-p}}{p(n-1)!}f^{(n)}(a+\theta h)$ is called the *remainder after n terms* and is denoted by R_n. There are three particular forms of R_n depending on the values of p ranging from 1 to n.

- *Schlomilch-Roche's form of remainder:* The general form of

$$R_n = \frac{h^n(1-\theta)^{n-p}}{p(n-1)!} f^{(n)}(a+\theta h)$$

  is known as the Schlomilch-Roche's form of remainder.
- *Cauchy's form of remainder:* For $p = 1$,

$$R_n = \frac{h^n(1-\theta)^{n-1}}{(n-1)!} f^{(n)}(a+\theta h)$$

  is called the Cauchy's form of remainder.
- *Lagrange's form of remainder:* For $p = n$,

$$R_n = \frac{h^n}{n!} f^{(n)}(a+\theta h)$$

  is called the Lagrange's form of remainder.

The above theorem is also true on the interval $(a, a+x]$ for any value $0 < x \leq h$. Choosing $a = 0$, we obtain:

Theorem 1.107 (Maclaurin's Theorem) *Let a function $f\colon [0, h] \to \mathbb{R}$ be such that:*

(a) $f^{(n-1)}$ is continuous on $[0, h]$.
(b) $f^{(n-1)}$ is differentiable on $(0, h)$.

Then for any $x \in (0, h]$, there exists a real number $\theta \in (0, 1)$ such that

$$f(x) = f(0) + xf'(0) + \frac{x^2}{2!}f''(0) + \cdots + \frac{x^{n-1}}{(n-1)!}f^{(n-1)}(0) + \frac{x^n(1-\theta)^{n-p}}{p(n-1)!}f^{(n)}(\theta x),$$

where $p \leq n$ is a positive integer.

Observing the fact that h can be negative as well, we arrive at:

Theorem 1.108 (General Form of Taylor's Theorem) *Let $a \in \mathbb{R}$. Let a real function f defined on some neighbourhood $N(a)$ of a be such that $f^{(n-1)}$ is differentiable on $N(a)$.*

Then for any x in the deleted neighbourhood $\hat{N}(a)$, there exists a real number $\theta \in (0, 1)$ such that

$$f(x) = f(a) + (x-a)f'(a) + \frac{(x-a)^2}{2!}f''(a) + \cdots + \frac{(x-a)^{n-1}}{(n-1)!}f^{(n-1)}(a) + \frac{(x-a)^n(1-\theta)^{n-p}}{p(n-1)!}f^{(n)}(a+\theta(x-a)),$$

where $p \leq n$ is a positive integer.

In the general form of Taylor's theorem, let us agree to write for any $x \in \hat{N}(a)$,

$$f(x) = P_n(x) + R_n(x),$$

where

$$P_n(x) = f(a) + (x-a)f'(a) + \frac{(x-a)^2}{2!}f''(a) + \cdots + \frac{(x-a)^{n-1}}{(n-1)!}f^{(n-1)}(a)$$

is a polynomial of degree $n-1$ and

$$R_n(x) = \frac{(x-a)^n(1-\theta)^{n-p}}{p(n-1)!}f^{(n)}(a+\theta(x-a))$$

is the remainder after n terms. $P_n(x)$ is called the nth *Taylor polynomial* of f about the point a. Observe that for every order of differentiation $k = 0, \ldots, n-1$,

$$P_n^{(k)}(a) = f^{(k)}(a),$$

where $f^{(k)}(a)$ denotes the kth derivative of f at a, and $f^{(0)}(a) = f(a)$. Now, if f happens to be a smooth (infinitely differentiable) function on $N(a)$, then the Taylor polynomial $P_n(x)$ exists for every positive integer n. In this case, a good point of interest is to investigate the behaviour of $P_n(x)$ as $n \to \infty$, i.e. the convergence of the infinite series

$$f(a) + (x-a)f'(a) + \frac{(x-a)^2}{2!}f''(a) + \frac{(x-a)^3}{3!}f'''(a) + \cdots. \tag{$*$}$$

As for every n, $f(x) = P_n(x) + R_n(x)$, an NASC for the convergence of $\{P_n(x)\}$ to the function $f(x)$ is $\lim_{n\to\infty} R_n(x) = 0$. Let $\lim_{n\to\infty} R_n(x) = 0$ hold for every point x from a subset $A \subset \hat{N}(a)$. Then, for each point $x \in A$, we will have

$$f(x) = f(a) + (x-a)f'(a) + \frac{(x-a)^2}{2!}f''(a) + \frac{(x-a)^3}{3!}f'''(a) + \cdots. \tag{$**$}$$

Observe that (**) is trivially satisfied for $x = a$. The right-hand side is called the *Taylor's infinite series* for the function f about the point a; the region of convergence of the series being $A \cup \{a\}$.

The series centred at $a = 0$ is called the *Maclaurin series* and takes the form

$$f(0) + xf'(0) + \frac{x^2}{2!}f''(0) + \frac{x^3}{3!}f'''(0) + \cdots .$$

Maxima and Minima

Definition 1.109 Let $f: [a, b] \to \mathbb{R}$ and $c \in [a, b]$. c is called:

(a) A *point of local maximum* for f if there exists a neighbourhood $N(c)$ of c so that for each point $x \in N(c) \cap [a, b]$, one has $f(x) \le f(c)$
(b) A *point of local minimum* for f if there exists a neighbourhood $N(c)$ of c so that for each point $x \in N(c) \cap [a, b]$, one has $f(x) \ge f(c)$
(c) A *point of local extremum* for f if c is a point of either local maximum or local minimum

Theorem 1.110 *Let $f: [a, b] \to \mathbb{R}$ and $c \in (a, b)$. If c is a point of local extremum for f, then either $f'(c) = 0$ or f is not differentiable at c.*

Theorem 1.111 *Let $f: [a, b] \to \mathbb{R}$ and $c \in (a, b)$. If $f'(c) = f''(c) = \cdots = f^{(n-1)}(c) = 0$ and $f^{(n)}(c) \neq 0$, then f has:*

i) No extremum at c if n is odd
ii) A local extremum at c if n is even. The extremum in this case is a maximum if $f^{(n)}(c) < 0$ and is a minimum if $f^{(n)}(c) > 0$

where $f^{(k)}$ denotes the kth derivative of f.

l'Hôpital's Rules

Let us agree to denote by $\mathbb{R}^*$, the set of all real numbers added with two ideal numbers ∞ and $-\infty$. If f is a function and c is a limit point of its domain, then we say $\lim_{x \to c} f(x)$ exists in $\mathbb{R}^*$ if

$$\lim_{x \to c} f(x) = l \ (l \in \mathbb{R}), \quad \text{or } \infty, \quad \text{or } -\infty.$$

Theorem 1.112 (l'Hôpital's Rule: Case $\frac{0}{0}$)

i) Let $c \in \mathbb{R}$. Let the functions f and g be defined and differentiable on some deleted neighbourhood $\hat{N}(c)$ of c. Let

$$\lim_{x \to c} f(x) = 0 = \lim_{x \to c} g(x) \quad \text{and} \quad g(x) \neq 0,\ g'(x) \neq 0 \text{ on } \hat{N}(c).$$

Then

$$\lim_{x\to c}\frac{f(x)}{g(x)}=\lim_{x\to c}\frac{f'(x)}{g'(x)}$$

whenever $\lim_{x\to c}\frac{f'(x)}{g'(x)}$ *exists in* $\mathbb{R}^*$.

ii) Let $c>0$, *and the functions* f *and* g *be differentiable on* $[c,\infty)$. *Let*

$$\lim_{x\to\infty}f(x)=0=\lim_{x\to\infty}g(x),\quad \textit{and}\quad g(x)\neq 0,\ g'(x)\neq 0 \textit{ on } (c,\infty).$$

Then

$$\lim_{x\to\infty}\frac{f(x)}{g(x)}=\lim_{x\to\infty}\frac{f'(x)}{g'(x)}$$

whenever $\lim_{x\to\infty}\frac{f'(x)}{g'(x)}$ *exists in* $\mathbb{R}^*$.

iii) Let $c<0$ *and the functions* f *and* g *be differentiable on* $(-\infty,c]$. *Let*

$$\lim_{x\to-\infty}f(x)=0=\lim_{x\to-\infty}g(x),\quad \textit{and}\quad g(x)\neq 0,\ g'(x)\neq 0 \textit{ on } (-\infty,c).$$

Then

$$\lim_{x\to-\infty}\frac{f(x)}{g(x)}=\lim_{x\to-\infty}\frac{f'(x)}{g'(x)}$$

whenever $\lim_{x\to-\infty}\frac{f'(x)}{g'(x)}$ *exists in* $\mathbb{R}^*$.

Theorem 1.113 (l'Hôpital's Rule: Case $\frac{\infty}{\infty}$)

i) Let $c\in\mathbb{R}$. *Let the functions* f *and* g *be defined and differentiable on some deleted neighbourhood* $\hat{N}(c)$ *of* c. *Let*

$$\lim_{x\to c}f(x)=\infty=\lim_{x\to c}g(x)\quad \textit{and}\quad g(x)\neq 0,\ g'(x)\neq 0 \textit{ on } \hat{N}(c).$$

Then

$$\lim_{x\to c}\frac{f(x)}{g(x)}=\lim_{x\to c}\frac{f'(x)}{g'(x)}$$

whenever $\lim_{x\to c}\frac{f'(x)}{g'(x)}$ *exists in* $\mathbb{R}^*$.

ii) Let $c>0$ *and the functions* f *and* g *be differentiable on* $[c,\infty)$. *Let*

$$\lim_{x\to\infty}f(x)=\infty=\lim_{x\to\infty}g(x),\quad \textit{and}\quad g(x)\neq 0,\ g'(x)\neq 0 \textit{ on } (c,\infty).$$

Then

$$\lim_{x\to\infty}\frac{f(x)}{g(x)}=\lim_{x\to\infty}\frac{f'(x)}{g'(x)}$$

whenever $\lim_{x\to\infty}\frac{f'(x)}{g'(x)}$ *exists in* $\mathbb{R}^*$.

iii) Let $c<0$ *and the functions* f *and* g *be differentiable on* $(-\infty,c]$. *Let*

$$\lim_{x\to-\infty}f(x)=\infty=\lim_{x\to-\infty}g(x),\quad and\quad g(x)\neq 0,\ g'(x)\neq 0\ on\ (-\infty,c).$$

Then

$$\lim_{x\to-\infty}\frac{f(x)}{g(x)}=\lim_{x\to-\infty}\frac{f'(x)}{g'(x)}$$

whenever $\lim_{x\to-\infty}\frac{f'(x)}{g'(x)}$ *exists in* $\mathbb{R}^*$.

1.2.8 Riemann Integration

Since the time of Archimedes, there had been a longstanding belief that:

> For any given function, the area under its graph, bounded by two ordinates and the x-axis, could be approximated by the total area of suitably constructed rectangles, provided the width of each rectangle became progressively smaller (Fig. 1.2).

The German mathematician Bernhard Riemann (1826–1866) sought to examine this idea in the latter half of the 19th century. He approached the problem by formulating it within the following framework.

Riemann's Formulation

Definition 1.114 Let $[a,b]$ be a closed and bounded interval and $f\colon[a,b]\to\mathbb{R}$ be a function:

(a) A *partition P* of $[a,b]$ is a collection of points $\{x_0,x_1,x_2,\ldots,x_n\}$ from $[a,b]$ such that $a=x_0<x_1<x_2<\cdots<x_{n-1}<x_n=b$.

We will use $\mathcal{P}[a,b]$ to denote the collection of *all* partitions of $[a,b]$.

(b) Let $P=\{a=x_0<x_1<\cdots<x_{n-1}<x_n=b\}$ be a partition of $[a,b]$. P endowed with an arbitrary choice of tags ξ_r in the rth subinterval $[x_{r-1},x_r]$, $r=1,2,\ldots,n$, is called a *tagged partition* of $[a,b]$.

We will use $\tilde{\mathcal{P}}[a,b]$ to denote the collection of *all* tagged partitions of $[a,b]$.

(c) Let $P=\{a=x_0<x_1<\cdots<x_{n-1}<x_n=b\}$ be a (tagged) partition of $[a,b]$. Then highest length of a subinterval is called the *norm* or *mesh size* of P

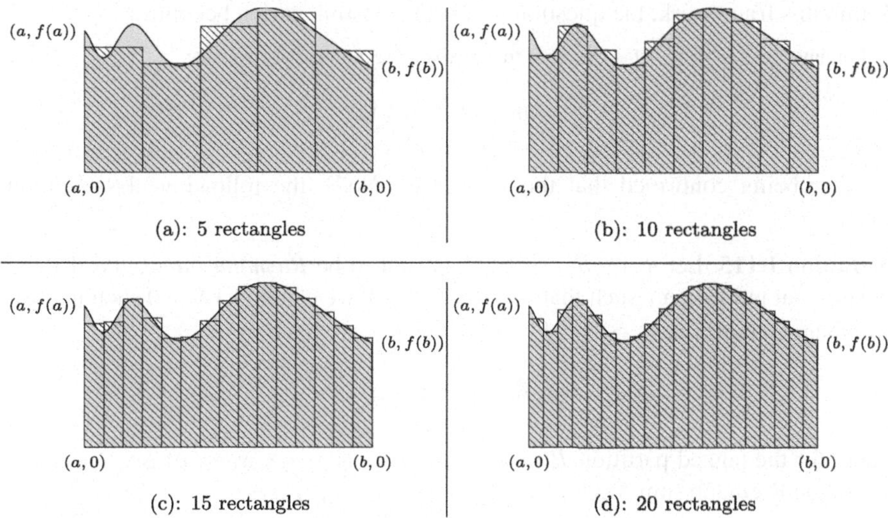

Fig. 1.2 Approximation of area with an increasing number of rectangles

and is denoted by $\|P\|$, i.e.

$$\|P\| := \max\{x_r - x_{r-1} : 1 \le r \le n\}.$$

(d) Let $\bar{P} = \{a = x_0 < x_1 < \cdots < x_n = b;\ \xi_1, \ldots, \xi_n\}$ be a tagged partition of $[a, b]$. The quantity

$$S(\bar{P}, f) = \sum_{r=1}^{n} (x_r - x_{r-1}) f(\xi_r)$$

is called the *Riemann sum* of the function f w.r.t. the tagged partition $\bar{P}$ of $[a, b]$.

(e) Let $f : [a, b] \to \mathbb{R}$. A real number A_f is called the *limit* of $S(\bar{P}, f)$ as $\|\bar{P}\|$ goes to 0, i.e.

$$\lim_{\|\bar{P}\| \to 0} S(\bar{P}, f) = A_f,$$

if for every $\epsilon > 0$, we can find a $\delta > 0$ such that for every tagged partition $\bar{P} \in \bar{\mathcal{P}}[a, b]$ with $\|\bar{P}\| < \delta$, we will have

$$|S(\bar{P}, f) - A_f| < \epsilon.$$

Within this framework, the question to test the existing notion becomes:

For any function $f:[a,b] \to \mathbb{R}$, does there exist a real number A_f so that

$$\lim_{\|\bar{P}\| \to 0} S(\bar{P}, f) = A_f?$$

After being confirmed that the answer is "No"[3], the following classification arises:

Definition 1.115 Let $f:[a,b] \to \mathbb{R}$. f is said to be *Riemann integrable* if there exists a real number A_f such that for every $\epsilon > 0$, we can find a $\delta > 0$ such that for every tagged partition $\bar{P} \in \bar{\mathcal{P}}[a,b]$ with $\|\bar{P}\| < \delta$, we have

$$|S(\bar{P}, f) - A_f| < \epsilon,$$

where for the tagged partition $\bar{P} = \{a = x_0 < x_1 < \cdots < x_n = b;\ \xi_1, \ldots, \xi_n\}$ of $[a,b]$, the Riemann sum $S(\bar{P}, f) := \sum_{r=1}^{n}(x_r - x_{r-1})f(\xi_r)$.

Proposition 1.116 *Let $f:[a,b] \to \mathbb{R}$ be Riemann integrable. Then f must be bounded.*

Darboux's Contribution

A couple of decades later, a French mathematician, viz., Jean Gaston Darboux (1842–1917), sought to simplify Riemann's formulation process. After Riemann had established boundedness as a necessary condition for integrability, Darboux worked exclusively with the class of bounded functions. In the process, he developed the following framework.

Definition 1.117 Let $[a,b]$ be a closed and bounded interval and $f:[a,b] \to \mathbb{R}$ be a bounded function:

(a) Let $P = \{a = x_0 < x_1 < \cdots < x_n = b\}$ be any partition of $[a,b]$. The *upper Darboux sum* or simply the *upper sum* of f w.r.t. the partition P is defined as the quantity

$$U(P, f) = \sum_{r=1}^{n} M_r (x_r - x_{r-1}),$$

where $M_r = \sup\{f(x) : x \in [x_{r-1}, x_r]\}$.

[3] The Dirichlet function, given by $f:[0,1] \to \mathbb{R}$ defined as $f(x) = \begin{cases} 1, & \text{if } x \in [0,1] \cap \mathbb{Q} \\ 0, & \text{if } x \in [0,1] \setminus \mathbb{Q}, \end{cases}$ is a popular counter-example.

(b) Let $P = \{a = x_0 < x_1 < \cdots < x_n = b\}$ be any partition of $[a, b]$. The *lower Darboux sum* or simply the *lower sum* of f w.r.t. the partition P is the quantity

$$L(P, f) = \sum_{r=1}^{n} m_r(x_r - x_{r-1}),$$

where $m_r = \inf\{f(x) : x \in [x_{r-1}, x_r]\}$.

(c) Let $P = \{a = x_0 < x_1 < \cdots < x_n = b\}$ be a partition of $[a, b]$. A partition Q is called a *refinement* of P if P is a proper subset of Q, i.e. Q is obtained by adjoining finitely many more points to P.

Lemma 1.118 *Let $f: [a, b] \to \mathbb{R}$ be bounded and $P \in \mathcal{P}[a, b]$. If Q is a refinement of P, then*

$$L(P, f) \le L(Q, f) \le U(Q, f) \le U(P, f).$$

Corollary 1.118.1 *Let $f: [a, b] \to \mathbb{R}$ be bounded and P, Q be any two partitions of $[a, b]$. Then*

$$L(P, f) \le U(Q, f) \quad \text{and} \quad L(Q, f) \le U(P, f).$$

Definition 1.119 (a) Let $f: [a, b] \to \mathbb{R}$ be bounded. The *lower integral* of f in $[a, b]$ is defined as the quantity

$$\underline{\int_a^b} f = \sup\{L(P, f) : P \in \mathcal{P}[a, b]\}.$$

(b) Let $f: [a, b] \to \mathbb{R}$ be bounded. The *upper integral* of f in $[a, b]$ is defined as the quantity

$$\overline{\int_a^b} f = \inf\{U(P, f) : P \in \mathcal{P}[a, b]\}.$$

Remark If $\int_a^b f$ denotes the actual area under the graph of the function f bounded by the ordinates $x = a$ and $x = b$, then

$$\underline{\int_a^b} f \le \int_a^b f \le \overline{\int_a^b} f.$$

Definition 1.120 Let $f: [a, b] \to \mathbb{R}$ be bounded. f is said to be *integrable in the sense of Darboux* if

$$\underline{\int_a^b} f = \overline{\int_a^b} f.$$

Theorem 1.121 *Let $f:[a,b] \to \mathbb{R}$ be bounded. Then f is Darboux integrable if and only if for every $\epsilon > 0$, there exists a partition P of $[a,b]$ such that*

$$U(P,f) - L(P,f) < \epsilon.$$

Theorem 1.122 (Darboux) *Let $f:[a,b] \to \mathbb{R}$ be bounded. Then for every $\epsilon > 0$, there exists $\delta > 0$ such that for every partition P of $[a,b]$ with $\|P\| < \delta$,*

$$\underline{\int_a^b} f - \epsilon < L(P,f) \le U(P,f) < \overline{\int_a^b} f + \epsilon.$$

Corollary 1.122.1 *Let $f:[a,b] \to \mathbb{R}$ be bounded. If $\{P_n\}$ is a sequence of partitions of $[a,b]$ such that $\lim \|P_n\| = 0$, then*

$$\lim_{n\to\infty} L(P_n, f) = \underline{\int_a^b} f \quad \text{and} \quad \lim_{n\to\infty} U(P_n, f) = \overline{\int_a^b} f.$$

Corollary 1.122.2 *Let $f:[a,b] \to \mathbb{R}$ be bounded. Then f is Darboux integrable if and only if for every $\epsilon > 0$, there exists a $\delta > 0$ such that for every partition P of $[a,b]$ with $\|P\| < \delta$, $U(P,f) - L(P,f) < \epsilon$.*

Theorem 1.123 (Equivalence of Two Definitions of Integrability) *Let $[a,b]$ be a closed and bounded interval and $f:[a,b] \to \mathbb{R}$ be bounded. Then f is Riemann integrable if and only if it is Darboux integrable.*

Despite Darboux's significant contributions, it remains customary to refer to this particular integration technique by Riemann's name. The set of all Riemann integrable functions defined on $[a,b]$ is denoted by $\mathcal{R}[a,b]$.

Properties and Consequences

Definition 1.124 A set $A \subset \mathbb{R}$ is said to be a *set of measure zero* if for each $\epsilon > 0$, there exists a countable collection $\{I_n\}$ of open intervals such that

$$S \subset \bigcup_n I_n \quad \text{and} \quad \sum_n |I_n| < \epsilon,$$

where $|I_n|$ denotes the length of the interval I_n.

Examples

1.124.1. A finite set is of measure zero.
1.124.2. An enumerable set is of measure zero.
1.124.3. If A is a bounded infinite subset of $\mathbb{R}$ having a finite number of limit points, then A is of measure zero.
1.124.4. Cantor set (see Appendix D) is a set of measure zero.

Theorem 1.125 (Lebesgue) *Let $f: [a, b] \to \mathbb{R}$ be a bounded function. Then f is integrable if and only if the set of points of discontinuity of f is of measure zero.*

Corollary 1.125.1 *A continuous function is integrable.*

Theorem 1.126 *Let $f, g: [a, b] \to \mathbb{R}$ be integrable. Let $\alpha \in \mathbb{R}$ be a chosen constant. Then:*

i) *$f + \alpha g$ is integrable on $[a, b]$ and*

$$\int_a^b (f + \alpha g) = \int_a^b f + \alpha \int_a^b g.$$

ii) *$|f|$ is integrable on $[a, b]$.*
iii) *fg is integrable on $[a, b]$.*
iv) *$\frac{f}{g}$ is integrable on $[a, b]$, provided for some $k > 0$, $g(x) \geq k$ holds for every point $x \in [a, b]$.*

Theorem 1.127 *Let $I = [a, b]$ and $f: I \to \mathbb{R}$ be integrable. Let $f(I) \subset [c, d]$ and $\varphi: [c, d] \to \mathbb{R}$ be continuous. Then the composite function $\varphi \circ f$ is integrable on $[a, b]$.*

Theorem 1.128 *Let $[a, b]$ be a closed and bounded interval and $c \in (a, b)$. Let $f: [a, b] \to \mathbb{R}$. Then f is integrable on $[a, b]$ if and only if f is integrable on $[a, c]$ and integrable on $[c, b]$. In either cases,*

$$\int_a^b f = \int_a^c f + \int_c^b f.$$

Theorem 1.129 *Let $f: [a, b] \to \mathbb{R}$ be integrable:*

i) *Let $m = \inf\{f(x) : x \in [a, b]\}$ and $M = \sup\{f(x) : x \in [a, b]\}$. Then*

$$m(b - a) \leq \int_a^b f \leq M(b - a).$$

ii) *If for every point $x \in [a, b]$, $f(x) \geq 0$, then $\int_a^b f \geq 0$.*
iii) *If for every point $x \in [a, b]$, $f(x) \geq 0$, and there is a point $c \in (a, b)$ so that f is continuous at c and $f(c) > 0$, then $\int_a^b f > 0$.*
iv) $\displaystyle\int_a^b |f| \geq \left|\int_a^b f\right|.$

Fundamental Theorem

Let $f: [a, b] \to \mathbb{R}$ be integrable on $[a, b]$. Then for any point $x \in [a, b]$, f is integrable on $[a, x]$ (Theorem 1.128). Define $F: [a, b] \to \mathbb{R}$ by

$$F(x) = \int_a^x f.$$

This function will be of tremendous importance in this portion.

Theorem 1.130 *Let $f\colon [a, b] \to \mathbb{R}$ be integrable. For $x \in [a, b]$, define*

$$F(x) = \int_a^x f :$$

i) Then F is continuous on $[a, b]$.
ii) If f is continuous, then F is differentiable and $F' = f$ on $[a, b]$.

Definition 1.131 Let I be an interval and $f\colon I \to \mathbb{R}$. A function $\varphi\colon I \to \mathbb{R}$ is called an *antiderivative* of f if for each point $x \in I$, $\varphi'(x) = f(x)$.

Remark If φ is an antiderivative of f and $\alpha \in \mathbb{R}$ be a constant, then $\varphi + \alpha$ is another antiderivative of f. Theorem 1.130 guarantees that if f is continuous, then it possesses an antiderivative. However, continuity is not a necessary condition for existence of an antiderivative. See Exercises 1.33 for a counter-example.

Theorem 1.132 (Fundamental Theorem of Integral Calculus) *Let $f\colon [a, b] \to \mathbb{R}$ be integrable on $[a, b]$, and it possesses an antiderivative φ on $[a, b]$. Then*

$$\int_a^b f = \varphi(b) - \varphi(b).$$

Mean Value Theorems

Theorem 1.133 (First Mean Value Theorem) *If $f, g\colon [a, b] \to \mathbb{R}$ be both integrable on $[a, b]$ and g maintains the same sign over entire $[a, b]$, then there is a number $\mu \in [m, M]$ so that*

$$\int_a^b fg = \mu \int_a^b g,$$

where $m = \inf\{f(x) : x \in [a, b]\}$ and $M = \sup\{f(x) : x \in [a, b]\}$.

Theorem 1.134 (Second Mean Value Theorem: Bonnet's Form) *Let $f, g\colon [a, b] \to \mathbb{R}$ be both integrable on $[a, b]$, and let f be monotone decreasing and non-negative on $[a, b]$. Then there exists a point $\xi \in [a, b]$ so that*

$$\int_a^b fg = f(a) \int_a^\xi g.$$

Theorem 1.135 (Second Mean Value Theorem: Weierstrass' Form) *Let $f, g: [a, b] \to \mathbb{R}$ be both integrable on $[a, b]$, and let f be monotone on $[a, b]$. Then there exists a point $\xi \in [a, b]$ so that*

$$\int_a^b fg = f(a) \int_a^\xi g + f(b) \int_\xi^b g.$$

Definition of Logarithmic and Exponential Functions

Definition 1.136 The *logarithmic function* L (or log) is defined by

$$L(x) = \int_1^x \frac{1}{t}\, \mathrm{d}t, \qquad \text{for } x > 0.$$

Proposition 1.137 *The logarithmic function L satisfies the following properties:*

i) $L(1) = 0$.
ii) $L(x) \lesseqgtr 0$ *according as* $x \lesseqgtr 1$.
iii) For any $x, y > 0$, $L(xy) = L(x) + L(y)$.
iv) For any real number α, $L(x^\alpha) = \alpha L(x)$.
v) L *is strictly increasing on* $(0, \infty)$, *and* $\lim_{x\to\infty} L(x) = \infty$ *and* $\lim_{x\to 0+} L(x) = -\infty$.
vi) L *is continuous on* $(0, \infty)$.
vii) $L'(x) = \frac{1}{x}$, *where* L' *is the derivative of* L.
viii) For $x > -1$ *and* $x \neq 0$, $\frac{x}{1+x} < L(1 + x) < x$. *As a corollary, we have*

$$\lim_{x\to 0} \frac{L(1 + x)}{x} = 1.$$

ix) L *is a bijective map from* $(0, \infty)$ *to* $(-\infty, \infty)$.

Definition 1.138 The unique real number x satisfying $L(x) = 1$ is denoted by e, i.e. e is defined by the equation

$$1 = \int_1^e \frac{1}{t}\, \mathrm{d}t.$$

Proposition 1.139 *e satisfies the following properties:*

i) $e = \lim_{n\to\infty} \left(1 + \frac{1}{n}\right)^n$.
ii) $2 < e < 3$.

As the logarithmic function is bijective from $(0, \infty)$ to $(-\infty, \infty)$, it admits an inverse function. We define:

Definition 1.140 The inverse of the logarithmic function defined from $(-\infty, \infty)$ to $(0, \infty)$ is called the *exponential function* and is denoted by $E(x)$.

Remark By definition, for any $x \in \mathbb{R}$, $(L \circ E)(x) = x$, and for any $x > 0$, $(E \circ L)(x) = x$. The monotonicity and continuity of the logarithmic function is reciprocated for the exponential function as well. $E(x)$ is continuous and monotonically increasing on $\mathbb{R}$. Also, $E(1) = e$.

Proposition 1.141 *The exponential function E enjoys the following properties:*

i) $E(0) = 1$.
ii) *For any* $x, y \in \mathbb{R}$, $E(x + y) = E(x)E(y)$.
iii) *For any* $\alpha \in \mathbb{R}$, $E(\alpha x) = [E(x)]^{\alpha}$.
iv) $\displaystyle\lim_{x\to 0} \frac{E(x) - 1}{x} = 1$.
v) $E'(x) = E(x)$, *where* E' *is the derivative of* E.

Remark Choose $x = 1$ in Theorem 1.141(*iii*). Then for any $\alpha \in \mathbb{R}$, you obtain

$$E(\alpha) = [E(1)]^{\alpha} = e^{\alpha}.$$

1.2.9 Sequence and Series of Functions

Let D be a non-empty subset of $\mathbb{R}$. For every $n \in \mathbb{N}$, let $f_n: D \to \mathbb{R}$. Then $\{f_n\}$ is a *sequence of functions* on D. The notion of convergence of a sequence $\{f_n\}$ of functions on D is rationalised in the following way:

For each point $x \in D$, the sequence $\{f_n\}_n$ of functions, evaluated at x, generates a sequence $\{f_n(x)\}$ of real numbers. In general, these sequences behave differently for different values of x. If for every point $x \in D$, the sequence $\{f_n(x)\}$ is convergent, we call the sequence $\{f_n\}$ of functions is *pointwise convergent*.

Let $\{f_n\}$ be a pointwise convergent sequence of functions. Let for the point $x \in D$, the sequence $\{f_n(x)\}$ converges to the real number l_x. Define $f: D \to \mathbb{R}$ by $f(x) = l_x$. Then $f(x)$ is called the *limit function* of the sequence $\{f_n\}$. We record the above discussion in the following:

Definition 1.142 Let $(\emptyset \neq)\ D \subset \mathbb{R}$ and $\{f_n\}$ be a sequence of functions on D. $\{f_n\}$ is said to *converge pointwise* to a function $f: D \to \mathbb{R}$ if for any $\epsilon > 0$ chosen arbitrarily and for any point $x \in D$, there exists a positive integer $k_\epsilon(x)$ so that for every index $n \geq k_\epsilon(x)$,

$$|f_n(x) - f(x)| < \epsilon.$$

In pointwise convergence, the threshold integer $k_\epsilon(x)$ depends on both ϵ and x. In certain cases, this local convergence can be elevated to global convergence by eliminating the dependence of $k_\epsilon(x)$ on x. This results in a uniform number k_ϵ, which is valid for every point x. Consequently, we arrive at:

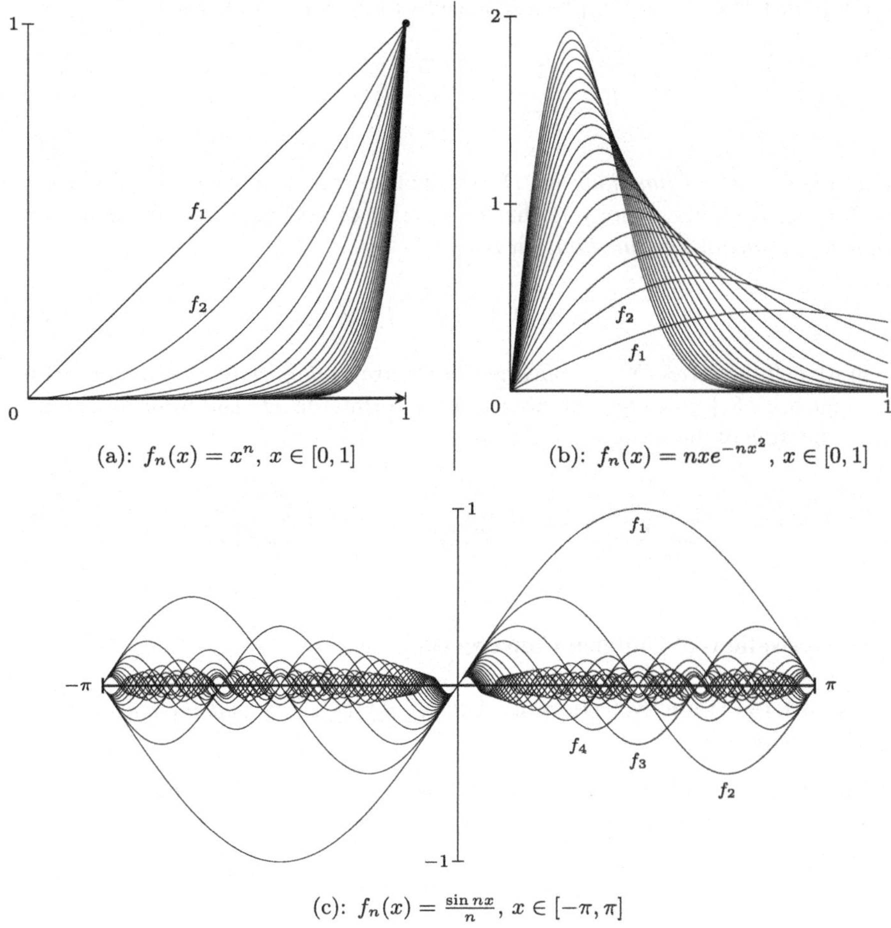

Fig. 1.3 Illustrations of sequence of functions

Definition 1.143 Let $(\varnothing \neq)\ D \subset \mathbb{R}$, and for each $n \in \mathbb{N}$, $f_n: D \to \mathbb{R}$. The sequence $\{f_n\}$ is said to *converge uniformly* to a function $f: D \to \mathbb{R}$ if for any $\epsilon > 0$ chosen arbitrarily, there exists a positive integer k_ϵ so that for every index $n \geq k_\epsilon$, the relation

$$|f_n(x) - f(x)| < \epsilon$$

holds for every point $x \in D$.

Example 1.143.1 In the examples illustrated in Fig. 1.3, Fig. 1.3a and b exhibits two non-uniformly convergent sequence of functions, whereas the sequence in Fig. 1.3c converges uniformly.

Let $(\emptyset \neq) \ D \subset \mathbb{R}$ and $\{f_n\}$ be a sequence of functions on D. Then

$$\sum f_n = \sum_{n=1}^{\infty} f_n = f_1 + f_2 + f_3 + \cdots$$

is called a *series of functions* on D. The notion of convergence for a series of functions is in direct analogy to that of a series of real numbers. We define the *sequence of partial sum functions* on D by

$$S_n = f_1 + f_2 + \cdots + f_n.$$

Then we call the series $\sum f_n$ *converges pointwise or uniformly* on D according as the sequence $\{S_n\}$ converges pointwise or uniformly on D. The limit function of $\{S_n\}$ is the *sum* of the series $\sum f_n$. Thus,

$$\sum f_n = \sum_{n=1}^{\infty} f_n = \lim_{n\to\infty} S_n = \lim_{n\to\infty} \Big(\sum_{k=1}^{n} f_k\Big).$$

Characterisations of Uniform Convergence

Theorem 1.144 (Cauchy Criterion) *Let $(\emptyset \neq) \ D \subset \mathbb{R}$, and for each $n \in \mathbb{N}$, $f_n \colon D \to \mathbb{R}$. Then:*

i) The sequence $\{f_n\}$ converges uniformly on D if and only if for each $\epsilon > 0$, there exists $k \in \mathbb{N}$ so that for any indices $m, n \geq k$, the relation

$$|f_m(x) - f_n(x)| < \epsilon$$

holds for every point $x \in D$.

ii) The series $\sum f_n$ converges uniformly on D if and only if for each $\epsilon > 0$, there exists $k \in \mathbb{N}$ such that for every index $n \geq k$, the relation

$$|f_{n+1}(x) + f_{n+2}(x) + \cdots + f_{n+p}(x)| < \epsilon$$

holds for each point $x \in D$ and any $p \in \mathbb{N}$.

Theorem 1.145 *Let $(\emptyset \neq) \ D \subset \mathbb{R}$, and for each $n \in \mathbb{N}$, $f_n \colon D \to \mathbb{R}$:*

i) Let $\{f_n\}$ converges pointwise to $f \colon D \to \mathbb{R}$. For every $n \in \mathbb{N}$, call

$$M_n = \sup\{|f_n(x) - f(x)| : x \in D\}.$$

Then f_n converges to f uniformly if and only if $\lim M_n = 0$.

ii) (Weierstrass M test): Let $\{M_n\}$ *be a sequence of positive real numbers such that for each n and for every* $x \in D$, $|f_n(x)| \leq M_n$. *If the series* $\sum M_n$ *is convergent, then* $\sum f_n$ *is uniformly convergent on D.*

Theorem 1.146 (Dini's Theorem) *Let K be a compact subset of* $\mathbb{R}$, *and for each* $n \in \mathbb{N}$, *let* $f_n\colon K \to \mathbb{R}$ *be continuous. Let* $\{f_n\}$ *be monotone on K and converge pointwise to* $f\colon K \to \mathbb{R}$, *i.e. either* $f_{n+1}(x) \geq f_n(x)$ *holds for every n and* $x \in D$, *or,* $f_{n+1}(x) \leq f_n(x)$ *holds for every n and* $x \in D$. *Then* $\{f_n\}$ *converges uniformly to* f.

Results and Consequences

Theorem 1.147 *Let* $(\varnothing \neq)\ D \subset \mathbb{R}$, *and for each* $n \in \mathbb{N}$, $f_n\colon D \to \mathbb{R}$:

i) Let $\{f_n\}$ *converge uniformly to* $f\colon D \to \mathbb{R}$. *Let* x_0 *be a limit point of D, and for each n, let* $\lim_{x\to x_0} f_n(x) = a_n$. *Then the sequence* $\{a_n\}$ *converges and* $\lim_{x\to x_0} f(x) = \lim_{n\to\infty} a_n$, *i.e.*

$$\lim_{x\to x_0}\left[\lim_{n\to\infty} f_n(x)\right] = \lim_{n\to\infty}\left[\lim_{x\to x_0} f_n(x)\right].$$

ii) Let $\sum f_n$ *converge uniformly to* $f\colon D \to \mathbb{R}$. *Let* x_0 *be a limit point of D, and for each n, let* $\lim_{x\to x_0} f_n(x) = a_n$. *Then the series* $\sum a_n$ *converges and* $\lim_{x\to x_0} f(x) = \sum a_n$, *i.e.*

$$\lim_{x\to x_0}\left[\sum_{n=1}^{\infty} f_n(x)\right] = \sum_{n=1}^{\infty}\left[\lim_{x\to x_0} f_n(x)\right].$$

Theorem 1.148 *Let* $(\varnothing \neq)\ D \subset \mathbb{R}$, *and for each* $n \in \mathbb{N}$, $f_n\colon D \to \mathbb{R}$:

i) Let $\{f_n\}$ *converge uniformly to* $f\colon D \to \mathbb{R}$. *If each* f_n *is bounded on D, then* f *is also bounded.*
ii) Let $\sum f_n$ *converge uniformly to* $f\colon D \to \mathbb{R}$. *If each* f_n *is bounded on D, then* f *is also bounded.*

Theorem 1.149 *Let* $(\varnothing \neq)\ D \subset \mathbb{R}$, *and for each* $n \in \mathbb{N}$, $f_n\colon D \to \mathbb{R}$:

i) Let $\{f_n\}$ *converge uniformly to* $f\colon D \to \mathbb{R}$. *If each* f_n *is continuous on D, then* f *is also continuous.*
ii) Let $\sum f_n$ *converge uniformly to* $f\colon D \to \mathbb{R}$. *If each* f_n *is continuous on D, then* f *is also continuous.*

Theorem 1.150 *Let* $[a, b]$ *be a closed and bounded interval, and for each* $n \in \mathbb{N}$, *let* $f_n\colon [a, b] \to \mathbb{R}$ *be a Riemann integrable function.*

i) *If the sequence* $\{f_n\}$ *converges uniformly to* $f:[a,b] \to \mathbb{R}$, *then* f *is also Riemann integrable and*

$$\int_a^b \left[\lim_{n\to\infty} f_n\right] = \int_a^b f = \lim_{n\to\infty} \int_a^b f_n.$$

ii) *If the series* $\sum f_n$ *converges uniformly to* $f:[a,b] \to \mathbb{R}$, *then* f *is also Riemann integrable and*

$$\int_a^b \left[\sum_{n=1}^{\infty} f_n\right] = \int_a^b f = \sum_{n=1}^{\infty} \int_a^b f_n.$$

Theorem 1.151 *Let* $[a,b]$ *be a closed and bounded interval, and for each* $n \in \mathbb{N}$, *let* $f_n:[a,b] \to \mathbb{R}$ *be a differentiable function:*

i) *If the sequence of derivatives* $\{f_n'\}$ *converges uniformly to* $g:[a,b] \to \mathbb{R}$ *and* $\{f_n\}$ *converges at least at one point* $x_0 \in [a,b]$, *then the sequence* $\{f_n\}$ *converges uniformly to a function* $f:[a,b] \to \mathbb{R}$ *such that* $f' = g$ *on* $[a,b]$.
ii) *If the series of derivatives* $\sum f_n'$ *converges uniformly to* $g:[a,b] \to \mathbb{R}$ *and* $\sum f_n$ *converges at least at one point* $x_0 \in [a,b]$, *then the series* $\sum f_n$ *converges uniformly to a function* $f:[a,b] \to \mathbb{R}$ *such that* $f' = g$ *on* $[a,b]$.

Remarks

(a) In case of uniform convergence of a series of functions, we can evaluate limits, or integrate, or differentiate the series *term by term*.
(b) A uniform convergent sequence of differentiable functions cannot guarantee the differentiability of the limit function.
(c) In the above theorems, the condition of uniform convergence is only sufficient, but not necessary. See Exercise 1.36 for counter-examples.

1.2.10 Power Series

A *power series* is a series of functions $\sum f_n(x)$ where $f_n(x) = a_{n-1}x^{n-1}$. Thus, a power series canonically takes the form

$$\sum_{n=0}^{\infty} a_n x^n = a_0 + a_1 x + a_2 x^2 + a_3 x^3 + \cdots, \tag{1.1}$$

where x and a_n's are all real numbers. (Strictly speaking, (1.1) is called a *real power series*.) The general form of a power series is

$$a_0 + a_1(x-\alpha) + a_2(x-\alpha)^2 + a_3(x-\alpha)^3 + \cdots. \tag{1.2}$$

It is called a *power series about* α. (1.2) reduces to the canonical form of (1.1) (which is a power series about 0) by the simple substitution $x' = x - \alpha$. In our discussion here, we will work with the canonical form of a power series. We will use the same notation $\sum_{n=0}^{\infty} a_n x^n$ or simply $\sum a_n x^n$ to denote both the power series and its sum, when it exists. Also, for brevity, we agree to use the notation $\mathbb{N}_0 := \mathbb{N} \cup \{0\}$.

Examples

1.154.1. The power series $1 + x + x^2 + x^3 + \cdots$ converges only for $x \in (-1, 1)$.
1.154.2. The power series $1 + x + \frac{x^2}{2!} + \frac{x^3}{3!} + \cdots$ converges for every $x \in \mathbb{R}$.
1.154.3. The power series $1 + x + 2!x^2 + 3!x^3 + \cdots$ converges only for $x = 0$.

Evidently, every power series converges at $x = 0$. It is a trivial point of convergence. Keeping this in mind, we set up the following:

Definition 1.152

(a) A power series is called *nowhere convergent* if it is convergent only at 0.
(b) A power series is called *everywhere convergent* if it is convergent for every real number.

There are power series which are neither everywhere convergent nor nowhere convergent (like one in Example 1.154.1). We will discover that for them, not any arbitrary subset of real numbers can serve as the exact set of points of convergence.

Proposition 1.153 *If a power series $\sum a_n x^n$ converges for $x = \alpha$, then it converges absolutely for every x with $|x| < |\alpha|$.*

Corollary 1.153.1 *If a power series $\sum a_n x^n$ diverges for $x = \beta$, then it diverges for every x with $|x| > |\beta|$.*

Theorem 1.154 *If the power series $\sum a_n x^n$ is neither nowhere convergent nor everywhere convergent, then there is some $R > 0$ such that the series converges absolutely for every x with $|x| < R$ and diverges for every x with $|x| > R$.*

Definition 1.155 Let $\sum a_n x^n$ be a power series. A positive real number R is called the *radius of convergence* of the series if the series converges absolutely for every x with $|x| < R$ and diverges for every x with $|x| > R$.

For a nowhere convergent power series, the radius of convergence is defined to be 0, whereas for an everywhere convergent power series, the radius of convergence is defined to be ∞.

The interval $(-R, R)$ is called the *interval of convergence* for the power series.

Remark There is no general behaviour of the power series $\sum a_n x^n$ at the points $\pm R$. It depends upon the coefficient sequence $\{a_n\}$.

Theorem 1.156 (Cauchy-Hadamard Theorem) *Let for the power series $\sum a_n x^n$, $\overline{\lim} |a_n|^{1/n} = \mu$. Then:*

i) If $\mu = 0$, the series is everywhere convergent.

ii) If $0 < \mu < \infty$*, the radius of convergence for the power series is* $\frac{1}{\mu}$.
iii) If $\mu = \infty$*, the series is nowhere convergent.*

Theorem 1.157 (Ratio Test) *Let for the power series* $\sum a_n x^n$, $\lim |\frac{a_{n+1}}{a_n}| = \mu$. *Then:*

i) If $\mu = 0$*, the series is everywhere convergent.*
ii) If $0 < \mu < \infty$*, the radius of convergence for the power series is* $\frac{1}{\mu}$.
iii) If $\mu = \infty$*, the series is nowhere convergent.*

Theorem 1.158 *Let* $\sum a_n x^n$ *be a power series with radius of convergence* $R > 0$. *Then the series converges uniformly on any closed and bounded interval* $[a, b] \subset (-R, R)$.

Remark The above theorem does not guarantee uniform convergence of the power series in its interval of convergence. See Exercise 1.38 for a counter-example.

Corollary 1.158.1 *A power series with a positive radius of convergence is continuous over its interval of convergence.*

Corollary 1.158.2 *A power series can be integrated term by term on any closed and bounded interval contained within the interval of convergence.*

Theorem 1.159 *Let* R (> 0) *be the radius of convergence of the power series* $\sum a_n x^n$. *Then:*

i) The radius of convergence of the power series

$$a_0 x + \frac{a_1}{2} x^2 + \frac{a_2}{3} x^3 + \cdots,$$

obtained by term-by-term integration, is also R.
ii) The radius of convergence of the power series

$$a_1 + 2a_2 x + 3a_3 x^2 + \cdots,$$

obtained by term-by-term differentiation, is also R.

Theorem 1.160 *A power series can be differentiated term by term within its radius of convergence.*

Corollary 1.160.1 (Uniqueness Property) *If the two power series* $\sum a_n x^n$ *and* $\sum b_n x^n$ *converge to the same function* f *on the same interval* $(-R, R)$*, then for each* $n \in \mathbb{N}_0$, $a_n = b_n$.

Exercises: 1.2

1.7. Prove the following properties of the absolute value function.

(a) For any $x, y \in \mathbb{R}$, $|xy| = |x||y|$.
(b) For any $x \in \mathbb{R}$, $|x|^2 = x^2$.
(c) For $a \geq 0$, $|x| \leq a$ if and only if $-a \leq x \leq a$.
(d) For any $x \in \mathbb{R}$, $-|x| \leq x \leq |x|$.
(e) Let $x \in \mathbb{R}$ be such that for any $\epsilon > 0$, $|x| < \epsilon$. Then $x = 0$.
(f) For any $x, y \in \mathbb{R}$, $\big||x| - |y|\big| \leq |x - y|$.

1.8. Prove the following corollaries from the Archimedean property (Proposition 1.41(*iv*)):

(a) For any $x \in \mathbb{R}$, there exists $n \in \mathbb{N}$ so that $n > x$.
(b) For any $\epsilon > 0$, there exists $n \in \mathbb{N}$ so that $0 < \frac{1}{n} < \epsilon$.
(c) For any $x \in \mathbb{R}$, there exists $m \in \mathbb{Z}$ so that $m - 1 \leq x < m$.

1.9. The completeness axiom is also known as the *lub property*. In this reference, the *glb property* for $\mathbb{R}$ is stated as:

Every non-empty bounded below subset of $\mathbb{R}$

has a greatest lower bound in $\mathbb{R}$.

Show that lub property holds if and only if glb property holds.

1.10. Let $A \subset \mathbb{R}$. Show that int A is the largest open set contained in A.

1.11. Let $A \subset \mathbb{R}$. Show that a point $x \in \mathbb{R}$ is a limit point of A if and only if every neighbourhood of x contains infinitely many points from A. Hence, conclude that a finite set does not have any limit point.

1.12. Prove the following facts about derived sets:

a. For $A \subset \mathbb{R}$, $(A')' \subset A'$.
b. For $A \subset B \subset \mathbb{R}$, show that $A' \subset B'$.
c. For $A, B \subset \mathbb{R}$, show that $(A \cup B)' = A' \cup B'$. Show by an example that this relation does not hold for union of infinitely many sets. In fact, if $\{A_n\}_n$ is a collection of subsets of $\mathbb{R}$, then show that

$$[A_1' \cup A_2' \cup A_3' \cup \ldots] \subset [A_1 \cup A_2 \cup A_3 \cup \ldots]'.$$

d. For $A, B \subset \mathbb{R}$, show that $(A \cap B)' \subset A' \cap B'$. Give an example where the inclusion is proper.
e. Let A and B be two subsets of $\mathbb{R}$ that differ by only finitely many elements. Show that $A' = B'$.
f. Let $A \subset \mathbb{R}$. For $\alpha, \beta \in \mathbb{R}$, define the set

$$B = \{\alpha x + \beta : x \in A\}.$$

We usually denote B by $\alpha A + \beta$. Show that when $\alpha \neq 0$,

$$(\alpha A + \beta)' = \alpha A' + \beta.$$

1.13. Show that a closed subset of a compact set is compact.

1.14. Let $\varnothing \neq A \subset \mathbb{R}$. For $x \in \mathbb{R}$, define the function $d(x, A)$ as

$$d(x, A) = \inf\{|x - a| : a \in A\}.$$

Prove the followings are equivalent:

(a) $x \in \bar{A}$.
(b) $d(x, A) = 0$.
(c) There is a sequence $\{a_n\}$ of points from A converging to x.

1.15. Show that limit of a convergent sequence is unique.

1.16. Let $\{x_n\} \subset \mathbb{R}$ be a bounded sequence. Show that l is a subsequential limit of $\{x_n\}$ if and only if every neighbourhood of l contains infinitely many terms of $\{x_n\}$.

1.17. Let $\{x_n\} \subset \mathbb{R}$ be a sequence. Consider the *tail* of $\{x_n\}$ as the set

$$T_n = \{x_n, x_{n+1}, x_{n+2}, \dots\}.$$

Consider the sequence $\{M_n\}$, where $M_n := \sup T_n$. Then $M_{n+1} \leq M_n$. If $\{x_n\}$ is bounded, show that $\inf_n M_n = \overline{\lim}\, x_n$, i.e. when $\{x_n\}$ is bounded, show that

$$\overline{\lim}\, x_n = \inf_n \sup\{x_n, x_{n+1}, x_{n+2}, \dots\}.$$

Construct and prove the analogous equivalent definition for $\underline{\lim}\, x_n$ as well.

1.18. Let $\{x_n\} \subset \mathbb{R}$ be a bounded sequence. Let $\overline{\lim}\, x_n = l$ and $\underline{\lim}\, x_n = m$. Show that for every $\epsilon > 0$:

(a) All but finitely many terms of $\{x_n\}$ are smaller than $l + \epsilon$.
(b) All but finitely many terms of $\{x_n\}$ are greater than $m - \epsilon$.

1.19. Let $\{x_n\} \subset \mathbb{R}$ be a sequence of positive numbers. Show that

$$\underline{\lim}\, \frac{x_{n+1}}{x_n} \leq \underline{\lim}\, \sqrt[n]{x_n} \leq \overline{\lim}\, \sqrt[n]{x_n} \leq \overline{\lim}\, \frac{x_{n+1}}{x_n}.$$

Hence, conclude that Cauchy's root test is stronger than D'Alembert's ratio test to check convergence of a series of positive numbers, in the sense that $\lim \frac{x_{n+1}}{x_n}$ may not exist, whereas $\lim \sqrt[n]{x_n}$ exists, but not conversely.

1.20. Show that a series $\sum x_n$ of real numbers is convergent if and only if for any $\epsilon > 0$, there is some $N \in \mathbb{N}$ so that for every $n \geq N$, the entire tail of the series

$$|x_{n+1} + x_{n+2} + \cdots| < \epsilon.$$

1.21. Show that limit of a function, if exists, is unique.

1.22. Let $(\emptyset \neq)\ D \subset \mathbb{R}$ and $f, g\colon D \to \mathbb{R}$. Let $c \in D'$, $\lim_{x\to c} f(x) = l$, and $\lim_{x\to c} g(x) = m$. Then prove that:

(a) For every real number α, $\lim_{x\to c} \alpha f(x) = \alpha l$.
(b) $\lim_{x\to c} (f(x) \pm g(x)) = l \pm m$.
(c) $\lim_{x\to c} (f(x)g(x)) = lm$.
(d) $\lim_{x\to c} \left(\frac{f(x)}{g(x)}\right) = \frac{l}{m}$, provided $g(x) \neq 0$ in some neighbourhood of c and $m \neq 0$.

1.23. Let $(\emptyset \neq)\ D \subset \mathbb{R}$ and $f, g, h\colon D \to \mathbb{R}$. Let $c \in D'$. Let for every point $x \in D \setminus \{c\}$, $f(x) \leq g(x) \leq h(x)$. If $\lim_{x\to c} f(x) = l = \lim_{x\to c} h(x)$, then $\lim_{x\to c} g(x)$ exists and is equal to l as well.

1.24. Prove the following limits without using L'Hôpital's rule:

(a) $\lim_{x\to a} \frac{x^n - a^n}{x-a} = na^{n-1}$.
(b) $\lim_{x\to 0} \frac{\sin x}{x} = 1$.
(c) $\lim_{x\to 0} \frac{e^x - 1}{x} = 1$.
(d) $\lim_{x\to 0} \frac{1}{x} \log(1+x) = 1$.
(e) $\lim_{x\to \infty} \left(1 + \frac{1}{x}\right)^x = e$.
(f) $\lim_{x\to 0} (1+x)^{\frac{1}{x}} = e$.

1.25. Let $(\emptyset \neq)\ D \subset \mathbb{R}$ and $f, g\colon D \to \mathbb{R}$ be continuous at $c \in D$. Let $\alpha \in \mathbb{R}$. Show that the functions αf, $f \pm g$, fg, and $\frac{f}{g}$ (provided $g(c) \neq 0$) are all continuous at c.

1.26. Let $f\colon \mathbb{R} \to \mathbb{R}$. Prove the followings are equivalent:

(a) f is continuous.
(b) For every open $G \subset \mathbb{R}$, $f^{-1}(G)$ is open.
(c) For every closed $F \subset \mathbb{R}$, $f^{-1}(F)$ is closed.

1.27. *Leibnitz theorem*: If f and g are two n times differentiable functions at a point c in their common domain, then the nth order derivative of their product function fg is given by

$$(fg)^{(n)}(c) = \sum_{k=0}^{n} {}^nC_k f^{(n-k)}(c) g^{(k)}(c),$$

where $f^{(r)}$ denotes the rth order derivative of f, $1 \leq r \leq n$, and $f^{(0)} = f$.

1.28. Let $f\colon [a,b] \to \mathbb{R}$ be such that $f' \equiv 0$ on $[a,b]$. Show that f is a constant.

1.29. A function $f\colon [a, b] \to \mathbb{R}$ is said to satisfy a *Lipschitz condition* of order α if there exists $M > 0$ so that the relation

$$|f(x) - f(y)| < M|x - y|^{\alpha}$$

holds for any pair of points $x, y \in [a, b]$. Show that if f satisfies a Lipschitz condition of order $\alpha > 1$ on $[a, b]$, then f must be constant.

1.30. Establish the Taylor's infinite series of the following functions in the specified domains:

(a) $e^x = 1 + x + \frac{x^2}{2!} + \frac{x^3}{3!} + \cdots + \frac{x^n}{n!} + \cdots, \quad x \in \mathbb{R}$.
(b) $\sin x = x - \frac{x^3}{3!} + \frac{x^5}{5!} - \cdots, \quad x \in \mathbb{R}$.
(c) $\cos x = 1 - \frac{x^2}{2!} + \frac{x^4}{4!} - \cdots, \quad x \in \mathbb{R}$.
(d) $(1 + x)^m = 1 + mx + \frac{m(m-1)}{2!}x^2 + \cdots + \frac{m(m-1)\ldots(m-r+1)}{r!}x^r + \cdots,$ $x \in (-1, 1)$.
(e) $\log(1 + x) = x - \frac{x^2}{2} + \frac{x^3}{3} - \cdots, \quad x \in (-1, 1]$.

1.31. Let $f\colon [a, b] \to \mathbb{R}$ be a monotone function. Show that f is integrable.

1.32. Let $f, g\colon [a, b] \to \mathbb{R}$ be bounded functions. Let $f(x) = g(x)$ hold for all but finitely many[4] points from $[a, b]$. If $f \in \mathcal{R}[a, b]$, show that $g \in \mathcal{R}[a, b]$ as well. Also, $\int_a^b f = \int_a^b g$.

1.33. Let $f\colon [-1, 1] \to \mathbb{R}$ be defined by

$$f(x) = \begin{cases} 2x \sin \frac{1}{x} - \cos \frac{1}{x}, & \text{when } x \neq 0 \\ 0, & \text{when } x = 0 \end{cases}.$$

Show that f possesses an antiderivative although it is not continuous on $[-1, 1]$.

1.34. Let $I = [a, b]$ and $J = [c, d]$ be closed and bounded intervals in $\mathbb{R}$. Let $f\colon I \to \mathbb{R}$ be continuous on I. Let $u, v\colon J \to \mathbb{R}$ be differentiable functions on J so that $u(I) \subset J$ and $v(I) \subset J$. If the function $g\colon J \to \mathbb{R}$ be defined by

$$g(x) = \int_{u(x)}^{v(x)} f, \qquad x \in J,$$

show that for every $x \in J$,

$$g'(x) = (f \circ v)(x)v'(x) - (f \circ u)(x)u'(x).$$

[4] The phrase "for all but finitely many" implies that the statement is true for all values of the variable, except only finitely many values.

1.35. Let $\alpha > 0$ be fixed. Show that:

(a) $\lim_{x\to\infty} \frac{L(x)}{x^\alpha} = 0$. (See Definition 1.136.)

(b) $\lim_{x\to\infty} \frac{x^\alpha}{E(x)} = 0$. (See Definition 1.140.)

[*Remark*: Above results show that $L(x)$ tends to ∞ slower than any positive power (however small) of x; and $E(x)$ tends to ∞ faster than any power (however big) of x.]

1.36. (a) For $n \in \mathbb{N}$, let $f_n: (0, 1) \to \mathbb{R}$ be defined by

$$f_n(x) = x^n.$$

Show that here the conclusions of Theorems 1.147(i) and 1.148(i) hold, although the convergence of $\{f_n\}$ is not uniform.

(b) For $n \in \mathbb{N}$, let $f_n: [0, 1] \to \mathbb{R}$ be defined by

$$f_n(x) = nxe^{-nx^2}.$$

Show that here the conclusions of Theorem 1.149(i) hold, although the convergence of $\{f_n\}$ is not uniform.

(c) For $n \in \mathbb{N}$, let $f_n: [0, 1] \to \mathbb{R}$ be defined by

$$f_n(x) = \frac{nx}{1 + n^2x^2}.$$

Show that here the conclusions of Theorem 1.150(i) hold, even when the convergence of $\{f_n\}$ is not uniform.

(d) For $n \in \mathbb{N}$, let $f_n: [0, 1] \to \mathbb{R}$ be defined by

$$f_n(x) = x - \frac{x^n}{n}.$$

Show that $\{f_n\}$ converges uniformly, even when the convergence of $\{f_n'\}$ is not uniform.

1.37. Use Exercise 1.19 to prove that Cauchy-Hadamard test (Theorem 1.156) is superior than the ratio test (Theorem 1.157) to determine the radius of convergence of a power series. Find an example where the ratio test fails, but Cauchy-Hadamard theorem successfully determines the radius of convergence of the series.

1.38. Show that Theorem 1.158 does not guarantee the uniform convergence of the power series in its interval of convergence by considering the counter-example

$$\sum a_n x^n = 1 + x + x^2 + x^3 + \cdots.$$

1.39. Let $f(x)$ be the sum of the power series $\sum a_n x^n$ in its interval of convergence I. If for every point $x \in I$, $f'(x) = f(x)$ and $f(0) = 1$, show that for each index n, $a_n = \frac{1}{n!}$.

1.40. Let $R > 0$ be fixed and let $f: (-R, R) \to \mathbb{R}$ be defined so that for some constant $M > 0$, $|f^{(n)}(x)| \leq M$ holds for each point $x \in (-R, R)$ and every index $n \in \mathbb{N}$. ($f^{(n)}$ is the nth order derivative of f.) Show that the Taylor's series

$$\sum_{n=0}^{\infty} \frac{f^{(n)}(0)}{n!} x^n$$

converges to $f(x)$ on $(-R, R)$.

Chapter 2
Basic Notions

The most natural way for mathematical theories to progress is by generalising existing concepts and structures. It is akin to searching for other planets in our solar system after realising we live on one. Though the planets are vastly different (for instance, only one currently sustains life), they all share certain intrinsic properties, such as their shape or the nature of their orbits around the Sun. It is through the discovery of these common features that we gain insight into the entire solar system. This process of broadening our perspective—from our own planet to the solar system and beyond to the galaxy—illustrates how science, along with our understanding of nature, expands.

In mathematics, the exact analogical process is known as "generalisation" or "abstraction" of concepts. From a pedagogical perspective, this process can be described to have the following structure:

- First, we examine a particular structure in a specific space (such as the analytical structure of the space of real numbers).
- Second, we identify other spaces that share similar structures (for example, the space of complex numbers).
- Third, we seek intrinsic properties common to all these spaces.
- Finally, we create a new abstract system that embodies these shared properties.

The original spaces now serve as examples of the newly generated system. As a result of this exercise, the discoveries and consequences of the generalised system can be applied to all the example spaces, eliminating the need to study them in their particular forms.

The natural extension of classical real analysis leads to the theories of metric spaces and topology. In real analysis, the key concepts are limits and continuity, both of which hinge on the fundamental idea of "distance" between points. To reconstruct classical analysis in an abstract setting, a natural starting point is to conceptualise the notion of distance for points in an abstract set. Consequently, we can study limits and continuity within this restructured framework, deepening our understanding of

S. Paul, *Metric Spaces*, University Texts in the Mathematical Sciences,
https://doi.org/10.1007/978-981-96-9259-0_2

the generalised structure. Moreover, we identify more spaces where such operations can be performed.

The motivation for this exercise to generalise concepts is not always straightforward. At times, it is driven by unsolved problems at a more elementary level. Paradoxically, a specific problem is sometimes easier to solve when generalised to a more abstract framework. For instance, procedures for solving polynomial equations of degree 4 were discovered by 1545. However, it took more than 275 years for mathematicians to realise that no general procedure exists for solving a quintic (degree 5) or higher degree polynomial equation. The proofs of this fact were provided by the Norwegian mathematician Niels Henrik Abel in 1826 and later by the French prodigy Évariste Galois in 1831, leading to their groundbreaking theory. The ideas used by both Abel and Galois laid the foundation for modern Abstract Algebra, a subject that originally emerged from the motivation to solve polynomial equations but has since become a central area in mathematics. A theory on generalisations of the algebraic structures was where the problem of solving polynomial equations found its resolution.

In our current context, the motivation cannot be attributed to a specific problem. The idea of an abstract space with metric properties was first addressed in 1906 by René Maurice Fréchet. This allowed mathematicians to study functions and sequences in a broader and more flexible way, which was crucial for the emerging field of functional analysis. At the time, functional analysis and rigorous modern analysis were still relatively new, and the abstract, axiomatic approach to mathematics was not as routine as it is today. Mathematicians were studying various spaces, primarily spaces of functions, each with its own notions of convergence, leading to ad hoc methods. Recognising the need for simplification and unification, Fréchet introduced metric spaces in his PhD dissertation by axiomatising the notion of distance, showing that many of these spaces were instances of metric spaces. By proving results axiomatically from the metric axioms, those results automatically applied to all instances. The term "Metric Space" was later coined by Felix Hausdorff in 1914, who, along with Stefan Banach, further refined and expanded the framework. Hausdorff also introduced topological spaces as a generalisation of metric spaces, marking a significant step forward in the abstraction and formalisation of mathematical spaces.

With this briefing, let us now delve deep into the subject. This first chapter consists of four sections. In the first two sections, we define a metric space and explore several examples. In the subsequent ones, we learn about pseudometrics and the concept of distance between sets in a metric space.

2.1 Definition of a Metric Space

We previously discussed the steps involved in abstracting an existing concept. For our current purpose, we have already completed the first two steps: We have studied real analysis in depth and recognised the existence of similar spaces. Now, as we move to the third step, aiming to redefine the notion of distance in an abstract setup, we examine the intrinsic properties of the concept of distance for real numbers that can be extended to other spaces. To do this, we analyse the Euclidean distance on the set of real numbers and eliminate any attributes unique to $\mathbb{R}$, ensuring that the remaining concepts can be applied to any abstract space.

The *Euclidean* distance between two real numbers x and y is defined as $|x - y|$. To generalise this, we examine the properties of the absolute value function and extend them to a more abstract setting. The objective is to develop a new concept in which the real numbers, equipped with $|\cdot|$, naturally emerge as a fundamental example.

For reference, we formally define the absolute value function on the set of real numbers.

Definition 2.1 The *absolute value* function $|\cdot|$ on $\mathbb{R}$ is defined as

$$|x| = \begin{cases} x, & \text{if } x \geq 0 \\ -x, & \text{if } x < 0 \end{cases}.$$

To begin the generalisation process, we start with a non-empty set X. We seek to define a function d on X that measures the distance between any two points in X. Thus, d must be defined on $X \times X$. The next question is: What values should d produce? In real analysis, the absolute value function $|\cdot|: \mathbb{R} \to \mathbb{R}$ returns a real number, allowing us to use the algebraic and order properties of $\mathbb{R}$ to conduct analysis. However, since X does not inherently possess any of the said structures, defining d as a function from $X \times X$ to X is not feasible. Instead, we comply ourselves by not deviating from the *real*-ity much further and decide to have $d: X \times X \to \mathbb{R}$.

We now start the process of setting up the general postulates for the distance function d.

- The first property of d is based on the fact that for any two points $x, y \in \mathbb{R}$, the absolute difference $|x - y|$ is always non-negative. Intuitively, "distance" cannot be negative, so we establish the rule:

$$\text{For any two points } x, y \in X, \quad d(x, y) \geq 0.$$

  Additionally, to ensure that d distinguishes points in X uniquely, we impose the condition:

$$\text{For points } x, y \in X, \quad d(x, y) = 0 \iff x = y.$$

- The second property of d reflects the symmetry of absolute differences in real numbers, i.e. for any $x, y \in \mathbb{R}$, we have $|x - y| = |y - x|$. This means that the distance from x to y is the same as the distance from y to x, which leads us to define d as a symmetric function:

$$\text{For any two points } x, y \in X, \quad d(x, y) = d(y, x).$$

- The third and most distinctive property of d as a distance function is the triangle inequality. This is based on the fact that for any two real numbers x and y, the inequality $|x + y| \leq |x| + |y|$ always holds (a proof of this will follow). Generalising this idea, we establish the rule:

$$\text{For any three points } x, y, z \in X, \quad d(x, z) \leq d(x, y) + d(y, z).$$

So finally, we record:

Definition 2.2 A non-empty set X is called a *metric space* if there is a *metric* or a *distance function* $d: X \times X \to \mathbb{R}$ such that:

(a) For any two points $x, y \in X$, $d(x, y) \geq 0$, and $d(x, y) = 0$ if and only if $x = y$. [*positivity*]
(b) For any two points $x, y \in X$, $d(x, y) = d(y, x)$. [*symmetry*]
(c) For any three points $x, y, z \in X$, $d(x, z) \leq d(x, y) + d(y, z)$. [*triangle inequality*]

An equivalent definition: A non-empty set X is called a *metric space* if there is a *metric* or a *distance function* $d: X \times X \to \mathbb{R}$ such that:

(a) For any two points $x, y \in X$, $d(x, y)$ is a finite real number.
(b) For any two points $x, y \in X$, $d(x, y) = 0$ if and only if $x = y$.
(c) For any three points $x, y, z \in X$, $d(y, z) \leq d(x, y) + d(x, z)$.

Exercises: 2.1

2.1. Show that Definition 2.2 is indeed equivalent to the definition of a metric space.

2.2 Classic Examples of Metric Spaces

In this section, we are going to see a number of examples of metric spaces. To check these spaces are indeed examples of metric spaces, we need a few inequalities at our disposal. Let us record them in the following subsubsection:

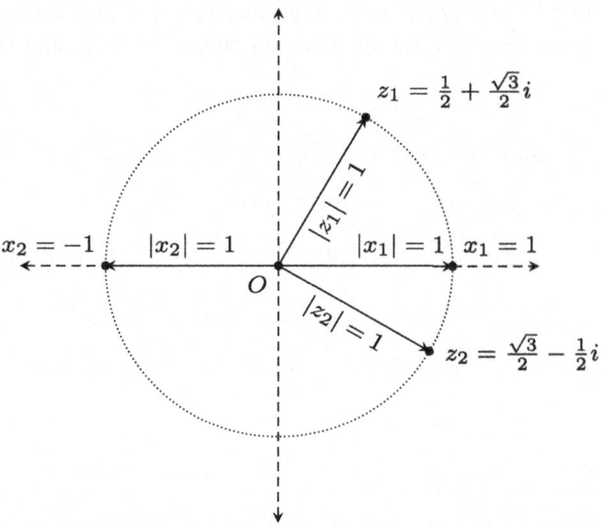

Fig. 2.1 Modulus function on $\mathbb{C}$ and on $\mathbb{R}$

Few Useful Inequalities

1. *Triangle inequalities:*

(a) *Over* $\mathbb{R}$: For any two real numbers x and y, we have

$$|x + y| \leq |x| + |y|.$$

Proof. Observe that for any real number x, $x \leq |x|$ and $-x \leq |x|$. Hence, if $|x + y| = x + y$, then

$$|x + y| = x + y \leq |x| + |y|.$$

Otherwise, if $|x + y| = -(x + y)$, then also

$$|x + y| = -(x + y) \leq |x| + |y|.$$

□

(b) *Over* $\mathbb{C}$: The *modulus* function $|\cdot|$ on $\mathbb{C}$ is defined as (Fig. 2.1)

$$|z| = \sqrt{x^2 + y^2}, \qquad \text{where } z = x + iy \text{ and } x, y \in \mathbb{R}.$$

Then for any two complex numbers z_1 and z_2,

$$|z_1 + z_2| \leq |z_1| + |z_2|.$$

Proof. If $\bar{z} = x - iy$ denotes the conjugate complex number of z, then $z\bar{z} = |z|^2$. Observe that for any two complex numbers z_1 and z_2,

$$
\begin{aligned}
|z_1 + z_2|^2 &= (z_1 + z_2)(\overline{z_1 + z_2}) \\
&= (z_1 + z_2)(\overline{z_1} + \overline{z_2}) \\
&= z_1\overline{z_1} + z_2\overline{z_2} + z_1\overline{z_2} + \overline{z_1}z_2 \\
&= |z_1|^2 + |z_2|^2 + 2\Re(z_1\overline{z_2}), \qquad [\Re(z) \text{ denotes the real part of } z] \\
&\leq |z_1|^2 + |z_2|^2 + 2|z_1\overline{z_2}|, \qquad [\because \forall z \in \mathbb{C},\ \Re(z) \leq |z|] \\
&= |z_1|^2 + |z_2|^2 + 2|z_1||z_2|, \qquad [\because \forall z \in \mathbb{C},\ |z| = |\bar{z}|] \\
&= (|z_1| + |z_2|)^2 .
\end{aligned}
$$

Hence, $|z_1 + z_2| \leq |z_1| + |z_2|$. □

Remark We use the same notation $|\cdot|$ to denote both the absolute value function on $\mathbb{R}$ and the modulus function on $\mathbb{C}$. Although the definitions of these two functions do not quite match in the naked eye, both of them essentially capture the *magnitude* of a point (in $\mathbb{R}$ or in $\mathbb{C}$) irrespective its direction w.r.t. the origin. In $\mathbb{R}$, both the numbers -1 and 1, despite sitting in the opposite directions w.r.t. the origin, have the same magnitude 1. Similarly both the points $\frac{1}{2} + \frac{\sqrt{3}}{2}i$ and $\frac{\sqrt{3}}{2} - \frac{1}{2}i$ have magnitude 1, although sitting in different directions w.r.t. origin. Actually, any complex number sitting on the unit circle has magnitude 1. And as a matter of fact, these two definitions comply with each other if we generalise the real numbers as complex numbers. For $x \in \mathbb{R}$, writing $x = x + 0i \in \mathbb{C}$, we see that

$$|x + 0i| = \sqrt{x^2 + 0} = |x|.$$

2. *Hölder's inequality:* If $\{a_k : k = 1, \ldots, n\}$ and $\{b_k : k = 1, \ldots, n\}$ are two sets of non-negative reals and p, q are two quantities greater than 1 such that $\frac{1}{p} + \frac{1}{q} = 1$, then

$$\sum_{k=1}^{n} a_k b_k \leq \left(\sum_{k=1}^{n} a_k^p\right)^{1/p} \left(\sum_{k=1}^{n} b_k^q\right)^{1/q} .$$

3. *Cauchy-Schwarz's inequality:* The special case when $p = q = 2$ in the Hölder's inequality is known as the Cauchy-Schwarz's inequality. Here, if $\{a_k : k = 1, \ldots, n\}$ and $\{b_k : k = 1, \ldots, n\}$ are two sets of non-negative reals, then

$$\left(\sum_{k=1}^{n} a_k b_k\right)^2 \leq \left(\sum_{k=1}^{n} a_k^2\right) \left(\sum_{k=1}^{n} b_k^2\right) .$$

Table 2.1 Classification of examples of metric spaces in this subsection

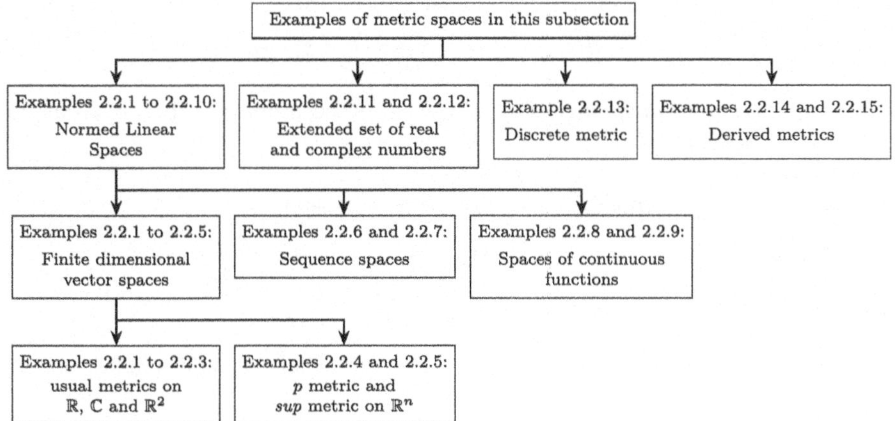

4. *Minkowski's inequality:* If $\{a_k : k = 1, \ldots, n\}$ and $\{b_k : k = 1, \ldots, n\}$ are two sets of non-negative reals and $p \geq 1$ is a real number, then

$$\left(\sum_{k=1}^{n}(a_k + b_k)^p\right)^{1/p} \leq \left(\sum_{k=1}^{n} a_k^p\right)^{1/p} + \left(\sum_{k=1}^{n} b_k^p\right)^{1/p}.$$

Remark The proofs of above three inequalities are included in Appendix A.

We will now explore 15 examples of metric spaces. The examples chosen here are quite classical in the sense that they keep appearing in the literature. Each has been rigorously proved with full details. However, an enthusiastic reader is not necessarily expected to focus extensively on these proofs but is instead encouraged to attempt constructing them independently. More important than the formal proofs is developing an awareness and appreciation about the existence of diverse spaces and the different notions of distance they illustrate.

A consistent approach has been followed in proving that each space satisfies the definition of a metric space, specifically by systematically verifying the conditions outlined in Definition 2.2. Additional interesting examples can be found in the exercises. A broad classification of the examples presented in this section is provided in Table 2.1.

Examples

2.2.1. $X = \mathbb{R}$; $d(x, y) := |x - y|, x, y \in \mathbb{R}$. Then (X, d) is a metric space.

Remark As we intended, this is the first example of a metric space. The proof of this fact, recorded here for mere formality, follows directly from the motivated construction of the distance function d for the general setup.

***Proof*.**

(a) *Positivity:* By definition of absolute value function, for any pair of points, $x, y \in \mathbb{R}$, $d(x, y) \geq 0$, and $d(x, y) = 0 \iff x = y$.

(b) *Symmetry:* Also for any pair of points $x, y \in \mathbb{R}$,

$$d(x, y) = |x - y| = |y - x| = d(y, x).$$

(c) *Triangle inequality:* Let $x, y, z \in \mathbb{R}$ be arbitrary. Then by triangle inequality of $\mathbb{R}$,

$$\begin{aligned} d(x, z) &= |x - z| = |(x - y) + (y - z)| \\ &\leq |x - y| + |y - z| = d(x, y) + d(y, z). \end{aligned}$$

□

2.2.2. $X = \mathbb{C}$; $d(z_1, z_2) := |z_1 - z_2|, z_1, z_2 \in \mathbb{C}$. Then (X, d) is a metric space.
Proof.

(a) *Positivity:* By definition of modulus function, for any pair of points, $z_1, z_2 \in \mathbb{C}$, $d(z_1, z_2) \geq 0$ and $d(z_1, z_2) = 0 \iff z_1 = z_2$.
(b) *Symmetry:* Also for any pair of points $z_1, z_2 \in \mathbb{C}$,

$$d(z_1, z_2) = |z_1 - z_2| = |z_2 - z_1| = d(z_2, z_1).$$

(c) *Triangle inequality:* Let $z_1, z_2, z_3 \in \mathbb{C}$ be arbitrary. Then by triangle inequality of $\mathbb{C}$,

$$|z_1 - z_3| = |(z_1 - z_2) + (z_2 - z_3)| \leq |z_1 - z_2| + |z_2 - z_3|.$$

Hence, $d(z_1, z_3) \leq d(z_1, z_2) + d(z_2, z_3)$. □

Remark The major convenience of $\mathbb{C}$ as an algebraic system is that it allows the extension of the concept of multiplication from the set of real numbers, making $\mathbb{C}$ an algebraically closed field[1]. The standard proof of the triangle inequality for $\mathbb{C}$ relies on the multiplication operation and its related properties. However, this dependency can be completely avoided by providing an independent proof of the triangle inequality. In doing so, we essentially interpret $\mathbb{C}$ as $\mathbb{R}^2$, viewed as a vector space over $\mathbb{R}$. In fact, the Euclidean metric in $\mathbb{R}^2$ can be defined in precisely the same way as in $\mathbb{C}$.

2.2.3. For points $\mathbf{x} = (x_1, x_2)$ and $\mathbf{y} = (y_1, y_2)$ from $\mathbb{R}^2$, define

$$d(\mathbf{x}, \mathbf{y}) = \sqrt{(x_1 - y_1)^2 + (x_2 - y_2)^2}.$$

Then d is a metric on $\mathbb{R}^2$.

[1] A field K is called algebraically closed if every polynomial of degree n over K has exactly n roots in K.

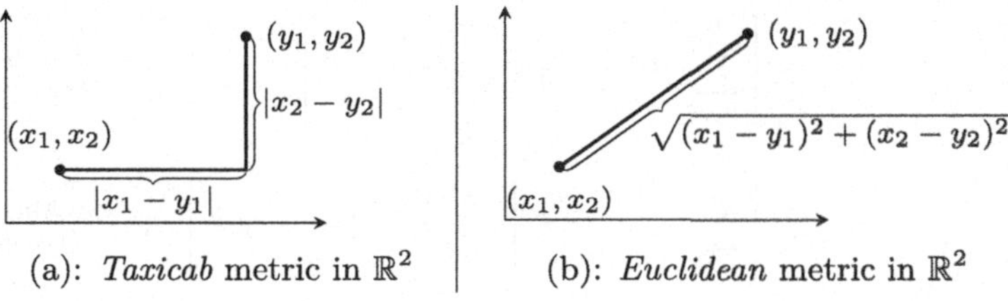

Fig. 2.2 d_p metric on $\mathbb{R}^2$ for $p = 1, 2$

Remark We skip the proof of the above fact. For, it is just a particular case of the general space $\mathbb{R}^n$ when $n = 2$. Also, we can tweak and generalise the metric and get a whole family of metrics in the following way:

2.2.4. Let $\mathbf{x} = (x_1, \ldots, x_n), \mathbf{y} = (y_1, \ldots, y_n) \in \mathbb{R}^n$. For $p \geq 1$, define

$$d_p(\mathbf{x}, \mathbf{y}) = \left[\sum_{k=1}^{n} |x_k - y_k|^p\right]^{\frac{1}{p}}.$$

Then d_p is a metric on $\mathbb{R}^n$ (Fig. 2.2).

Proof.

(a) *Positivity:* By definition of d_p for any pair of points $\mathbf{x}, \mathbf{y} \in \mathbb{R}^n$, $d_p(\mathbf{x}, \mathbf{y}) \geq 0$ and

$$\begin{aligned} & d_p(\mathbf{x}, \mathbf{y}) = 0 \\ \iff & \left[\sum_{k=1}^{n} |x_k - y_k|^p\right]^{\frac{1}{p}} = 0 \\ \iff & \forall k \in \{1, \ldots, n\},\ x_k = y_k \\ \iff & \mathbf{x} = \mathbf{y}. \end{aligned}$$

(b) *Symmetry:* For any pair of points $\mathbf{x}, \mathbf{y} \in \mathbb{R}^n$, $d_p(\mathbf{x}, \mathbf{y}) = d_p(\mathbf{y}, \mathbf{x})$ holds trivially.

(c) *Triangle inequality:* Let $\mathbf{x}, \mathbf{y}, \mathbf{z} \in \mathbb{R}^n$ be arbitrary. Then

$$\begin{aligned} d_p(\mathbf{x}, \mathbf{z}) &= \left[\sum_{k=1}^{n} |x_k - z_k|^p\right]^{\frac{1}{p}} \\ &= \left[\sum_{k=1}^{n} |(x_k - y_k) + (y_k - z_k)|^p\right]^{\frac{1}{p}} \end{aligned}$$

$$\leq \left[\sum_{k=1}^{n} (|x_k - y_k| + |y_k - z_k|)^p\right]^{\frac{1}{p}}, \qquad \text{[by triangle inequality on } \mathbb{R}]$$

$$\leq \left[\sum_{k=1}^{n} |x_k - y_k|^p\right]^{\frac{1}{p}} + \left[\sum_{k=1}^{n} |y_k - z_k|^p\right]^{\frac{1}{p}}, \qquad \text{[by Minkowski's inequality]}$$

$$= d_p(\mathbf{x}, \mathbf{y}) + d_p(\mathbf{y}, \mathbf{z}).$$

□

Remarks

(a) For $p = 1$, this metric is called the *taxicab* metric. In this case, the distance between two points $\mathbf{x}$ and $\mathbf{y}$

$$d_1(\mathbf{x}, \mathbf{y}) = |x_1 - y_1| + \cdots + |x_n - y_n|$$

is the sum of the differences in each coordinate.

(b) For $p = 2$, this metric is called the *Euclidean* metric on $\mathbb{R}^n$ and is the usual metric there. In this case, the distance between two points $\mathbf{x}$ and $\mathbf{y}$

$$d_2(\mathbf{x}, \mathbf{y}) = \sqrt{(x_1 - y_1)^2 + \cdots + (x_n - y_n)^2}$$

is the linear distance between them.

2.2.5. For $\mathbf{x} = (x_1, \ldots, x_n), \mathbf{y} = (y_1, \ldots, y_n) \in \mathbb{R}^n$; define

$$d_\infty(\mathbf{x}, \mathbf{y}) := \max\{|x_k - y_k| : 1 \leq k \leq n\}.$$

Then d_∞ is a metric on $\mathbb{R}^n$ (Fig. 2.3).

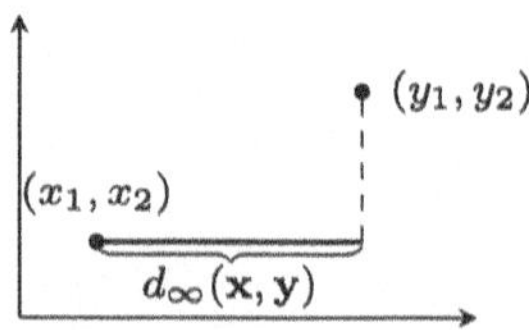

Fig. 2.3 d_∞ metric on $\mathbb{R}^2$

***Proof*.**

(a) *Positivity:* By definition of d_∞ for any pair of points $\mathbf{x}, \mathbf{y} \in \mathbb{R}^n$, $d_\infty(\mathbf{x}, \mathbf{y}) \geq 0$ and

$$\begin{aligned} & d_\infty(\mathbf{x}, \mathbf{y}) = 0 \\ \iff & \max_{1\leq k\leq n} \{|x_k - y_k|\} = 0 \\ \iff & \forall k \in \{1, \dots, n\}, \ x_k = y_k \\ \iff & \mathbf{x} = \mathbf{y}. \end{aligned}$$

(b) *Symmetry:* For any pair of points $\mathbf{x}, \mathbf{y} \in \mathbb{R}^n$, $d_\infty(\mathbf{x}, \mathbf{y}) = d_\infty(\mathbf{y}, \mathbf{x})$ holds trivially.

(c) *Triangle inequality:* Let $\mathbf{x}, \mathbf{y}, \mathbf{z} \in \mathbb{R}^n$ be arbitrary. If $d_\infty(\mathbf{x}, \mathbf{z}) = |x_j - z_j|$ for some $j \in \{1, \dots, n\}$, then by triangle inequality of real numbers,

$$\begin{aligned} d_\infty(\mathbf{x}, \mathbf{z}) &= |x_j - z_j| \\ &\leq |x_j - y_j| + |y_j - z_j| \\ &\leq \max\{|x_k - y_k| : 1 \leq k \leq n\} + \max\{|y_k - z_k| : 1 \leq k \leq n\} \\ &= d_\infty(\mathbf{x}, \mathbf{y}) + d_\infty(\mathbf{y}, \mathbf{z}). \end{aligned}$$

□

2.2.6. Let $1 \leq p < \infty$. Call

$$\ell_p := \left\{ \{x_k\}_{k=1}^{\infty} : x_k \in \mathbb{R} \text{ or } \mathbb{C} \text{ and } \sum_{k=1}^{\infty} |x_k|^p < \infty \right\}.$$

Then for elements $\mathbf{x} = \{x_k\}$ and $\mathbf{y} = \{y_k\}$ from ℓ_p, the map

$$d_p(\mathbf{x}, \mathbf{y}) := \left[\sum_{k=1}^{\infty} |x_k - y_k|^p \right]^{\frac{1}{p}}$$

is a metric on ℓ_p.

***Proof*.** We start by showing d_p is well-defined. As both $\sum |x_k|^p$ and $\sum |y_k|^p$ are finite, we have

$$d_p(\mathbf{x}, \mathbf{y}) = \left[\sum_{k=1}^{\infty} |x_k - y_k|^p \right]^{\frac{1}{p}} = \left[\lim_{n\to\infty} \sum_{k=1}^{n} |x_k - y_k|^p \right]^{\frac{1}{p}}$$

$$= \lim_{n\to\infty} \left[\sum_{k=1}^{n} |x_k - y_k|^p\right]^{\frac{1}{p}}. \tag{$*$}$$

But by Minkowski's inequality, for each $n \in \mathbb{N}$, we have

$$\left[\sum_{k=1}^{n} |x_k - y_k|^p\right]^{\frac{1}{p}} \leq \left[\sum_{k=1}^{n} |x_k|^p\right]^{\frac{1}{p}} + \left[\sum_{k=1}^{n} |y_k|^p\right]^{\frac{1}{p}}. \tag{$**$}$$

Hence from $(*)$, we get

$$\begin{aligned} d_p(\mathbf{x}, \mathbf{y}) &= \lim_{n\to\infty} \left[\sum_{k=1}^{n} |x_k - y_k|^p\right]^{\frac{1}{p}} \\ &\leq \lim_{n\to\infty} \left[\sum_{k=1}^{n} |x_k|^p\right]^{\frac{1}{p}} + \lim_{n\to\infty} \left[\sum_{k=1}^{n} |y_k|^p\right]^{\frac{1}{p}}, \qquad \text{([by (**)])} \\ &< \infty. \qquad \left[\text{as both } \textstyle\sum_1^\infty |x_k|^p < \infty \text{ and } \sum_1^\infty |y_k|^p < \infty\right]. \end{aligned}$$

Hence, d_p is well-defined.

(a) *Positivity:* By definition of d_p, for any $\mathbf{x}, \mathbf{y} \in \ell_p$, $d_p(\mathbf{x}, \mathbf{y}) \geq 0$ and

$$\begin{aligned} & d_p(\mathbf{x}, \mathbf{y}) = 0 \\ \iff & \left[\sum_{k=1}^{\infty} |x_k - y_k|^p\right]^{\frac{1}{p}} = 0 \\ \iff & \forall k \in \mathbb{N},\ x_k = y_k \\ \iff & \mathbf{x} = \mathbf{y}. \end{aligned}$$

(b) *Symmetry:* For any pair of points $\mathbf{x}, \mathbf{y} \in \ell_p$, $d_p(\mathbf{x}, \mathbf{y}) = d_p(\mathbf{y}, \mathbf{x})$ holds trivially.

(c) *Triangle inequality:* Let $\mathbf{x}, \mathbf{y}, \mathbf{z} \in \ell_p$ be arbitrary. Then

$$\begin{aligned} d_p(\mathbf{x}, \mathbf{z}) &= \left[\sum_{k=1}^{\infty} |x_k - z_k|^p\right]^{\frac{1}{p}} \\ &= \left[\lim_{n\to\infty} \sum_{k=1}^{n} |(x_k - y_k) + (y_k - z_k)|^p\right]^{\frac{1}{p}} \end{aligned}$$

$$\leq \lim_{n\to\infty}\left[\sum_{k=1}^{n}(|x_k - y_k| + |y_k - z_k|)^p\right]^{\frac{1}{p}}, \quad \text{[by triangle inequality on } \mathbb{R}\text{]}$$

$$\leq \lim_{n\to\infty}\left[\left[\sum_{k=1}^{n}|x_k - y_k|^p\right]^{\frac{1}{p}} + \left[\sum_{k=1}^{n}|y_k - z_k|^p\right]^{\frac{1}{p}}\right], \quad \text{[by Minkowski's inequality]}$$

$$\leq \lim_{n\to\infty}\left[\sum_{k=1}^{n}|x_k - y_k|^p\right]^{\frac{1}{p}} + \lim_{n\to\infty}\left[\sum_{k=1}^{n}|y_k - z_k|^p\right]^{\frac{1}{p}}$$

$$= d_p(\mathbf{x}, \mathbf{y}) + d_p(\mathbf{y}, \mathbf{z}). \quad \text{[as both limits exist finitely]}$$

□

2.2.7. Let ℓ_∞ denote the set of all bounded sequences of real or complex numbers. Then for $\mathbf{x} = \{x_k\}, \mathbf{y} = \{y_k\} \in \ell_\infty$, the map

$$d_\infty(\mathbf{x}, \mathbf{y}) := \sup\{|x_k - y_k| : k \in \mathbb{N}\}$$

is a metric on ℓ_∞.

Proof. Here also, we start by showing d is well-defined. For any two elements $\mathbf{x} = \{x_k\}, \mathbf{y} = \{y_k\}$ from ℓ_∞, as both the sequences $\{x_k\}$ and $\{y_k\}$ are bounded, we obtain positive reals M_x and M_y such that for each index k, $|x_k| \leq M_x$ and $|y_k| \leq M_y$. Then by triangle inequality of $\mathbb{R}$ or $\mathbb{C}$,

$$\forall k \in \mathbb{N}, \quad |x_k - y_k| \leq |x_k| + |y_k| \leq M_x + M_y,$$

i.e. the set $\{|x_k - y_k| : k \in \mathbb{N}\}$ is a non-empty bounded above subset of non-negative real numbers. Hence by completeness axiom, $d_\infty(\mathbf{x}, \mathbf{y}) = \sup\{|x_k - y_k| : k \in \mathbb{N}\}$ exists finitely.

(a) *Positivity:* By definition, for any two elements $\mathbf{x}, \mathbf{y} \in \ell_\infty$, $d_\infty(\mathbf{x}, \mathbf{y}) \geq 0$, and $d_\infty(\mathbf{x}, \mathbf{y}) = 0 \iff x_k = y_k$ holds for every $k \in \mathbb{N}$, i.e. $\mathbf{x} = \mathbf{y}$.
(b) *Symmetry:* Follows trivially from the definition.
(c) *Triangle inequality:* For $\mathbf{x}, \mathbf{y}, \mathbf{z} \in \ell_\infty$, we have for every n,

$$|x_k - z_k| \leq |x_k - y_k| + |y_k - z_k|.$$

$$\begin{aligned}
\text{Hence } d_\infty(\mathbf{x}, \mathbf{z}) &= \sup\{|x_k - z_k| : k \in \mathbb{N}\} \\
&\leq \sup\{|x_k - y_k| + |y_k - z_k| : k \in \mathbb{N}\} \\
&\leq \sup\{|x_k - y_k| : k \in \mathbb{N}\} + \sup\{|y_k - z_k| : k \in \mathbb{N}\} \\
&\leq d_\infty(\mathbf{x}, \mathbf{y}) + d_\infty(\mathbf{y}, \mathbf{z}).
\end{aligned}$$

□

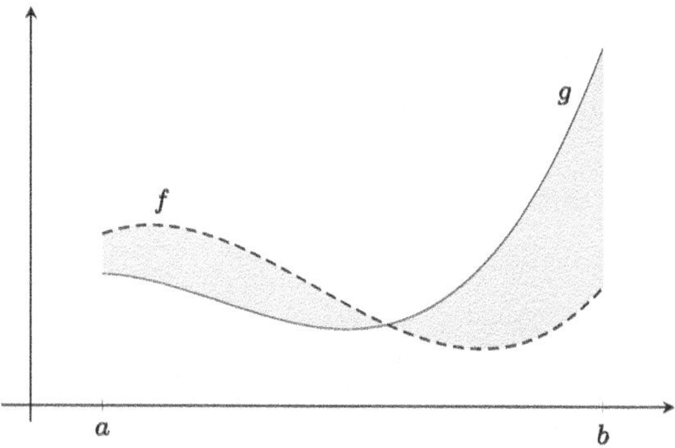

Fig. 2.4 $d_1(f, g)$ is the area of the shaded region

2.2.8. Let $\mathcal{C}[a, b]$ denote the set of all real-valued continuous functions defined on the closed and bounded interval $[a, b]$ of $\mathbb{R}$. For $f, g \in \mathcal{C}[a, b]$, let

$$d_1(f, g) := \int_a^b |f - g|.$$

Then d_1 is a metric on $\mathcal{C}[a, b]$, known as the *integral metric*. We denote the pair $(\mathcal{C}[a, b], d_1)$ by $\mathcal{C}_{\text{int}}[a, b]$ (Fig. 2.4).

***Proof*.** Clearly, as both f and g are continuous, so is $|f - g|$, and hence d_1 is well-defined.

(a) *Positivity:* For any two continuous functions $f, g \in \mathcal{C}[a, b]$, $d_1(f, g) \geq 0$ and $f = g$ implies $d_1(f, g) = 0$. Now from the result in the footnote[2], we have $d_1(f, g) = 0$ iff $f = g$.

(b) *Symmetry:* For any two continuous functions $f, g \in \mathcal{C}[a, b]$,

$$d_1(f, g) := \int_a^b |f - g| = \int_a^b |g - f| = d_1(g, f).$$

[2] If for $f \in \mathcal{C}[a, b]$, $\int_a^b |f| = 0$, then $f = 0$ on $[a, b]$. Because otherwise, let $f(\xi) \neq 0$ for some $\xi \in [a, b]$. Then $|f(\xi)| = \alpha > 0$, say. Since $|f|$ is also continuous on $[a, b]$, $|f(\xi)| = \alpha > 0$ implies we find $\delta > 0$ such that $|f(x)| > \frac{\alpha}{2} > 0$ for every $x \in (\xi - \delta, \xi + \delta) \cap [a, b]$. When $(\xi - \delta, \xi + \delta) \subset [a, b]$, we obtain

$$\int_a^b |f| \geq \int_{\xi-\delta}^{\xi+\delta} |f| > \int_{\xi-\delta}^{\xi+\delta} \frac{\alpha}{2} = \alpha\delta > 0,$$

a contradiction. Similar arguments lead to the same contradiction even when $(\xi - \delta, \xi + \delta)$ is not entirely contained in $[a, b]$.

(c) *Triangle inequality*: For continuous functions $f, g, h \in \mathcal{C}[a, b]$,

$$\begin{aligned} d_1(f, g) &= \int_a^b |f - g| \\ &= \int_a^b |f - h + h - g| \\ &\leq \int_a^b |f - h| + \int_a^b |h - g| \\ &\leq d_1(f, h) + d_1(h, g). \end{aligned}$$

□

2.2.9. On $\mathcal{C}[a, b]$, define for $f, g \in \mathcal{C}[a, b]$,

$$d_\infty(f, g) := \sup_{x \in [a,b]} |f(x) - g(x)|.$$

Then d_∞ is a metric on $\mathcal{C}[a, b]$, known as the *sup metric* or the *uniform metric*. We denote the pair $(\mathcal{C}[a, b], d_\infty)$ by $\mathcal{C}_{\text{sup}}[a, b]$ (Fig. 2.5).
***Proof*.** Since a continuous function defined over a closed and bounded interval $[a, b]$ is bounded, the metric d_∞ is well-defined.

(a) *Positivity:* For any two continuous functions $f, g \in \mathcal{C}[a, b]$, $d(f, g) \geq 0$ and

$$\begin{aligned} d_\infty(f, g) = 0 &\iff \sup_{x \in [a,b]} |f(x) - g(x)| = 0 \\ &\iff \forall x \in [a, b],\ f(x) = g(x). \end{aligned}$$

(b) *Symmetry:* Follows trivially from the definition just like in the previous example.
(c) *Triangle inequality*: A continuous function over a closed and bounded interval $[a, b]$ attains its bounds. Now for continuous functions $f, g, h \in \mathcal{C}[a, b]$, if $\sup_{x \in [a,b]} |f(x) - g(x)| = |f(\xi) - g(\xi)|$ for some $\xi \in [a, b]$, then we have

$$\begin{aligned} d_\infty(f, g) &= \sup_{x \in [a,b]} |f(x) - g(x)| \\ &= |f(\xi) - g(\xi)| \\ &\leq |f(\xi) - h(\xi)| + |h(\xi) - g(\xi)| \\ &\leq d_\infty(f, h) + d_\infty(h, g). \end{aligned}$$

□

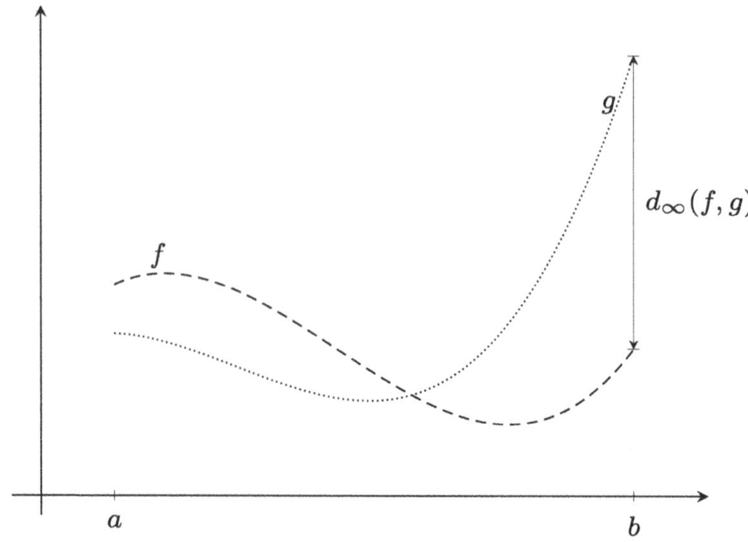

Fig. 2.5 $d_\infty(f, g)$ is the highest difference between f and g

2.2.10. Let V be a vector space over the field of $\mathbb{R}$ or $\mathbb{C}$. A real-valued function $\|\cdot\|: V \to \mathbb{R}$ is called a *norm* on V if:

i) For every vector $\mathbf{x} \in V$, $\|\mathbf{x}\| \geq 0$ and $\|\mathbf{x}\| = 0$ if and only if $\mathbf{x} = \mathbf{0}$.
ii) For every scalar $\alpha \in \mathbb{R}$ or $\mathbb{C}$ and for every vector $\mathbf{x} \in V$, $\|\alpha\mathbf{x}\| = |\alpha|\,\|\mathbf{x}\|$.
iii) For every vectors $\mathbf{x}, \mathbf{y} \in V$, $\|\mathbf{x}+\mathbf{y}\| \leq \|\mathbf{x}\| + \|\mathbf{y}\|$.

A vector space equipped with a norm is called a *normed linear space* (abbreviated as "NLS").
Now, let V be an NLS and define $d: V \times V \to \mathbb{R}$ by

$$d(\mathbf{x}, \mathbf{y}) = \|\mathbf{x} - \mathbf{y}\|.$$

Then d is a metric on V.
***Proof*.**

(a) *Positivity:* From the first property of a norm, for any pair of points $\mathbf{x}, \mathbf{y} \in V$, $d(\mathbf{x}, \mathbf{y}) \geq 0$, and $d(\mathbf{x}, \mathbf{y}) = 0 \iff \mathbf{x} = \mathbf{y}$.
(b) *Symmetry:* From the second property of a norm, choosing $\alpha = -1$, we see that $\|-\mathbf{x}\| = \|\mathbf{x}\|$. Hence for any pair of points $\mathbf{x}, \mathbf{y} \in V$,

$$d(\mathbf{x}, \mathbf{y}) = \|\mathbf{x} - \mathbf{y}\| = \|-(\mathbf{x} - \mathbf{y})\| = \|\mathbf{y} - \mathbf{x}\| = d(\mathbf{y}, \mathbf{x}).$$

(c) *Triangle inequality:* From the third property of a norm, we get for $\mathbf{x}, \mathbf{y}, \mathbf{z} \in V$

$$\begin{aligned} d(\mathbf{x}, \mathbf{z}) &= \|\mathbf{x} - \mathbf{z}\| \\ &= \|(\mathbf{x} - \mathbf{y}) + (\mathbf{y} - \mathbf{z})\| \\ &\leq \|\mathbf{x} - \mathbf{y}\| + \|\mathbf{y} - \mathbf{z}\| \\ &= d(\mathbf{x}, \mathbf{y}) + d(\mathbf{y}, \mathbf{z}). \end{aligned}$$

□

Remarks

(a) A metric thus defined via a norm is called "metric *induced* by a norm".
(b) In the previous examples, the following metrics are induced by norms:

- In Example 2.2.1, the metric is induced by the norm

$$\|x\| := |x|, \qquad x \in \mathbb{R}.$$

  It is the *Euclidean* norm on $\mathbb{R}$.
- In Example 2.2.2, the metric is induced by the norm

$$\|z\| := |z|, \qquad z \in \mathbb{C}.$$

  It is the *Euclidean* norm on $\mathbb{C}$.
- In Example 2.2.3, the metric is induced by the norm

$$\|\mathbf{x}\| := \sqrt{x_1^2 + x_2^2}, \qquad \mathbf{x} = (x_1, x_2) \in \mathbb{R}^2.$$

  It is the *Euclidean* norm on $\mathbb{R}^2$.
- In Example 2.2.4, the metric is induced by the norm

$$\|\mathbf{x}\|_p := \left[\sum_{k=1}^{n} |x_k|^p\right]^{\frac{1}{p}}, \qquad \mathbf{x} = (x_1, \ldots, x_n) \in \mathbb{R}^n.$$

  It is generally called the p-norm on $\mathbb{R}^n$. For $p = 2$, it is the *Euclidean* norm on $\mathbb{R}^n$.
- In Example 2.2.5, the metric is induced by the norm

$$\|\mathbf{x}\|_\infty := \max\{|x_k| : 1 \leq k \leq n\}, \qquad \mathbf{x} = (x_1, \ldots, x_n) \in \mathbb{R}^n.$$

  It is known as the *sup* norm on $\mathbb{R}^n$.

- In Example 2.2.6, the metric is induced by the norm

$$\|\mathbf{x}\|_p := \left[\sum_{k=1}^{\infty}|x_k|^p\right]^{\frac{1}{p}}, \qquad \mathbf{x} = \{x_k\} \in \ell_p.$$

- In Example 2.2.7, the metric is induced by the norm

$$\|\mathbf{x}\|_\infty := \sup\{|x_k| : k \in \mathbb{N}\}, \qquad \mathbf{x} = \{x_k\} \in \ell_\infty.$$

- In Example 2.2.8, the metric is induced by the norm

$$\|f\|_1 := \int_a^b |f|, \qquad f \in \mathcal{C}[a, b].$$

- In Example 2.2.9, the metric is induced by the norm

$$\|f\|_\infty := \sup\{|f(x)| : x \in [a, b]\}, \qquad f \in \mathcal{C}[a, b].$$

  It is known as the *sup* norm on $\mathcal{C}[a, b]$.

(c) A common tendency is to replace the metric notation "d" by the norm notation "$\|\cdot\|$". We will use both d and $\|\cdot\|$ interchangeably.

2.2.11. Denote by $\mathbb{R}^*$, the set of *extended real numbers* comprising of the real numbers together with two ideal elements ∞ and $-\infty$ with the property that for every real x, $-\infty < x < \infty$. Let $f : \mathbb{R}^* \to \mathbb{R}$ be defined by

$$f(x) = \begin{cases} \frac{x}{1+|x|}, \text{ if } x \in \mathbb{R} \\ 1, \text{ if } x = \infty \\ -1, \text{ if } x = -\infty \end{cases}.$$

Define for $x, y \in \mathbb{R}^*$,

$$d(x, y) = |f(x) - f(y)|.$$

Then d is a metric on $\mathbb{R}^*$.
Proof.

(a) *Positivity:* By definition, for $x, y \in \mathbb{R}^*$, $d(x, y) \geq 0$. It is left as a small exercise to check that f is an injective map, i.e. for $x, y \in \mathbb{R}^*$, $f(x) = f(y) \iff x = y$. Therefore, $d(x, y) = 0 \iff x = y$.
(b) *Symmetry:* Symmetry of d follows trivially.
(c) *Triangle inequality:* To see the triangle inequality, for any three points $x, y, z \in \mathbb{R}^*$, as f is a real-valued function, we have

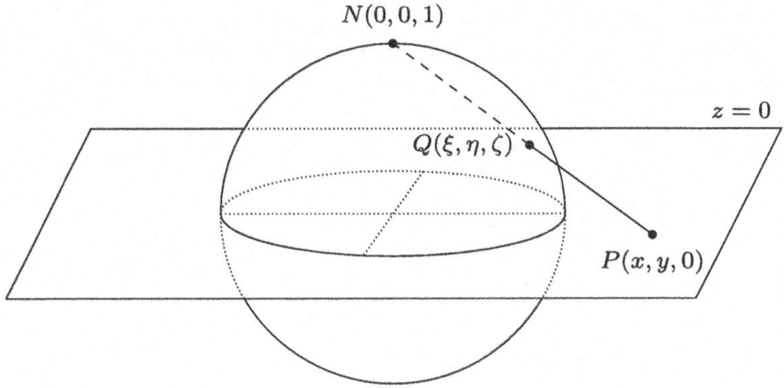

Fig. 2.6 Stereographic projection

$$\begin{aligned} d(x, z) &= |f(x) - f(z)| \\ &= |(f(x) - f(y)) + (f(y) - f(z))| \\ &\leq |f(x) - f(y)| + |f(y) - f(z)| \\ &= d(x, y) + d(y, z). \end{aligned}$$

□

2.2.12. The *extended set of complex numbers* comprises all complex numbers with an ideal number ∞ and is usually denoted by $\mathbb{C}^* = \mathbb{C} \cup \{\infty\}$. One representation of $\mathbb{C}^*$ is given by the *Riemann sphere* $x^2+y^2+z^2 = 1$ whose stereographic projection onto the plane $z = 0$, taking the point $(0, 0, 1)$ as the vertex of projection, is the extended complex plane (Fig. 2.6). The details about this are worked out in Appendix B. Let $z = x + iy$ be any complex number. We identify z as the point $P(x, y, 0)$ on the xy-plane in $\mathbb{R}^3$. Let $Q(\xi, \eta, \zeta)$ be that point on the Riemann sphere whose stereographic projection is P. In this association, let z_1, z_2 be two points on the complex plane and Q_1, Q_2 be the corresponding points on the Riemann sphere, respectively. Then from the details worked out in Appendix B, we know that the chordal distance[3] between Q_1 and Q_2 in terms of z_1 and z_2 is given by

$$\frac{2|z_1 - z_2|}{\sqrt{1 + |z_1|^2}\sqrt{1 + |z_2|^2}}.$$

[3] Chordal distance between Q_1 and Q_2 is the length of the line segment Q_1Q_2 as a chord of the sphere.

Define a distance d between z_1 and z_2 in terms of the chordal distance between Q_1 and Q_2. This way, extending our definition for the extended complex numbers, we have

$$d(z_1, z_2) = \begin{cases} \dfrac{2|z_1 - z_2|}{\sqrt{1+|z_1|^2}\sqrt{1+|z_2|^2}}, & \text{if } z_1, z_2 \in \mathbb{C} \\ \dfrac{2}{\sqrt{1+|z_1|^2}}, & \text{if } z_1 \in \mathbb{C} \text{ and } z_2 = \infty \end{cases}.$$

Then d is a metric on $\mathbb{C}^*$.

Proof.

(a) *Positivity:* By definition, for every $z_1, z_2 \in \mathbb{C}^*$, $d(z_1, z_2) \geq 0$. Also, for $z_1, z_2 \in \mathbb{C}$, $d(z_1, z_2) = 0 \iff z_1 = z_2$. Now,

$$\begin{aligned} d(z_1, \infty) = 0 &\iff \frac{2}{\sqrt{1+|z_1|^2}} = 0 \\ &\iff |z_1| = \infty \iff z_1 = \infty. \end{aligned}$$

(b) *Symmetry:* Symmetry of d follows trivially.

(c) *Triangle inequality:* First let us prove the following inequality. For $z_1, z_2 \in \mathbb{C}$,

$$\begin{aligned} (1 + z_1 z_2)(1 + \overline{z_1 z_2}) &= 1 + z_1 z_2 + \overline{z_1 z_2} + |z_1|^2 |z_2|^2 \\ &= 1 + 2\Re(z_1 z_2) + |z_1|^2 |z_2|^2 \\ &\leq 1 + 2|z_1||z_2| + |z_1|^2 |z_2|^2 \\ &\leq 1 + |z_1|^2 + |z_2|^2 + |z_1|^2 |z_2|^2 \\ &= (1 + |z_1|^2)(1 + |z_2|^2). \end{aligned}$$

As $|1 + z_1 z_2| = |1 + \overline{z_1 z_2}|$, from here we get

$$|1 + z_1 z_2| \leq \sqrt{(1 + |z_1|^2)(1 + |z_2|^2)}. \qquad (*)$$

Now from the identity

$$(z_1 - z_2)(1 + z_3 \overline{z_3}) = (z_1 - z_3)(1 + z_2 \overline{z_3}) + (z_3 - z_2)(1 + z_1 \overline{z_3})$$

we get

$$|z_1 - z_2|(1 + |z_3|^2) \leq |z_1 - z_3||1 + z_2 \overline{z_3}| + |z_3 - z_2||1 + z_1 \overline{z_3}|$$

$$\leq |z_1 - z_3|\sqrt{(1+|z_2|^2)(1+|z_3|^2)}$$
$$+ |z_3 - z_2|\sqrt{(1+|z_1|^2)(1+|z_3|^2)} \qquad (**)$$

using $(*)$. Now dividing both sides of $(**)$ by $\sqrt{(1+|z_1|^2)(1+|z_2|^2)(1+|z_3|^2)}$, we get

$$d(z_1, z_2) \leq d(z_1, z_3) + d(z_3, z_2),$$

when all of $z_1, z_2, z_3 \in \mathbb{C}$. The other cases are left as exercise. □

Remark If the arguments in the proof of the triangle inequality in the previous example seem artificial, an alternative approach can be taken: First, demonstrate that the given metric corresponds to the chordal distance between the associated points on the Riemann sphere; second, since these points lie in $\mathbb{R}^3$ and the chordal distance is essentially the standard Euclidean distance, the triangle inequality follows naturally. The first step requires some calculations, which, although intricate, are inherently intuitive, as illustrated in Appendix B.

2.2.13. Let X be any non-empty set. Define $d\colon X \times X \to \mathbb{R}$ by

$$d(x, y) = \begin{cases} 1, & \text{if } x \neq y \\ 0, & \text{if } x = y \end{cases}.$$

Then (X, d) is a metric space. This is called the *discrete* metric.
Proof.

(a) *Positivity:* By the given definition, for any pair of points $x, y \in X$, $d(x, y) \geq 0$, and $d(x, y) = 0 \iff x = y$.
(b) *Symmetry:* For any pair of points $x, y \in X$, $d(x, y) = d(y, x)$.
(c) *Triangle inequality:* Let $x, y, z \in X$ be arbitrary. If $x = z$, then $0 = d(x, z) \leq d(x, y) + d(y, z)$ holds trivially. If $x \neq z$, then LHS $= d(x, z) = 1$. In that case, y cannot be equal to both x and z. Hence RHS $= d(x, y) + d(y, z)$ is at least 1. Hence $d(x, z) \leq d(x, y) + d(y, z)$.

□

Remark This is the first example of an abstract metric space. If X has three elements, this metric places them at the vertices of an equilateral triangle with unit side; if $|X| = 4$, then the elements are situated at the vertices of a regular tetrahedron with unit side.

2.2.14. Let (X, d) be a metric space. For $x, y \in X$, define

$$d'(x, y) = \frac{d(x, y)}{1 + d(x, y)}.$$

Then d' is another metric on X.

***Proof*.**

(a) *Positivity:* Since d is a metric, both the positivity conditions hold true for d' trivially.
(b) *Symmetry:* Also follows trivially from the symmetry property of d.
(c) *Triangle inequality:* From the triangle inequality of the metric d, we have for points $x, y, z \in X$, $d(x, z) \leq d(x, y) + d(y, z)$. Hence

$$\begin{aligned}
d'(x, y) + d'(y, z) &= \frac{d(x, y)}{1 + d(x, y)} + \frac{d(y, z)}{1 + d(y, z)} \\
&\geq \frac{d(x, y)}{1 + d(x, y) + d(y, z)} + \frac{d(y, z)}{1 + d(x, y) + d(y, z)} \\
&\qquad ([\because d(x, y), d(y, z) \geq 0]) \\
&= \frac{d(x, y) + d(y, z)}{1 + d(x, y) + d(y, z)} \\
&= \frac{1}{1 + \frac{1}{d(x,y)+d(y,z)}} \\
&\geq \frac{1}{1 + \frac{1}{d(x,z)}} \\
&\qquad ([\because d(x, z) \leq d(x, y) + d(y, z)]) \\
&= \frac{d(x, z)}{1 + d(x, z)} \\
&= d'(x, z).
\end{aligned}$$

□

2.2.15. Let (X, d_X) and (Y, d_Y) be metric spaces. Then

$$d((x_1, y_1), (x_2, y_2)) := \max\{d_X(x_1, x_2), d_Y(y_1, y_2)\}$$

is a metric on the set $X \times Y$. It is called as the *product metric* on $X \times Y$.
***Proof*.**

(a) *Positivity:* For $(x_1, y_1), (x_2, y_2) \in X \times Y$, by the positivity of the metrics d_X and d_Y, it follows trivially that $d((x_1, y_1), (x_2, y_2)) \geq 0$ and $d((x_1, y_1), (x_2, y_2)) = 0 \iff d_X(x_1, x_2) = 0$ and $d_Y(y_1, y_2) = 0$, i.e. $(x_1, y_1) = (x_2, y_2)$.
(b) *Symmetry:* Also follows trivially from the symmetry of d_X and d_Y.
(c) *Triangle inequality:* Let (x_k, y_k), $k = 1, 2, 3$ be arbitrary. WLOG, assume $d((x_1, y_1), (x_3, y_3)) = d_X(x_1, x_3)$. Then

$$\begin{aligned}
d((x_1, y_1), (x_3, y_3)) &= d_X(x_1, x_3) \\
&\leq d_X(x_1, x_2) + d_X(x_2, x_3) \\
&\leq d((x_1, y_1), (x_2, y_2)) + d((x_2, y_2), (x_3, y_3)).
\end{aligned}$$

□

Exercises: 2.2

2.2. For any two points $(x_1, y_1), (x_2, y_2)$ from $\mathbb{R}^2$, define

$$d((x_1, y_1), (x_2, y_2)) = \begin{cases} |x_1 - x_2|, \text{ if } y_1 = y_2 \\ |x_1| + |y_1 - y_2| + |x_2|, \text{ otherwise} \end{cases}.$$

Show that d is a metric on $\mathbb{R}^2$.

2.3. For positive integers m and n, define

$$d(m, n) = \left| \frac{1}{m} - \frac{1}{n} \right|.$$

Show that d is a metric on the set $\mathbb{N}$ of all positive integers.

2.4. On the set of extended real numbers, define for points $x, y \in \mathbb{R}^*$,

$$d(x, y) := |\tan^{-1} x - \tan^{-1} y|,$$

where $-\frac{\pi}{2} < \tan^{-1} x < \frac{\pi}{2}$ for $x \in \mathbb{R}$ and $\tan^{-1}(\infty) = \frac{\pi}{2}$, $\tan^{-1}(-\infty) = -\frac{\pi}{2}$. Show that d is a metric on $\mathbb{R}^*$.

2.5. Show that the map

$$d(x, y) := \left| \frac{x}{1 + \sqrt{1 + x^2}} - \frac{y}{1 + \sqrt{1 + y^2}} \right|, \qquad x, y \in \mathbb{R}$$

is a metric on $\mathbb{R}$. If $d(x, \pm\infty)$ and $d(-\infty, \infty)$ are defined via limits, will it still be a metric on the extended real line?

2.6. Let X be the set of all sequences of real or complex numbers. For two sequences $\mathbf{x} = \{x_n\}, \mathbf{y} = \{y_n\} \in X$, let

$$d(\mathbf{x}, \mathbf{y}) := \begin{cases} 0, & \text{if } \mathbf{x} = \mathbf{y} \\ \frac{1}{\min\{n : x_n \neq y_n\}}, & \text{if } \mathbf{x} \neq \mathbf{y} \end{cases}.$$

Show that d is a metric on X.

2.7. Let X be the set of all sequences of real or complex numbers. If $\sum a_n$ is a convergent series of positive reals, then for two sequences $\mathbf{x} = \{x_n\}, \mathbf{y} = \{y_n\} \in X$, let

$$d(\mathbf{x}, \mathbf{y}) := \sum_{n=1}^{\infty} \left(a_n \cdot \frac{|x_n - y_n|}{1 + |x_n - y_n|} \right).$$

Show that d is a metric on X.

2.8. Denote by $\mathcal{B}[a, b]$, the set of all bounded real-valued functions defined over the closed and bounded interval $[a, b] \subset \mathbb{R}$. For $f, g \in \mathcal{B}[a, b]$, define

$$d(f, g) := \sup\{|f(x) - g(x)| : x \in [a, b]\}.$$

Show that d is a metric on $\mathcal{B}[a, b]$.

2.9. Let $\mathcal{C}^1[a, b]$ denote the set of all continuously differentiable real-valued functions defined on the closed and bounded interval $[a, b]$. For $f, g \in \mathcal{C}^1[a, b]$, let

$$d(f, g) = |f(a) - g(a)| + \sup\{|f'(x) - g'(x)| : x \in [a, b]\}.$$

Show that d is a metric on $\mathcal{C}^1[a, b]$.

2.10. Fix $p \geq 1$. On the set $\mathcal{C}[a, b]$ of all real-valued continuous functions defined on the closed and bounded interval $[a, b]$, define the map d_p by

$$d_p(f, g) = \left[\int_a^b |f - g|^p\right]^{\frac{1}{p}}, \qquad f, g \in \mathcal{C}[a, b].$$

Show that d_p is a metric on $\mathcal{C}[a, b]$ for each value of $p \geq 1$.

2.11. Give an example of a metric on $\mathbb{R}^n$ which is not induced by a norm.

2.12. Let (X, d_X) and (Y, d_Y) be two metric spaces. Show that

$$d_p((x_1, y_1), (x_2, y_2)) := \Big[[d_X(x_1, x_2)]^p + [d_Y(y_1, y_2)]^p\Big]^{1/p}$$

is a metric on $X \times Y$ for any value of $p \geq 1$. Generalise the result for product of n spaces.

2.13. Let (X, d) be a metric space. Show that $\min\{1, d\}$ is also a metric on X.

2.14. Let (X, d) be a metric space. If $f : X \to X$ be a one-to-one map, then show that

$$d'(x, y) := d(f(x), f(y)), \qquad x, y \in X$$

is also a metric on X.

2.15. Let (X, d) be a metric space. Is $\sqrt{d}$ a metric on X? Is d^2 a metric on X?

2.16. Let d_1 and d_2 be two metrics on a non-empty set X. Show that $d_1 + d_2$ is also a metric on X. Can you provide more metrics on X using both d_1 and d_2?

2.17. Let (X, d) be a metric space. For which real value of k, will $a)$ $d + k$, $b)$ kd be metrics on X?

2.3 Pseudometric

A good number of important spaces lack a little on being a metric space. A corresponding small relaxation without compromising the basic nature of a "distance" function helps to include these organic examples into our study. These are called pseudometric spaces. Also there exists a natural way to morph them to a metric space as shown in Miscellaneous result 2.5.

Definition 2.3 Let X be a non-empty set. A map $d\colon X \times X \to \mathbb{R}$ is called a *pseudometric* if:

(a) For any two points $x, y \in X$, $d(x, y) \geq 0$ and $d(x, x) = 0$.
(b) For any two points $x, y \in X$, $d(x, y) = d(y, x)$.
(c) For any three points $x, y, z \in X$, $d(x, y) \leq d(x, z) + d(z, y)$.

Remarks

(a) A pseudometric d can be zero for distinct points x and y.
(b) Clearly, a metric is always a pseudometric, whereas the converse is not always true. Hence an example of a metric space is always an example of a pseudometric space as well. Below, we provide examples of pseudometric spaces which are *not* metric spaces.

Examples

2.3.1. The map $d(x, y) := |x^2 - y^2|$, $x, y \in \mathbb{R}$ is a pseudometric on $\mathbb{R}$.
Proof. To check d is a pseudometric is easy. To see that it is not a metric, observe that $d(1, -1) = 0$. □

2.3.2. For arbitrary points (x_1, y_1) and (x_2, y_2) from $\mathbb{R}^2$, define

$$d((x_1, y_1), (x_2, y_2)) = |x_1 - x_2|.$$

Then d is a pseudometric on $\mathbb{R}^2$ which is not a metric. The latter can be verified by observing $d((1, 2), (1, 3)) = 0$.

2.3.3. Denote by $\mathcal{R}[a, b]$, the set of all real-valued Riemann integrable functions defined on the closed and bounded interval $[a, b]$. For $f, g \in \mathcal{R}[a, b]$, define

$$d(f, g) := \int_a^b |f - g|.$$

Then d is a pseudometric on $\mathcal{R}[a, b]$.
Proof. The proof to show d is a pseudometric will go along the lines of that of Example 2.2.8, with the exception that in this case, $d(f, g) = 0$ will not necessarily imply $f = g$. Obtain a counter-example by choosing $[a, b] = [0, 1]$, and

$$f(x) = \begin{cases} 0, \text{ when } x \in (0, 1) \\ 1, \text{ when } x \in \{0, 1\} \end{cases}; \qquad g(x) = 0, \; x \in [0, 1].$$

Then $d(f, g) = 0$, but $f \neq g$. □

Exercises: 2.3

2.18. Let $\{d_n\}_{n \in \mathbb{N}}$ be a sequence of metrics on a set X. Let for every pair of points $x, y \in X$, $\{d_n(x, y)\}$ be a bounded sequence of real numbers. Define

$$d(x, y) = \overline{\lim_{n \to \infty}} d_n(x, y).$$

Show that d is a pseudometric on X. Give an example to thus produce a pseudometric which is not a metric.

2.4 Diameter of a Metric Space

If you need to place a pizza inside a polygonal box, the box must be at least as long as the pizza's diameter. The diameter of a circle (or a disc, like a pizza) is the length of its longest chord, providing an estimate of the disc's size and, consequently, the required size of the container. We appropriate this idea in a general metric space as follows:

Definition 2.4 Let (X, d) be a metric space and A be a non-empty subset of X. Consider the set

$$S := \{d(x, y) : x, y \in A\}.$$

If S is bounded above, $\sup S$ exists and is called as *diameter* of the set A, denoted by $\operatorname{diam}(A)$. If S is unbounded above, it is customary to write $\operatorname{diam}(A) = \infty$.

Remarks

(a) In some books, $\operatorname{diam}(A)$ is denoted by the notation $\delta(A)$.
(b) As a strange consequence, the diameter of the empty set is $\delta(\varnothing) = -\infty$[4]!

[4] $\operatorname{diam}(\varnothing) = \sup \varnothing =$ least upper bound of $\varnothing$. Since every real number is an upper bound of $\varnothing$, $\sup \varnothing = -\infty$, which strictly speaking is not counter-intuitive. What is counter-intuitive is this fact together with the fact that $\inf \varnothing = \infty$, which is true for the exact same reason!

Definition 2.5 Let (X, d) be a metric space and $(\varnothing \neq)A \subset X$. If diam$(A)$ exists finitely, A is called a *bounded set*. If diam(X) is finite, then X is a *bounded space*. Otherwise they are *unbounded*.

Example 2.5.1 Consider $\mathbb{R}$ with the usual metric. Then $\mathbb{R}$ is unbounded. But if we choose the metric $d(x, y) = \frac{|x-y|}{1+|x-y|}$ (as in Example 2.2.14), then $\forall x, y \in \mathbb{R}$, $d(x, y) < 1$. Hence, $\mathbb{R}$ is bounded w.r.t. the metric d. In this case, diam$(\mathbb{R}) = 1$.

Remark This example shows that a set can be both bounded or unbounded depending on the underlying metric defined on it.

Definition 2.6 Let (X, d) be a metric space and A, B be two non-empty subsets of X. Then:

- The *distance between the sets A and B* is defined as

$$d(A, B) := \inf\{d(a, b) : a \in A, b \in B\}.$$

- If $A = \{a\}$ is a singleton set, then $d(a, B) := \inf\{d(a, b) : b \in B\}$ is the *distance of the set B from the point a*.

Remark $d(A, B) = 0$ does not always imply $A \cap B \neq \varnothing$. For example, let the metric space be $\mathbb{R}$ with usual metric and consider $A = (-1, 0)$, $B = (0, 1)$.

Miscellaneous Results

2.1. In a metric space (X, d), for any $x, y, z \in X$, show that

$$|d(x, z) - d(y, z)| \leq d(x, y).$$

Proof. Let (X, d) be a metric space and $x, y, z \in X$ be arbitrary. Then by triangle inequality of d, we have

$$d(x, z) \leq d(x, y) + d(y, z) \implies d(x, z) - d(y, z) \leq d(x, y). \qquad (*)$$

Interchanging x and y in $(*)$, we get

$$d(y, z) - d(x, z) \leq d(y, x) = d(y, x). \qquad (**)$$

Hence, combining $(*)$ and $(**)$, we have

$$|d(x, z) - d(y, z)| \leq d(x, y).$$

□

2.2. In Example 2.2.4, will d_p be a metric on $\mathbb{R}^n$ if p is a real number in $(0, 1)$?
Proof. No. It will not satisfy the triangle inequality. Choose $\mathbf{x} = (-1, 0, \ldots, 0)$, $\mathbf{y} = (0, 0, \ldots, 0)$ and $\mathbf{z} = (1, 0, \ldots, 0)$. Then, $d_p(\mathbf{x}, \mathbf{z}) = 2^{1/p} > 2$, as $0 < p < 1$. But $d_p(\mathbf{x}, \mathbf{y}) = 1 = d_p(\mathbf{y}, \mathbf{z})$. □

2.3. Let (X, d) be a metric space. Denote by $\mathcal{P}(X)$, the *power set* of X, i.e. the set of all subsets of X. Define for $A, B \in \mathcal{P}(X)$,

$$D(A, B) := \begin{cases} d(A, B), \text{ if both } A, B \neq \varnothing, \\ \quad 0, \text{ if either } A \text{ or } B \text{ is } \varnothing \end{cases}.$$

Show that D is a pseudometric on $\mathcal{P}(X)$.
Proof. Remember for non-empty subsets $A, B \subset X$,

$$d(A, B) = \inf\{d(a, b) : a \in A, b \in B\} :$$

(a) *Positivity*: Follows trivially. Also for $A \in \mathcal{P}(X)$, $D(A, A) = 0$.
(b) *Symmetry*: Also follows trivially from the symmetry of the metric d.
(c) *Triangle inequality*: Let A, B and C be three arbitrary non-empty subsets of X. Then for arbitrary elements a, b, c from A, B and C, respectively, we have by the triangle inequality of d,

$$\begin{aligned} & d(a, b) \leq d(a, c) + d(c, b) \\ \Longrightarrow \quad & \inf_{a \in A} d(a, b) \leq \inf_{a \in A} d(a, c) + d(c, b) \\ \Longrightarrow \quad & d(b, A) \leq d(c, A) + d(c, b) \\ \Longrightarrow \quad & \inf_{b \in B} d(b, A) \leq d(c, A) + \inf_{b \in B} d(c, b) \\ \Longrightarrow \quad & d(B, A) \leq d(c, A) + d(c, B) \\ \Longrightarrow \quad & d(B, A) \leq d(C, A) + d(C, B), \\ & \qquad\qquad ([\text{since } c \in C \text{ is arbitrary}]) \\ \Longrightarrow \quad & D(A, B) \leq D(A, C) + D(C, B). \end{aligned}$$

Also, if either A or B is $\varnothing$, the inequality $D(A, B) \leq D(A, C) + D(C, B)$ is satisfied trivially. □

Remark To see that D is only a pseudometric on $\mathcal{P}(X)$ and not a metric, observe that if $A \subsetneq B \subseteq X$, then $D(A, B) = 0$. In particular, $D(A, X) = 0$ for every $A \in \mathcal{P}(X)$.

2.4. Denote by $\mathcal{C}(\mathbb{R})$, the set of all real-valued continuous functions defined over $\mathbb{R}$. Let n be a natural number. For $f, g \in \mathcal{C}(\mathbb{R})$, define

$$d_n(f, g) := \sup_{x \in [-n, n]} |f(x) - g(x)|.$$

Call

$$d(f,g) := \sum_{n\in\mathbb{N}} \frac{1}{2^n} \cdot \frac{d_n(f,g)}{1+d_n(f,g)}.$$

Show that d is a metric on $\mathcal{C}(\mathbb{R})$.

Proof. Along the lines of the proof in Example 2.2.9, we can easily see that for each $n \in \mathbb{N}$, d_n is a pseudometric on $\mathcal{C}(\mathbb{R})$. Hence by the arguments in the proof of Example 2.2.14, $\frac{d_n(f,g)}{1+d_n(f,g)}$ is also a pseudometric for every natural number n. Since $\forall f, g \in \mathcal{C}(\mathbb{R})$, $\frac{d_n(f,g)}{1+d_n(f,g)} < 1$, by the comparison test of a series of positive reals, we conclude that $\sum \frac{1}{2^n} \frac{d_n(f,g)}{1+d_n(f,g)}$ is convergent. Hence d is well-defined.

Now we see that d is indeed a metric:

(a) *Positivity*: Clearly for $f, g \in \mathcal{C}(\mathbb{R})$, $d(f,g) \geq 0$. Now $d(f,g) = 0$ holds iff

$$\begin{aligned}\forall n \in \mathbb{N},\ d_n(f,g) = 0 &\iff \forall n,\ f(x) = g(x),\ \text{on } [-n,n]\\ &\iff f = g \text{ on } \mathbb{R}.\end{aligned}$$

(b) *Symmetry*: Follows trivially as for each n, $d_n(f,g) = d_n(g,f)$.

(c) *Triangle inequality*: Since for each n, $\frac{d_n(f,g)}{1+d_n(f,g)}$ is a pseudometric, then for arbitrary $f, g, h \in \mathcal{C}(X)$, we have

$$\begin{aligned}&\frac{d_n(f,h)}{1+d_n(f,h)} \leq \frac{d_n(f,g)}{1+d_n(f,g)} + \frac{d_n(g,h)}{1+d_n(g,h)}\\ \Longrightarrow\ &\frac{1}{2^n}\frac{d_n(f,h)}{1+d_n(f,h)} \leq \frac{1}{2^n}\frac{d_n(f,g)}{1+d_n(f,g)} + \frac{1}{2^n}\frac{d_n(g,h)}{1+d_n(g,h)}\\ \Longrightarrow\ &\sum_{n\in\mathbb{N}}\frac{1}{2^n}\frac{d_n(f,h)}{1+d_n(f,h)} \leq \sum_{n\in\mathbb{N}}\frac{1}{2^n}\frac{d_n(f,g)}{1+d_n(f,g)} + \sum_{n\in\mathbb{N}}\frac{1}{2^n}\frac{d_n(g,h)}{1+d_n(g,h)},\\ &\text{([since all limits exist finitely])}\\ \Longrightarrow\ &d(f,h) \leq d(f,g) + d(g,h).\end{aligned}$$

□

2.5. Let d be a pseudometric on a non-empty set X. Define a relation $\sim$ on X by

$$x \sim y \text{ if and only if } d(x,y) = 0,\ x, y \in X:$$

(a) Show that $\sim$ is an equivalence relation on X.

(b) Consider the set of all equivalence classes $\widetilde{X} := X/\!\sim$. Define $\widetilde{d}$ on $\widetilde{X} \times \widetilde{X}$ by

$$\widetilde{d}([x],[y]) = d(x,y),$$

where $[x], [y] \in \widetilde{X}$ are equivalence classes containing x and y, respectively. Show that $\widetilde{d}$ is a well-defined metric on $\widetilde{X}$.

***Proof*.**

(a) Since d is a pseudometric on X, $\forall x \in X$, $d(x, x) = 0$. Hence, $\sim$ is reflexive. Again by symmetry of d, if $d(x, y) = 0$, then $d(y, x) = 0$ as well. Hence d is symmetric. Finally if for $x, y, z \in X$, $d(x, y) = 0$ and $d(y, z) = 0$, then

$$d(x, z) \leq d(x, y) + d(y, z) = 0.$$

But as d is a pseudometric, $d(x, z) \geq 0$, and therefore $d(x, z) = 0$. Hence d is transitive, and therefore, $\sim$ is an equivalence relation on X.

(b) Let $[x], [y] \in \widetilde{X} := X/\sim$. We have defined $\widetilde{d}([x], [y]) = d(x, y)$. Now let $[x] = [x']$ and $[y] = [y']$. To show $\widetilde{d}$ is well-defined is to show $d(x, y) = d(x', y')$.

Now $[x] = [x']$ and $[y] = [y']$ hold iff $x \sim x'$ and $y \sim y'$. Hence

$$d(x', y') \leq d(x', x) + d(x, y) + d(y, y') = d(x, y). \tag{$*$}$$

Interchanging $x \leftrightarrow x'$ and $y \leftrightarrow y'$, from ($*$), we get $d(x, y) \leq d(x', y')$ as well. Hence $d(x, y) = d(x', y')$, i.e. $\widetilde{d}$ is well-defined.

Next to check whether $\widetilde{d}$ is a metric on $X/\sim$. From the definition of $\widetilde{d}$, it is clear that for any $[x], [y] \in X/\sim$, $\widetilde{d}([x], [y]) \geq 0$. Also $\widetilde{d}([x], [y]) = 0$ holds true iff $d(x, y) = 0$, i.e. $x \sim y$, i.e. $[x] = [y]$. Hence, positivity condition holds for $\widetilde{d}$. Symmetry of $\widetilde{d}$ follows trivially from the symmetry of d. And finally, the triangle inequality for $\widetilde{d}$ also follows trivially from the triangle inequality of the pseudometric d. □

2.6. Let (X, d) be a metric space and $A, B \subset X$. Then

$$\operatorname{diam}(A \cup B) \leq \operatorname{diam}(A) + \operatorname{diam}(B) + d(A, B).$$

***Proof*.** Let $a, a' \in A$ and $b, b' \in B$ be arbitrary. Then

$$\begin{aligned} d(a, b) &\leq d(a, a') + d(a', b') + d(b', b) \\ &\leq \operatorname{diam}(A) + d(a', b') + \operatorname{diam}(B). \end{aligned} \tag{$*$}$$

As $d(A, B) = \inf\{d(a, b) : a \in A,\ b \in B\}$, by definition of infimum of a set, for every $\epsilon > 0$, we find $a' \in A$ and $b' \in B$ such that $d(a', b') < d(A, B) + \epsilon$. Since $\epsilon > 0$ is arbitrary, using ($*$), we get

$$d(a, b) \leq \operatorname{diam}(A) + d(A, B) + \operatorname{diam}(B). \tag{$**$}$$

Now, let $x, y \in A \cup B$ be arbitrary. If both $x, y \in A$, then $d(x, y) \leq \text{diam}(A) \leq \text{diam}(A) + \text{diam}(B) + d(A, B)$. Else if, both $x, y \in B$, similarly, $d(x, y) \leq \text{diam}(B) \leq \text{diam}(A) + \text{diam}(B) + d(A, B)$. And finally if $x \in A$ and $y \in B$, since they are arbitrary, from $(**)$, we get

$$\text{diam}(A \cup B) = \sup_{x,y \in A \cup B} d(x, y) \leq \text{diam}(A) + \text{diam}(B) + d(A, B).$$

□

Exercises: 2.4

2.19. Go through Miscellaneous result 2.5. Find the equivalence classes thus formed in Examples 2.3.1, to 2.3.3 and Miscellaneous result 2.3.

Chapter 3
Topology of Metric Spaces

In Mathematics, topology is defined as the collection of all open sets in a space where it makes sense to talk about them. But over the course of last century, the study of topology has standardised itself by including discussions about the preservation and propagation of spatial properties like compactness and connectedness through continuous functions. Following that convention, the name of this entire book should have been the "Topology of Metric Spaces". However, adhering to the literal meaning of the term, we will break from this convention and use this title only for the current chapter. In this chapter, we will explore open sets, closed sets, subspaces, sequences, and related concepts, following a similar approach as in the previous chapter—generalising these ideas, originally introduced in real analysis, to the abstract setting of metric spaces.

3.1 Open Sets

In real analysis, the ϵ-neighbourhood of a point x is defined as the set $(x-\epsilon, x+\epsilon)$. In a metric space, the analogical concept is called an "open ball".

Definition 3.1 Let (X, d) be a metric space. Let $x \in X$ and $r > 0$. The subset

$$B_d(x; r) := \{y \in X : d(x, y) < r\}$$

is called the *open ball* with centre at x and radius r with respect to the metric d.

Remarks

(a) An open ball is always non-empty as it contains the centre.
(b) The term *ball* in the name "open ball" comes from the visualisation of $B(x; r)$ in our regular three-dimensional space where it looks like a solid ball. Also it

S. Paul, *Metric Spaces*, University Texts in the Mathematical Sciences,
https://doi.org/10.1007/978-981-96-9259-0_3

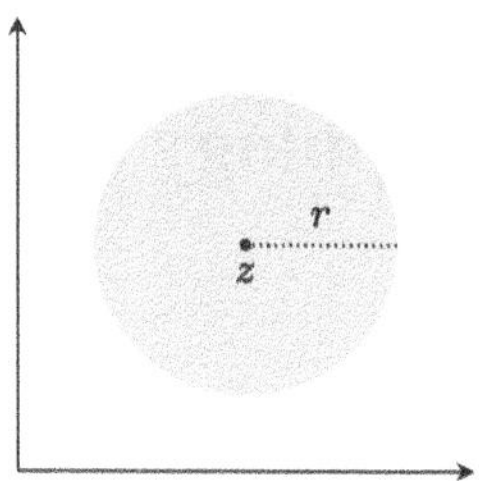

Fig. 3.1 $B(z; r)$ in $\mathbb{C}$

must be noted that some authors like to call it a *sphere*. Both "ball" and "sphere" are standard terminologies to denote the above-mentioned set.

(c) We often tend to omit the "d" in the notation $B_d(x; r)$ and simply write $B(x; r)$ when there is no confusion over the metric d.

(d) There is a reason we are using a "semicolon" in the notation $B(x; r)$ to separate the centre and the radius. The notation $(\cdot, \cdot)$ has already got too many meanings associated to it. (For example, an ordered pair, or, an open interval in $\mathbb{R}$.) Our endeavour here is to not impose another meaning, hence, reducing the chance of any potential confusion in future. Nevertheless, you will find other authors using a "comma" as per convention.

Examples

3.1.1. Let $X = \mathbb{R}$ with the usual metric. Then the open ball $B(x; r) = (x-r, x+r)$.

3.1.2. Let $X = [0, 1]$ with the usual metric. Then the open ball $B(0; 1) = [0, 1)$. This may come as a bit of shock at first. But one must realise that points from outside the space are completely irrelevant. The entire metric space is the *universal set* and does serve as our "Universe" in the context. Anything outside that simply does not exist!

3.1.3. Let $X = \mathbb{C}$ with usual metric. Then $B(z; r)$ is the open disc centred at z with radius r (Fig. 3.1).

3.1.4. Let X be a discrete space. Then for any $x \in X$, $B(x; r)$ is the singleton set $\{x\}$ for $r \leq 1$ and is the entire space X for $r > 1$.

3.1.5. Consider Example 2.2.4 and Example 2.2.5. Choose $X = \mathbb{R}^2$. Then for different values of $1 \leq p \leq \infty$, the open unit balls $B_{d_p}(\mathbf{0}; 1)$ centred at $\mathbf{0} = (0, 0)$, denoted briefly by $B_p(\mathbf{0}; 1)$, take different shapes as shown by shaded regions in Fig. 3.2.

3.1.6. Consider Example 2.2.9. The open unit ball around a function $f \in \mathcal{C}_{\text{sup}}[a, b]$ is the set of all functions situated at less than 1 distance from f. In Fig. 3.3, $g \in B(f; 1)$.

3.1.7. Consider $\mathbb{R}^2$ with the metric $d(\mathbf{x}, \mathbf{y}) := \frac{\|\mathbf{x}-\mathbf{y}\|_2}{1+\|\mathbf{x}-\mathbf{y}\|_2}$, as in Example 2.2.14, where $\mathbf{x} = (x_1, x_2)$ and $\mathbf{y} = (y_1, y_2)$. Here, denoting the origin by $\mathbf{0} = (0, 0)$, we have

$$B_d\left(\mathbf{0}; \tfrac{1}{2}\right) = \{\mathbf{x} \in \mathbb{R}^2 : \|\mathbf{x}\|_2 < 1\},$$

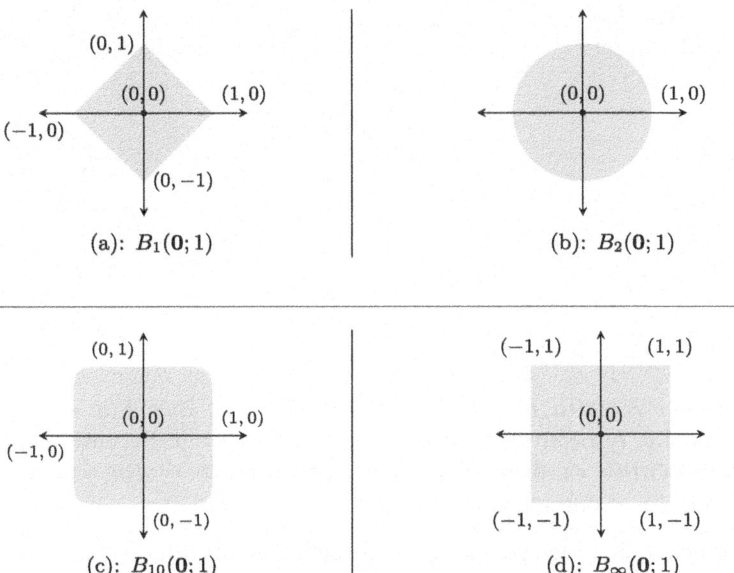

Fig. 3.2 Open unit balls in $\mathbb{R}^2$ for $p = 1, 2, 10, \infty$

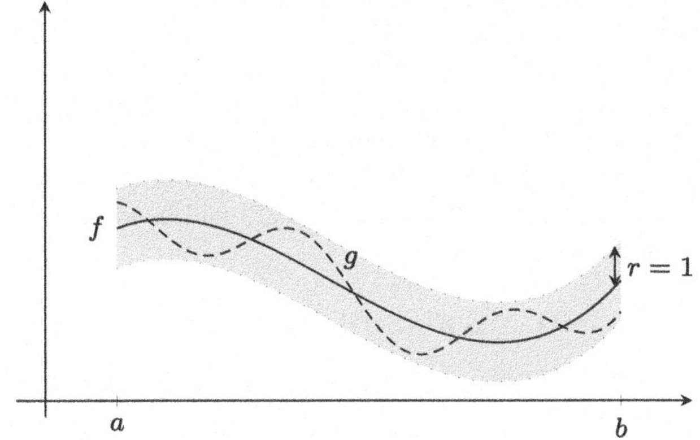

Fig. 3.3 $B(f; 1)$ in the metric space $\mathcal{C}_{\text{sup}}[a, b]$

$$B_d\left(\mathbf{0}; \tfrac{2}{3}\right) = \{\mathbf{x} \in \mathbb{R}^2 : \|\mathbf{x}\|_2 < 2\}, \text{ and}$$

$$B_d\,(\mathbf{0}; 1) = \mathbb{R}^2.$$

Following is an important property enjoyed by all metric spaces.

Proposition 3.2 (***Hausdorff Property***) *Let a and b be two distinct points in a metric space (X, d). Then there exist disjoint open balls $B(a)$ and $B(b)$ containing a and b, respectively.*

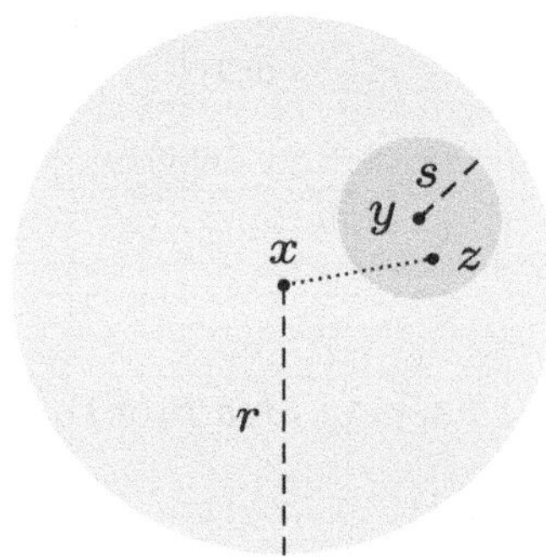

Fig. 3.4 $B(y; s) \subset B(x; r)$ for $0 < s < r - d(x, y)$

Proof Since $a \neq b$, $d(a, b) > 0$. Choose $\epsilon = \frac{1}{2}d(a, b)$. Then $B(a; \epsilon)$ and $B(b; \epsilon)$ contain a and b, respectively, and $B(a; \epsilon) \cap B(b; \epsilon) = \varnothing$. Because otherwise, if $y \in B(a; \epsilon) \cap B(b; \epsilon)$, then by the choice of ϵ, $d(a, b) \leq d(a, y) + d(y, b) < \epsilon + \epsilon = d(a, b)$, a contradiction. ■

As you probably have guessed already, we define an open set in a metric space in the following way.

Definition 3.3 A subset G of a metric space X is said to be *open* if for every point $x \in G$, we can find $r > 0$ such that the ball $B(x; r) \subset G$.

Examples

3.3.1. In a metric space (X, d), the open ball $B(x; r)$ is an open set.
Proof. Let $y \in B(x; r)$ be arbitrary. We want to find $s > 0$ such that $B(y; s) \subset B(x; r)$ (Fig. 3.4). To find s, let us work backwards. Choose $z \in B(y; s)$ arbitrarily. Then $B(y; s) \subset B(x; r)$ will be implied if $d(x, z) < r$. But by triangle inequality,

$$d(x, z) \leq d(x, y) + d(y, z) < d(x, y) + s.$$

Hence $d(x, z) < r$ will be satisfied if we choose s so that $d(x, y) + s < r$, i.e. $0 < s < r - d(x, y)$. Hence for this choice of s, $z \in B(x; r)$. Since $z \in B(y; s)$ is arbitrary, $B(y; s) \subset B(x; r)$. Again as $y \in B(x; r)$ is arbitrary, $B(x; r)$ is open. □

3.3.2. In a metric space (X, d), both of $\varnothing$ and X are open sets.
Proof. To show $\varnothing$ is open, we need to show every point in $\varnothing$ is the centre of an open ball contained in $\varnothing$. But there is no point in $\varnothing$. Hence, the condition gets satisfied vacuously.

X is open trivially, because every open ball around every point in X is again there in X. [Remember X is the universal set as discussed in Example 3.1.2.] □

3.3.3. In a discrete metric space X, every subset of X is open.

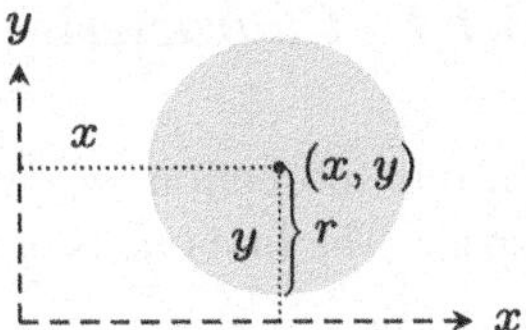

Fig. 3.5 $B((x, y); r) \subset S$ for $0 < r < \min\{x, y\}$

Proof. Let $A \subset X$. If A is empty, it is open. If not, then $\forall x \in A$, $B\left(x; \frac{1}{2}\right) = \{x\} \subset A$. Hence A is open. □

3.3.4. In $\mathbb{R}^2$ with usual metric, the set

$$S := \{(x, y) \in \mathbb{R}^2 : x > 0, y > 0\}$$

is open.

Proof. Let $(x, y) \in S$ be arbitrary. Then as shown in Fig. 3.2b, an open ball around (x, y) is a circular disc centred at (x, y). Now, to have the open ball at (x, y) entirely inside S, choose $0 < r < \min\{x, y\}$ (Fig. 3.5). Then $B((x, y); r) \subset S$. □

3.3.5. Consider $\mathcal{C}_{\sup}[a, b]$. Let

$$A := \{f \in \mathcal{C}[a, b] : f(a) \neq 0\}.$$

Then A is open in $\mathcal{C}_{\sup}[a, b]$.

Proof. Choose $f \in A$ arbitrarily. Call $\mu := |f(a)| > 0$. Consider the open ball $B\left(f; \frac{\mu}{2}\right)$. If $g \in B\left(f; \frac{\mu}{2}\right)$, then $\sup\{|g(x) - f(x)| : x \in [a, b]\} < \frac{\mu}{2}$. Hence $|g(a) - f(a)| < \frac{\mu}{2}$, i.e. $f(a) - \frac{\mu}{2} < g(a) < f(a) + \frac{\mu}{2}$. But this implies

$$g(a) \text{ is } \begin{cases} > f(a) - \frac{\mu}{2} > 0, \text{ if } f(a) > 0 \\ < f(a) + \frac{\mu}{2} < 0, \text{ if } f(a) < 0 \end{cases}.$$

Hence in both cases, $g(a) \neq 0$, i.e. $g \in A$. Since $g \in B\left(f; \frac{\mu}{2}\right)$ is arbitrary, $B\left(f; \frac{\mu}{2}\right) \subset A$. Hence A is open. □

3.3.6. In a metric space (X, d), let $A \subset X$ be finite. Then $X \setminus A$ is an open subset of X.

Proof. If $A = \varnothing$ or X, we are done by Example 3.3.2. Otherwise, let $A := \{a_1, \ldots, a_m\}$ and choose $x \in X \setminus A$ arbitrarily. For $k = 1, \ldots, m$, call $\delta_k := d(x, a_k)$. Choose $\delta = \min\{\delta_k : k = 1, \ldots, m\} > 0$. Then $B(x; \delta) \cap A = \varnothing$, and hence $B(x; \delta) \subset X \setminus A$. Since $x \in X \setminus A$ is arbitrary, $X \setminus A$ is open. □

3.1.1 Characterisations of Open Sets

In this subsection, we will discuss about the structure of an open set in a metric space. The first one is in terms of *interior* of a set. We start with the following:

Definition 3.4 Let A be a subset of a metric space X. A point $x \in A$ is called an *interior point* of A if there is an open ball around x contained in A, i.e. there exists some $r > 0$ such that $B(x; r) \subset A$.

The set of all interior points of A is called the *interior* of A and is denoted by A° or $\operatorname{int} A$.

Remark Clearly, for every subset $A \subset X$, $\operatorname{int} A \subset A$.

The following characterisation is immediate:

Theorem 3.5 *In a metric space (X, d), a set $G \subset X$ is open iff every point $x \in G$ is an interior point of G.*

Examples

3.5.1. In the metric space of real numbers with usual metric, $\operatorname{int}(0, 1) = (0, 1)$ and $\operatorname{int}[1, 2] = (1, 2)$. Therefore, $(0, 1)$ is an open set in $\mathbb{R}$, whereas $[1, 2]$ is not.

3.5.2. Again in $\mathbb{R}$ with usual metric, $\operatorname{int} \mathbb{Q} = \varnothing$. Because for every $q \in \mathbb{Q}$ and for every $\delta > 0$ (however small), the open ball $B(q; \delta)$ contains irrational numbers. Hence no point is an interior point of $\mathbb{Q}$.

3.5.3. In the metric space $\mathcal{C}_{\sup}[a, b]$, for an arbitrarily fixed real number α, consider the set

$$A_\alpha = \{f \in \mathcal{C}[a, b] : \inf f > \alpha\}.$$

Then of course $A_\alpha \neq \varnothing$, as the constant function $f(x) = \alpha + 1$ is there in A_α. Now, for $f \in A_\alpha$ chosen arbitrarily, consider a real number r such that $0 < r < \inf f - \alpha$. Choose a function $g \in B(f; r)$ arbitrarily. Then $d_\infty(f, g) = \sup|f - g| < r$ implies that for each point

$$x \in [a, b], \quad f(x) - r < g(x) < f(x) + r. \tag{$*$}$$

(See Fig. 3.3.) Since both f and g are continuous functions over a closed and bounded interval $[a, b]$, both $\inf f$ and $\inf g$ exist finitely. Moreover by $(*)$, $\inf g \geq \inf f - r > \alpha$. Hence $g \in A_\alpha$. Since $g \in B(f; r)$ is arbitrary, $B(f; r) \subset A_\alpha$. Since $f \in A_\alpha$ is arbitrary, $\operatorname{int} A_\alpha = A_\alpha$. Therefore, for each α arbitrary, the set A_α is open.

Proposition 3.6 *For a subset A of a metric space (X, d), $\operatorname{int} A$ is the largest open set contained in A.*

Proof We first prove that $\operatorname{int} A$ is an open set. If $\operatorname{int} A = \varnothing$, we are done. Otherwise, choose $x \in \operatorname{int} A$ arbitrarily. Then for some $r > 0$, $B(x; r) \subset A$. From the proof of

Example 3.3.1, we conclude that for each point $y \in B(x; r)$, there is some $s > 0$ such that $B(y; s) \subset B(x; r) \subset A$. Hence y is also an interior point of A. Since $y \in B(x; r)$ is arbitrary, $B(x; r) \subset \operatorname{int} A$. Hence $\operatorname{int} A$ is an open set.

To see $\operatorname{int} A$ is the largest open set, let $G \subset A$ be open. Then by Theorem 3.5, every point of G is an interior point of G, hence an interior point of A as well. Therefore, $G = \operatorname{int} G \subset \operatorname{int} A$, i.e. $\operatorname{int} A$ is the largest open set contained in A. ■

The second characterisation is in terms of open balls.

Theorem 3.7 *Let (X, d) be a metric space. A set $G \subset X$ is open in X if and only if G is the union of a collection of open balls.*

Proof Let $G \subset X$ be open. If $G = \varnothing$, then it is the union of an empty collection of open balls. If not, choose $x \in G$ arbitrarily. As G is open, for some $\delta_x > 0$ we have $B(x; \delta_x) \subset G$. Then clearly, $G = \bigcup_{x \in G} B(x; \delta_x)$.

Conversely, let $G \subset X$ be the union of a collection $\mathcal{G}$ of open balls. If $\mathcal{G} = \varnothing$, G is also $\varnothing$ and we are done. Otherwise, choose $x \in G$ arbitrarily. Then $x \in B(a; \delta)$ for some open ball $B(a; \delta) \in \mathcal{G}$. Then by Example 3.3.1, we must have some $\delta_x > 0$ such that $B(x; \delta_x) \subset B(a; \delta) \subset G$. Hence x is an interior point of G. Since $x \in G$ is arbitrary, by Theorem 3.5, G is open. ■

Structure of Open Sets in $\mathbb{R}$ with Usual Metric

In the set of all real numbers with the usual metric, we can say a bit more than the general setup in Theorem 3.7. We start by recalling a couple of facts.

Definition 3.8 A non-empty subset $I \subset \mathbb{R}$ is called an *interval* if for any two points $x < y$ from I and for every real number z with $x < z < y$, we have $z \in I$.

Proposition 3.9 *Let $I \subset \mathbb{R}$ be an interval. Then I is one of the following forms:*

- $(-\infty, \infty) := \{x \in \mathbb{R} : -\infty < x < \infty\}$;
- $(-\infty, b) := \{x \in \mathbb{R} : -\infty < x < b\}$;
- $(-\infty, b] := \{x \in \mathbb{R} : -\infty < x \leq b\}$;
- $(a, \infty) := \{x \in \mathbb{R} : a < x < \infty\}$;
- $[a, \infty) := \{x \in \mathbb{R} : a \leq x < \infty\}$;
- $(a, b) := \{x \in \mathbb{R} : a < x < b\}$;
- $(a, b] := \{x \in \mathbb{R} : a < x \leq b\}$;
- $[a, b) := \{x \in \mathbb{R} : a \leq x < b\}$;
- $[a, b] := \{x \in \mathbb{R} : a \leq x \leq b\}$.

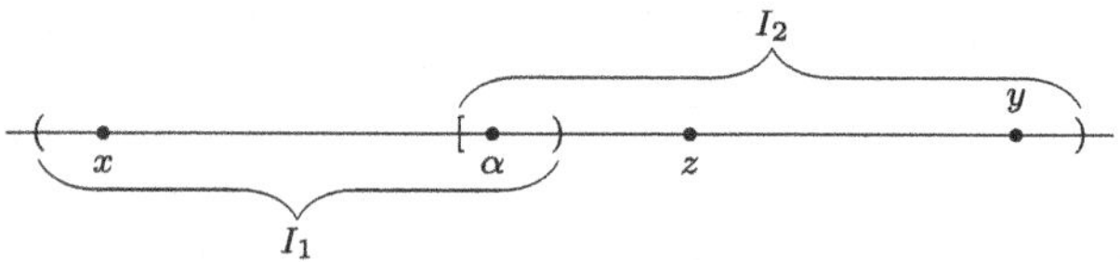

Fig. 3.6 Union of two intersecting intervals is an interval

We now have the following structure theorem for the open sets in $\mathbb{R}$ (with usual metric).

Theorem 3.10 *Let $G \subset \mathbb{R}$ be open. Then G is a disjoint union of countably many open intervals in $\mathbb{R}$.*

The following lemma will reduce our efforts proving the above theorem.

Lemma 3.11 *If I_1 and I_2 are two intersecting intervals in $\mathbb{R}$, then $I = I_1 \cup I_2$ is also an interval.*

Proof Choose arbitrary points $x < y$ from I. If both x and y come from the same interval I_1 or I_2, then any real number z with $x < z < y$ will also belong to the same interval I_1 or I_2, respectively. So WLOG, let us assume $x \in I_1$, $y \in I_2$, and z be any real number between x and y. We need to show $z \in I$ (Fig. 3.6).

Let $\alpha \in I_1 \cap I_2$. If $x < z \leq \alpha$, then $z \in I_1 \subset I$; again, if $\alpha < z < y$, then $z \in I_2 \subset I$. Hence I is an interval. ■

Now let us see:

Proof of Theorem 3.10 If $G = \varnothing$, it is a union of an empty collection of open intervals. So assume $G \neq \varnothing$. Define a relation $\sim$ on G by

$$x \sim y \iff x \text{ and } y \text{ are in the same open interval in } G.$$

We will show that $\sim$ is an equivalence relation.

As G is open, for every point $x \in G$, there is some $\delta_x > 0$ such that $(x - \delta_x, x + \delta_x) \subset G$. Hence for each point $x \in G$, $x \sim x$. Therefore, $\sim$ is reflexive. Next, it is trivially true that $\sim$ is symmetric. To see that $\sim$ is transitive as well, let for $x, y, z \in G$, $x \sim y$ and $y \sim z$. Let $x, y \in I_1$ and $y, z \in I_2$, where I_1 and I_2 are two open intervals in G. Then $y \in I_1 \cap I_2$, and hence by the last lemma, $I = I_1 \cup I_2$ is an interval, containing all of x, y, and z. Therefore $x \sim z$, i.e. $\sim$ is transitive. Consequently, $\sim$ is an equivalence relation.

The $\sim$-equivalence classes force a partition on G. We will show that:

i) Each equivalence class is an open interval.
ii) There are countably many equivalence classes.

Let J be an $\sim$-equivalence class. To show J is open, choose $x \in J$ arbitrarily. Then, as $x \in G$ and G is open, for some $\delta > 0$, $(x - \delta, x + \delta) \subset G$ as well. But this implies that each point in $(x - \delta, x + \delta)$ is in the same open interval as with x, i.e. each point in $(x-\delta, x+\delta)$ is $\sim$-related to x. Hence $(x-\delta, x+\delta) \subset J$. Since $x \in J$ is arbitrary, J is open.

Since J is open, J contains at least two distinct elements. To see J is an interval, let $x, y \in J$ and $z \in \mathbb{R}$ be such that $x < z < y$. Now $x, y \in J$ implies $x \sim y$, i.e. there is an open interval J_1 such that $x, y \in J_1$. But then $x < z < y$ implies that $z \in J_1$ as well, i.e. $x \sim z$ (and $y \sim z$ as well). Therefore $z \in J$, and this shows J is an interval. Consequently, G is the union of a collection of open intervals.

Lastly, we will show that this collection of open intervals is countable. Let $\mathcal{G} = \{J_\lambda : \lambda \in \Lambda\}$ be the collection of distinct equivalence classes (open intervals) thus formed. To see $\mathcal{G}$ is countable, choose a rational number r_λ from each J_λ (thanks to density property of rational numbers!). Now, for $\alpha, \beta \in \Lambda, \alpha \neq \beta$ implies $J_\alpha \cap J_\beta = \emptyset$. Hence $r_\alpha \neq r_\beta$, i.e. the map from $\Lambda \to \mathbb{Q}$ given by $\lambda \mapsto r_\lambda$ is one to one. Since $\mathbb{Q}$ is countable, Λ is also so and so is $\mathcal{G}$. ■

Next we will see the properties of a collection of open sets in a metric space.

Proposition 3.12 *In a metric space, union of an arbitrary collection of open sets is open.*

Proof Let (X, d) be a metric space and $\mathcal{G} = \{G_\lambda : \lambda \in \Lambda\}$ be a collection of open sets in X. We want to show $G = \bigcup_{\lambda \in \Lambda} G_\lambda$ is also open.

Assuming G is non-empty, choose $x \in G$ arbitrarily. Then $x \in G_\alpha$ for some open set $G_\alpha \in \mathcal{G}$. Since G_α is open, for some $\delta > 0$ we have $B(x; \delta) \subset G_\alpha$. Then $B(x; \delta) \subset \bigcup_{\lambda \in \Lambda} G_\lambda = G$, and as $x \in G$ is arbitrary, G is open. ■

Proposition 3.13 *In a metric space, intersection of a finite collection of open sets is open.*

Proof Let (X, d) be a metric space and $\mathcal{G} := \{G_1, \ldots, G_m\}$ be a finite collection of open sets in X. We want to show $G = \bigcap_{k=1}^{m} G_k$ is also open.

If $G = \emptyset$, we are done. Otherwise, choose $x \in G$ arbitrarily. Then $x \in G_k$ for every $k \in \{1, \ldots, m\}$. Since each G_k is open for some $\delta_k > 0$, $B(x; \delta_k) \subset G_k$. Choose $\delta = \min\{\delta_k : k = 1, \ldots, m\} > 0$. Then for each k, $B(x; \delta) \subset B(x; \delta_k) \subset G_k$. Hence $B(x; \delta) \subset \bigcap_{k=1}^{m} G_k = G$. Since $x \in G$ is arbitrary, G is open in X. ■

Remark Intersection of an infinite collection of open sets is not open in general. Consider the following counter-example: Let $G_n = \left(-\frac{1}{n}, \frac{1}{n}\right)$ and consider $\{G_n : n \in \mathbb{N}\}$. Then all G_n's are open, but $\bigcap_{n \in \mathbb{N}} G_n = \{0\}$ is not open.

3.1.2 Topology and Equivalent Metrics

The facts about a collection of open sets have been explored in Propositions 3.13 and 3.12 and in Example 3.3.2. We gather them together in the following:

Theorem 3.14 *Let (X, d) be a metric space and $\mathcal{T}$ denote the collection of all open sets in X. Then $\mathcal{T}$ has the following properties:*

i) $\varnothing, X \in \mathcal{T}$.
ii) Arbitrary union of members from $\mathcal{T}$ *lies in* $\mathcal{T}$.
iii) Finite intersection of members from $\mathcal{T}$ *lies in* $\mathcal{T}$.

This theorem is one of the most instructive theorems in this chapter. It tells us how does a collection of open sets behave in a metric space. In our quest of abstractifying the study of analysis, this very fact will become the stepping stone for further elevation. We will start with an abstract collection $\mathcal{T}$ of subsets of a non-empty set X and call those subsets *open* if the collection satisfies the criteria laid in the above theorem. The collection will then be called a *topology* on X.

Owing to its importance, let us record the definition of topology formally:

Definition 3.15 Let X be a non-empty set and $\mathcal{T}$ be a collection of subsets of X. $\mathcal{T}$ is called a *topology* on X if:

(a) $\varnothing, X \in \mathcal{T}$.
(b) Arbitrary union of members from $\mathcal{T}$ again lies in $\mathcal{T}$.
(c) Finite intersection of members from $\mathcal{T}$ again lies in $\mathcal{T}$.

The pair $(X, \mathcal{T})$ is called a *topological space*.

Every metric space is a topological space, but the reverse is not always true. Defining a metric on a non-empty set X determines a topology on X because the metric defines open balls, which in turn generate the open sets (see Theorem 3.7). Consequently, it is natural to expect that different metrics on the same set may give rise to different collections of open sets. For example, in the discrete metric on $\mathbb{R}$, every singleton subset is open, whereas this is not the case in the usual metric. This leads us to the following:

Definition 3.16 Let (X, d) be a metric space. The collection $\mathcal{T}$ of all open sets w.r.t. the metric d in X is called the *topology generated by the metric* $\boldsymbol{d}$.

But we stumble upon another problem with this definition. Consider the following:

Example 3.16.1 Consider the metric $d(m, n) = \left|\frac{1}{m} - \frac{1}{n}\right|$ on the set $\mathbb{N}$ of all positive integers, as seen in Exercise 2.3. For a positive integer n, fixed arbitrarily, consider the open ball $B(n; \epsilon)$ for $\epsilon = \frac{1}{2n^2}$. Then

$$\begin{aligned} B\left(n; \frac{1}{2n^2}\right) &= \left\{m \in \mathbb{N} : \left|\frac{1}{m} - \frac{1}{n}\right| < \frac{1}{2n^2}\right\} \\ &= \left\{m \in \mathbb{N} : \frac{1}{n} - \frac{1}{2n^2} < \frac{1}{m} < \frac{1}{n} + \frac{1}{2n^2}\right\} \\ &= \left\{m \in \mathbb{N} : \frac{2n-1}{2n^2} < \frac{1}{m} < \frac{2n+1}{2n^2}\right\} \\ &= \left\{m \in \mathbb{N} : \frac{2n^2}{2n+1} < m < \frac{2n^2}{2n-1}\right\} \end{aligned}$$

$$= \left\{ m \in \mathbb{N} : n - \frac{n}{2n+1} < m < n + \frac{n}{2n-1} \right\}$$
$$= \{m \in \mathbb{N} : n - 1 < m < n + 1\}$$
$$= \{n\}.$$

Therefore, for this choice of ϵ, $B(n; \epsilon) = \{n\}$, i.e. each singleton set $\{n\}$ is open in $(\mathbb{N}, d)$. Clubbing this fact with Proposition 3.12 and Example 3.3.2, we can conclude that the topology generated by d on $\mathbb{N}$ comprises all possible subsets of $\mathbb{N}$.

On the other hand, by Example 3.3.3, the discrete metric also generates the same topology on $\mathbb{N}$.

Essentially, d and the discrete metric are two different metrics on the set $\mathbb{N}$ generating the same topology. This example establishes the need of the following:

Definition 3.17 Two metrics d and d' on a non-empty set X are said to be *equivalent metrics* if the topologies generated by d and d' are the same. This is equivalent to say that $G \subset X$ is open with respect to the metric d iff it is open with respect to the metric d'.

Using Theorem 3.7, we can say that two metrics on a given set X will be equivalent if and only if every open ball in one metric is also open in the other metric and vice versa. This fact is often recorded in the following more usable form:

Theorem 3.18 *Let X be a non-empty set and d, d' are two metrics on X. Then d and d' are equivalent if and only if*

for each point $x \in X$ and for any $r > 0$, $\exists r' > 0$ such that $B_{d'}(x; r') \subset B_d(x; r)$;

and

for each point $x \in X$ and for any $s' > 0$, $\exists s > 0$ such that $B_d(x; s) \subset B_{d'}(x; s')$.

Examples

3.18.1. Let (X, d) be a metric space. Refer Example 2.2.14. The metric $d' = \frac{d}{1+d}$ is equivalent to d.

***Proof*.** Let $x \in X$ and $r > 0$ be arbitrary. Then for $y \in X$,

$$d(x, y) < r \iff 1 + d(x, y) < 1 + r$$
$$\iff \frac{1}{1 + d(x, y)} > \frac{1}{1 + r}$$
$$\iff 1 - \frac{1}{1 + d(x, y)} < 1 - \frac{1}{1 + r}$$

$$\iff \frac{d(x, y)}{1 + d(x, y)} < \frac{r}{1 + r}$$

$$\iff d'(x, y) < \frac{r}{1 + r}.$$

For chosen $r > 0$ arbitrarily, choose $r' = \frac{r}{1+r}$. Then

$$B_d(x; r) = B_{d'}(x; r').$$

Again, for any chosen $0 < s' < 1$, choose $s = \frac{s'}{1-s'}$. Then

$$B_{d'}(x; s') = B_d(x; s).$$

For $s' \geq 1$, any $s > 0$ will produce

$$B_d(x; s) \subset B_{d'}(x; s') = X.$$

Hence by Theorem 3.18, d and d' are equivalent metrics on X. □

3.18.2. Let (X, d) be a metric space. Refer Exercise 2.13. The metric $d'' = \min\{1, d\}$ is equivalent to d.

Proof. Fix $x \in X$ arbitrarily. Then for arbitrary $r > 0$, choose $0 < r'' < \min\{1, r\}$. Then

$$B_{d''}(x; r'') = B_d(x; r'') \subset B_d(x; r).$$

Again, for arbitrary $s'' > 0$, choose $0 < s < \min\{1, s''\}$. Then

$$B_d(x; s) = B_{d''}(x; s) \subset B_{d''}(x; s'').$$

Hence by Theorem 3.18, d and d'' are equivalent metrics on X. □

Remark The above two examples show that every metric is equivalent to a bounded metric.

The following is a sufficient condition for equivalence of two metrics:

Theorem 3.19 *Let X be a non-empty set. Two metrics d and d' on X are equivalent if there are constants $K_1, K_2 > 0$ such that for any two points $x, y \in X$,*

$$K_1 d'(x, y) \leq d(x, y) \leq K_2 d'(x, y).$$

Proof We first observe that for $x, y \in X$ and $r, r' > 0$ arbitrary,

$$d(x, y) < r \implies d'(x, y) < \frac{r}{K_1} \quad \text{and} \quad d'(x, y) < r' \implies d(x, y) < K_2 r'.$$

Hence $B_d(x;r) \subset B_{d'}(x; \frac{r}{K_1})$ and $B_{d'}(x;r') \subset B_d(x; K_2r')$. Therefore by Theorem 3.18, d and d' are equivalent metrics. ■

Remark The fact that the above theorem is only sufficient but not a necessary condition is established by the following counter-example: From Example 3.18.1, we know that the metrics d and $d' = \frac{d}{1+d}$ are equivalent. Choose $X = \mathbb{R}$ and d to be the usual metric on $\mathbb{R}$. Now if we assume that there are constants $K_1, K_2 > 0$ such that for every pair of points $x, y \in \mathbb{R}$, we have

$$K_1 d'(x, y) \le d(x, y) \le K_2 d'(x, y),$$

then choosing points $x = 2K_2$ and $y = 0$, we get $d(2K_2, 0) = |2K_2 - 0| = 2K_2$ and $d'(2K_2, 0) = \frac{2K_2}{1+2K_2} < 1$. Hence $d(2K_2, 0) \le K_2 d'(2K_2, 0) < K_2 \cdot 1$, i.e. $2K_2 < K_2 \implies 2 < 1$, a contradiction.

Examples

3.19.1. Consider Examples 2.2.4 and 2.2.5. For $1 \le p \le \infty$, all metrics d_p are equivalent to one another.

Proof. Let $\{a_k : k = 1, \ldots, n\}$ be a set of non-negative reals. In Hölder's inequality, for each $k = 1, \ldots, n$, choose $b_k = 1$. Using the norm notations mentioned in the remarks on page 65, we have for $p > 1$,

$$\begin{aligned}\sum_{k=1}^{n} a_k &\le \left(\sum_{k=1}^{n} a_k^p\right)^{1/p} \times n^{1/q} \qquad \text{[this quantity is equal to } \|\mathbf{a}\|_p \times n^{\frac{1}{q}}\text{]}\\ &\le \left(\sum_{k=1}^{n} [\max_k\{a_k\}]^p\right)^{1/p} \times n^{1/q}\\ &= n^{1/p} \times \max_k\{a_k\} \times n^{1/q}\\ &= n\|\mathbf{a}\|_\infty.\end{aligned}$$

Therefore,

$$\begin{aligned} n^{\frac{1}{q}} \|\mathbf{a}\|_p &\le n\|\mathbf{a}\|_\infty \\ \text{or,} \quad \|\mathbf{a}\|_p &\le n^{\frac{1}{p}} \|\mathbf{a}\|_\infty. \end{aligned} \tag{$*$}$$

Clearly, $(*)$ is true for $p = 1$ as well. Again, for any $p \ge 1$,

$$\|\mathbf{a}\|_p = \left(\sum_{k=1}^{n} a_k^p\right)^{1/p} \ge \left[\left(\max_k\{a_k\}\right)^p\right]^{\frac{1}{p}} = \max_k\{a_k\} = \|\mathbf{a}\|_\infty. \tag{$**$}$$

Therefore, from (∗) and (∗∗), we can conclude that for points $\mathbf{x} = (x_1, \ldots, x_n), \mathbf{y} = (y_1, \ldots, y_n) \in \mathbb{R}^n$, and for $p \in [1, \infty)$,

$$d_\infty(\mathbf{x}, \mathbf{y}) \le d_p(\mathbf{x}, \mathbf{y}) \le n^{\frac{1}{p}} d_\infty(\mathbf{x}, \mathbf{y}).$$

Therefore, by Theorem 3.19, every d_p metric is equivalent to d_∞. □

Remark So in essence, we can say that all d_p metrics are equivalent to the standard d_2 metric on $\mathbb{R}^n$, which we interchangeably denote by the $\|\cdot\|$ notation.

3.19.2. In $\mathcal{C}[a, b]$, the metrics d_1 and d_∞ (as defined in Examples 2.2.8 and 2.2.9) are not equivalent.

Proof. Denote the constant function $f(x) = 0$ on $[a, b]$ by $\mathbf{0}$. We will prove that there cannot be a $\delta > 0$ so that the open ball $B_1(\mathbf{0}; \delta) \subset B_\infty(\mathbf{0}; 1)$. Now

$$B_1(\mathbf{0}; \delta) = \left\{ f \in \mathcal{C}[a, b] : \int_a^b |f| < \delta \right\}$$

and

$$B_\infty(\mathbf{0}; 1) = \left\{ f \in \mathcal{C}[a, b] : \sup_{x \in [a,b]} |f(x)| < 1 \right\}.$$

For brevity and geometric intuition, we will prove the fact for $\mathcal{C}[0, 1]$. $\mathcal{C}[0, 1]$ serves as an exact prototype of $\mathcal{C}[a, b]$ in general, and the arguments for the general case can be easily constructed from here.

Choose $\delta > 0$ arbitrarily. Our goal is to find a continuous function on $[0, 1]$ with area smaller than δ, but with height (absolute maximum value) more than or equal to 1. WLOG, assume $0 < \delta < 1$. Obtain the continuous function on $[0, 1]$ which erects an isosceles triangle on $[0, \delta]$ of height 1, and is 0 outside $[0, \delta]$. Then this function will have area $\frac{\delta}{2}$ under its curve but will remain outside of $B_\infty(\mathbf{0}; 1)$ ball (Fig. 3.7).

Let us now proceed formally. For $\delta' > 0$ chosen arbitrarily, fix $0 < \delta < \min\{\delta', 1\}$. Define $g\colon [0, 1] \to \mathbb{R}$ by

$$g(x) = \begin{cases} \dfrac{x}{\delta/2}, & \text{if } 0 \le x < \frac{\delta}{2} \\ 2 - \dfrac{x}{\delta/2}, & \text{if } \frac{\delta}{2} \le x \le \delta \\ 0, & \text{if } \delta < x \le 1 \end{cases}.$$

Then g is continuous, $g\left(\frac{\delta}{2}\right) = 1$, and

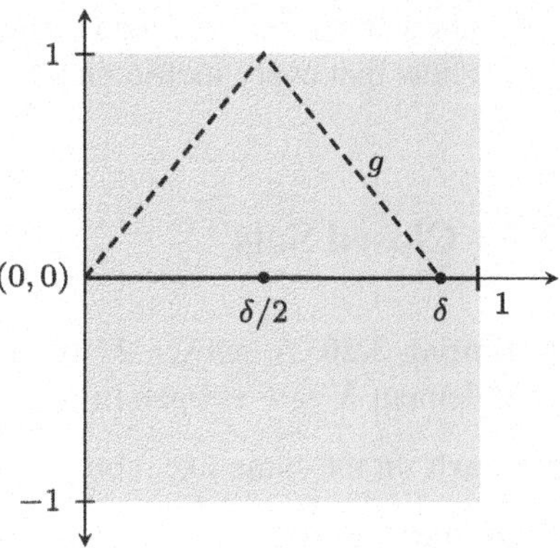

Fig. 3.7
$B_1(\mathbf{0}; \delta) \not\subset B_\infty(\mathbf{0}; 1)$

$$\int_0^1 |g| = \int_0^{\frac{\delta}{2}} \frac{x}{\delta/2}\, dx + \int_{\frac{\delta}{2}}^{\delta} \left(2 - \frac{x}{\delta/2}\right) dx = \frac{\delta}{2} < \delta < \delta'.$$

Hence g belongs to $B_1(\mathbf{0}; \delta')$ but does not belong to $B_\infty(\mathbf{0}; 1)$, i.e. $B_1(\mathbf{0}; \delta') \not\subset B_\infty(\mathbf{0}; 1)$. Since $\delta' > 0$ is arbitrary, by Theorem 3.18, d_1 and d_∞ on $\mathcal{C}[0, 1]$ are not equivalent. □

Exercises: 3.1

3.1. If in a metric space, $B(x_1; r_1) = B(x_2; r_2)$, can you always conclude $x_1 = x_2$ and $r_1 = r_2$?

3.2. If A and B are subsets of a metric space, show that $\operatorname{int}(A \cap B) = \operatorname{int} A \cap \operatorname{int} B$.

3.3. Let $\{(a_k, b_k) : k = 1, \ldots, n\}$ be a collection of open intervals in $\mathbb{R}$. Show that the product $\prod_k (a_k, b_k)$ is open in $\mathbb{R}^n$. Use this result to show that if $\{U_k : k = 1, \ldots, n\}$ is a collection of open sets in $\mathbb{R}$, then $\prod_k U_k$ is open in $\mathbb{R}^n$.

3.4. Let C denote the subset of the metric space ℓ_∞ (Example 2.2.7) given by

$$C = \left\{\mathbf{x} = \{x_k\}_k \in \ell_\infty : |x_k| < \frac{1}{k}\right\}.$$

Check whether C is open in ℓ_∞.

3.5. Consider the metrics as described in Exercise 2.12. Show that each metric is equivalent to the product metric (Example 2.2.15) on $X \times Y$.

3.6. Refer to Examples 2.2.2 and 2.2.12. On the set of all complex numbers $\mathbb{C}$, for $z_1, z_2 \in \mathbb{C}$, call

$$d(z_1, z_2) = |z_1 - z_2| \text{ and } \rho(z_1, z_2) = \frac{2|z_1 - z_2|}{\sqrt{1 + |z_1|^2}\sqrt{1 + |z_2|^2}}.$$

Show that d and ρ are equivalent metrics on $\mathbb{C}$.

3.7. Show that every metric on a finite set is equivalent to the discrete metric.

3.2 Closed Sets

Definition 3.20 A subset F of a metric space X is said to be *closed* if its complement $X \setminus F$ is open in X.

Remark In the same line, complement of a closed set is open.

Examples

3.20.1. In a metric space X, as both $\varnothing$ and X are open sets, their complements, X and $\varnothing$, respectively, are closed sets.

3.20.2. Let (X, d) be a metric space. Let $x \in X$ and $r > 0$. The subset

$$B_d[x; r] := \{y \in X : d(x, y) \leq r\}$$

is called the *closed ball* with centre at x and radius r with respect to the metric d. In a metric space, a closed ball is always closed.

Proof. Let $x \in X$ and $r > 0$ be arbitrary. If $B[x; r] = X$, then it is closed by the last example. If not, choose $y \in X \setminus B[x; r]$. Then $d(x, y) > r$. Choose δ such that $0 < \delta < d(x, y) - r$. Now consider the open ball $B(y; \delta)$. If $z \in B(y; \delta)$, then by triangle inequality of the metric d,

$$d(x, z) \geq d(x, y) - d(y, z) > d(x, y) - \delta = r.$$

Hence $z \notin B[x; r]$, i.e. $z \in X \setminus B[x; r]$ (Fig. 3.8). Therefore as $z \in B(y; \delta)$ is arbitrary, $B(y; \delta) \subset X \setminus B[x; r]$. Hence y is an interior point of $X \setminus B[x; r]$. Since y is arbitrary, $X \setminus B[x; r]$ is open, and hence $B[x; r]$ is closed. □

3.20.3. In a discrete metric space X, every subset of X is closed.

Proof. This follows trivially as in a discrete metric space, every subset of X is open as well (see Example 3.3.3). □

3.20.4. Consider $\mathcal{C}_{\text{sup}}[a, b]$. Let $B := \{f \in \mathcal{C}[a, b] : f(a) = 0\}$. Then B is closed in $\mathcal{C}_{\text{sup}}[a, b]$.

Proof. Follows trivially from Example 3.3.5. □

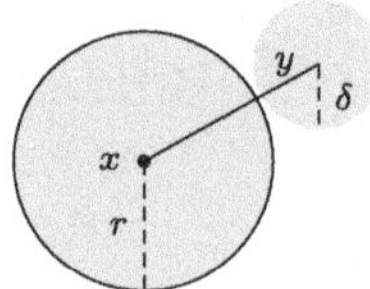

Fig. 3.8 $B[x; r]$ is closed

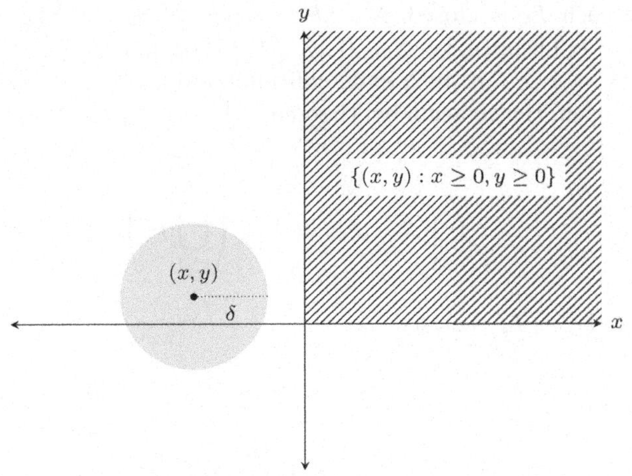

Fig. 3.9 $\{(x, y) \in \mathbb{R}^2 : x \geq 0, y \geq 0\}$ is closed

3.20.5. In a metric space, a finite set is always closed.
Proof. Follows trivially from Example 3.3.6. □

3.20.6. In $\mathbb{R}^2$ with usual metric, the set $S := \{(x, y) \in \mathbb{R}^2 : x \geq 0, y \geq 0\}$ is closed.
Proof. Let $(x, y) \in \mathbb{R}^2 \setminus S$ be arbitrary. Then (x, y) is in any one of the second, third or fourth quadrant in $\mathbb{R}^2$ and $x \neq 0$, $y \neq 0$ (see Fig. 3.9). Choose δ such that $0 < \delta < \min\{|x|, |y|\}$. Then it is an easy task to check that $B((x, y); \delta) \subset \mathbb{R}^2 \setminus S$. Hence $\mathbb{R}^2 \setminus S$ is open, i.e. S is closed. □

The results regarding a collection of closed sets are recorded in the following theorem. They follow from their counterpart results for open sets in Theorem 3.14 and De Morgan's laws.

Theorem 3.21 *Let (X, d) be a metric space. Let $\mathfrak{F}$ denote the set of all closed sets in X. Then $\mathfrak{F}$ has the following properties:*

i) $\varnothing, X \in \mathfrak{F}$.
ii) Arbitrary intersection of members of $\mathfrak{F}$ lies in $\mathfrak{F}$.
iii) Union of any finite number of members of $\mathfrak{F}$ lies in $\mathfrak{F}$.

Proof

i) Follows from Example 3.20.1.
ii) Let $\{F_\lambda : \lambda \in \Lambda\}$ be an arbitrary collection of closed sets. Then

$$X \setminus \left(\bigcap_{\lambda \in \Lambda} F_\lambda \right) = \bigcup_{\lambda \in \Lambda} (X \setminus F_\lambda).$$

But since each F_λ is closed, $X \setminus F_\lambda$ is open. Hence by Theorem 3.14(*ii*), $\bigcup_{\lambda \in \Lambda}(X \setminus F_\lambda)$ is open, i.e. $X \setminus \left(\bigcap_{\lambda \in \Lambda} F_\lambda\right)$ is open, i.e. $\bigcap_{\lambda \in \Lambda} F_\lambda$ is closed.

iii) Let $\{F_k : k = 1, \ldots, n\}$ be a finite collection of closed sets. Then for each k, $X \setminus F_k$ is open. Hence again by Theorem 3.14(*iii*), $\bigcap_k (X \setminus F_k)$ is open. But then

$$\bigcap_{k=1}^{n}(X \setminus F_k) = X \setminus \left(\bigcup_{k=1}^{n} F_k\right)$$

is open, i.e. $\bigcup_{k=1}^{n} F_k$ is closed.

■

Remarks

(a) The fact that the union of only a finite collection of closed sets may not be closed in general is shown by the following counter-example: Let $F_n = \left[\frac{1}{n}, 1 - \frac{1}{n}\right]$, $n \geq 3$. Then each F_n is closed but $\bigcup_{n \geq 3} F_n = (0, 1)$ which is not closed.

(b) Since closed sets are complements of open sets and vice versa, the collection of all closed sets in a space remains same under equivalent metrics.

3.2.1 Characterisations of Closed Sets

In this subsection, we will explore various criteria for a subset to be closed. For that, let us develop some terminologies first.

Definition 3.22 A *neighbourhood* $N(x)$ of a point x in a metric space is a set containing an open set U_x that contains x, i.e. $x \in U_x \subset N(x)$.

Example 3.22.1 For every positive $\delta > 0$, both $B(x; \delta)$ and $B[x; \delta]$ are neighbourhoods of x.

Remark Essentially in a metric space, we often bother ourself only with the symmetric open neighbourhood $B(x; \delta)$ around x. We call it the ***δ**-neighbourhood* of x. In this context, the set

$$\hat{B}(x; \delta) := B(x; \delta) \setminus \{x\}$$

is called the *deleted **δ**-neighbourhood* of x.

Definition 3.23 Let A be a subset of a metric space X. A point $x \in X$ is said to be a *limit point* of A if every neighbourhood of x contains a point from A other than x. The set of all limit points of A is called the *derived set* of A and is denoted by A'.

It is a good exercise to show that the above definition is equivalent to the following form.

Let A be a subset of a metric space X. A point $x \in X$ is said to be a *limit point* of A if for every $\epsilon > 0$, the deleted ϵ-neighbourhood of x contains a point from A, i.e. $\hat{B}(x; \epsilon) \cap A \neq \varnothing$.

In analysis, the symbol ϵ is almost always accompanied by the silent tag: *however small*. This case is no exception. A point x is a limit point of A if every neighbourhood of x, *however small*, contains a point of A other than x itself. As we consider progressively smaller positive values of ϵ, the point x increasingly appears to be an accumulation point of A, i.e. a cluster of points from A lies infinitesimally close to x.

A useful analogy is that of a bar magnet dropped into a jar of iron sand: The sand particles cling to the poles of the magnet, making the poles the limit points of the set of all iron sand particles. With this in mind, it is easy to visualise the following:

Proposition 3.24 *Let A be a subset of a metric space X. A point $x \in X$ is a limit point of A if and only if every neighbourhood of x (or equivalently, every ϵ-neighbourhood) contains infinitely many points from A.*

Proof of this fact is left as an exercise.

Remark As an immediate consequence of the above proposition, a finite set in a metric space cannot have a limit point.

Example 3.24.1 In the metric space $\mathbb{R}^n$, the derived set of $\mathbb{Q}^n$ is the entire $\mathbb{R}^n$.

Proof Let $\mathbf{x} = (x_1, \ldots, x_n) \in \mathbb{R}^n$ and $\epsilon > 0$ be arbitrary. Then by the density property of rational numbers, for every k^{th} coordinate, we will find rational numbers q_k, such that $0 < |x_k - q_k| < \frac{\epsilon}{\sqrt{n}}$. Call the point $\mathbf{q} = (q_1, \ldots, q_n) \in \mathbb{Q}^n$. Then

$$0 < \|\mathbf{x} - \mathbf{q}\| = \sqrt{\sum_{k=1}^{n} |x_k - q_k|^2} < \sqrt{n \cdot \frac{\epsilon^2}{n}} = \epsilon.$$

Therefore for every point $\mathbf{x} \in \mathbb{R}^n$, $\mathbf{x}$ is a limit point of $\mathbb{Q}^n$. Consequently, the derived set of $\mathbb{Q}^n$ is entire $\mathbb{R}^n$. □

A limit point may or may not belong to the set in general. The first characterisation of closed sets comes in terms of its limit points.

Theorem 3.25 *In a metric space, a set $A \subset X$ is closed iff A contains all its limit points, i.e. $A' \subset A$.*

Proof Let $A \subset X$ be closed in X. Then $X \setminus A$ is open. If $X \setminus A = \varnothing$, then $A = X$, and hence $A' \subset A$. Otherwise, choose $x \in X \setminus A$. Since $X \setminus A$ is open, $\exists \delta > 0$ such that $B(x; \delta) \subset X \setminus A$. Therefore, $B(x; \delta) \cap A = \varnothing$, i.e. x is not a limit point of A. So $x \notin A$ implies $x \notin A'$. Contrapositively, $x \in A'$ implies $x \in A$. Hence $A' \subset A$.

Conversely, let $A \subset X$ contain all its limit points. We need to show that A is closed, i.e. to show $X \setminus A$ is open. If $X \setminus A = \varnothing$, we are done. Otherwise, choose

$x \in X \setminus A$ arbitrarily. Now as A contains all of its limit points and $x \in X \setminus A$, x is not a limit point of A. Therefore, $\exists\delta > 0$ such that $\hat{B}(x; \delta) \cap A = \varnothing$, i.e. $B(x; \delta) \cap A = \varnothing$ (since $x \notin A$), i.e. $B(x; \delta) \subset X \setminus A$. But this shows that x is an interior point of $X \setminus A$. Since x is arbitrary, $X \setminus A$ is open. ■

The second characterisation of closed sets is in terms of its adherent points.

Definition 3.26 Let A be a subset of a metric space X. A point $x \in X$ is called an *adherent point* of A if every neighbourhood of x contains a point from A.

The collection of all adherent points of A is called the *closure* of A and is denoted by $\bar{A}$.

The above definition is equivalent to saying that:

> Let A be a subset of a metric space X. A point $x \in X$ is said to be an *adherent point* of A if for every $\epsilon > 0$, the ϵ-neighbourhood of x contains a point from A, i.e. $B(x; \epsilon) \cap A \neq \varnothing$.

Example 3.26.1 In the metric space $\mathbb{R}$, the set $A = \left\{\frac{1}{n} : n \in \mathbb{N}\right\}$ has only one limit point 0, but every point of A is an adherent point of A along with 0.

It is easy to see that an adherent point of a set is either a point from the set, or a limit point, or both. This simple fact is recorded in the following proposition with the proof left for the reader.

Proposition 3.27 *In a metric space (X, d), for a subset $A \subset X$, $\bar{A} = A \cup A'$.*

Theorem 3.28 *In a metric space (X, d), a set A is closed iff $\bar{A} = A$.*

Proof By Proposition 3.27, for any subset $A \subset X$, $A \subseteq \bar{A}$. If A is closed, i.e. $A' \subset A$, then

$$\bar{A} = A \cup A' \subseteq A \subseteq \bar{A}.$$

Hence $\bar{A} = A$.

Conversely, if $\bar{A} = A$, then since $\bar{A} = A \cup A'$, we have $A' \subset A$, i.e. A is closed. ■

Another justification of the above fact is given in the following:

Proposition 3.29 *For any subset A of a metric space X, $\bar{A}$ is the smallest closed set containing A.*

Proof To show $\bar{A}$ is closed, we show $X \setminus \bar{A}$ is open. If $X \setminus \bar{A} = \varnothing$, we are done. Otherwise, choose $x \in X \setminus \bar{A}$. Then $\exists\delta > 0$ such that $B(x; \delta) \cap A = \varnothing$. Hence $B(x; \delta) \subset X \setminus A$. Now, choose $y \in B(x; \delta)$ arbitrarily. Then by Example 3.3.1, $B(x; \delta)$ is also a neighbourhood of y, and since $B(x; \delta) \cap A = \varnothing$, y is not a limit point of A. As $y \in B(x; \delta)$ is arbitrary, $B(x; \delta) \cap A' = \varnothing$, i.e. $B(x; \delta) \subset X \setminus A'$. Consequently, $B(x; \delta) \subset X \setminus \bar{A}$, and hence $X \setminus \bar{A}$ is open, i.e. $\bar{A}$ is closed.

To show $\bar{A}$ is the smallest one, let $F \subset X$ be a closed set containing A. Then by Exercise 3.8(a), $A \subset F$ implies $A' \subset F'$. But F is closed implies $F' \subset F$. Hence

$$\bar{A} = A \cup A' \subset F \cup F' = \bar{F} = F.$$

■

Remarks

(a) In the light of the above proposition, Theorem 3.28 trivially reduces to

$$A \text{ is closed} \iff A \text{ is the smallest closed set containing itself.}$$

(b) A different way to devise the above framework is also there in existence. There, we *define* $\bar{A}$ to be the smallest closed set containing A as in the last proposition. Then as in the above remark, Theorem 3.28, which characterises closed sets in terms of its closure, follows trivially. Only thing that remains to show is $\bar{A} = A \cup A'$, which is left for the reader in Exercise 3.11.

Intuitively, an open set allows each of its points some "cushion"—a neighbourhood within which it can "stretch its arms" freely. In contrast, a closed set fully encloses its points, even if it means sacrificing these surrounding cushions for some of them. If a point lies infinitesimally close to the set, a closed set absorbs it, forming a complete whole.

A helpful analogy is to consider two fruits: one with its skin intact and the other peeled. The peeled fruit represents an open set, while the unpeeled fruit corresponds to a closed set. If we place them side by side and treat their union as the universal set, it becomes natural to see one as the complement of the other.

As consumers of fruit, we are concerned with two aspects: maximising the edible portion and minimising the space required for storage. In both cases, the fruit's skin plays a crucial role. When eating, we aim to peel the skin as thinly as possible, whereas for safekeeping at the minimum level, we want to add nothing extra to the skin. This analogy aligns well with the two previously introduced definitions and also naturally leads to a new one:

- The *interior* of a set (the largest open subset contained within it) corresponds to the consumable portion of the fruit.
- The *closure* of a set (the smallest closed set containing it) is the entire unpeeled fruit.
- The *boundary* of the set is like the fruit's skin—it consists of all points that separate the inside from the outside.

Thus, removing the skin yields an open set, while keeping it intact results in a closed set. This perspective offers another way to characterise open and closed sets. We now formalise this intuition starting with the following:

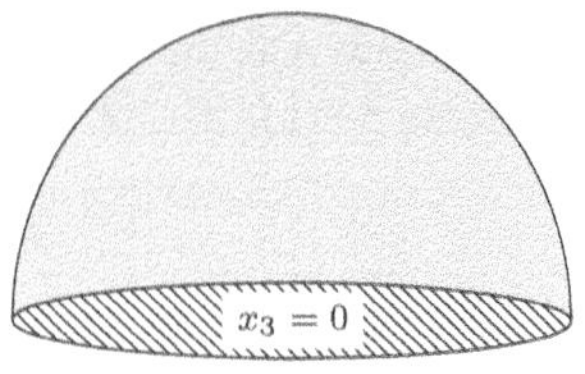

Fig. 3.10 Upper hemisphere in $\mathbb{R}^3$

Definition 3.30 Let A be a subset of a metric space X. The *boundary* of A (denoted by ∂A) is defined by the set

$$\partial A = \bar{A} \setminus A^\circ.$$

Examples

3.30.1. In $\mathbb{R}^3$, let $S = \{\mathbf{x} = (x_1, x_2, x_3) \in \mathbb{R}^3 : \|\mathbf{x}\| \leq 1, x_3 > 0\}$ be the upper hemisphere of the closed unit ball. Then $\bar{S} = \{\mathbf{x} \in \mathbb{R}^3 : \|\mathbf{x}\| \leq 1, x_3 \geq 0\}$ and $S^\circ = \{\mathbf{x} \in \mathbb{R}^3 : \|\mathbf{x}\| < 1, x_3 > 0\}$. Hence $\partial S = \{\mathbf{x} \in \mathbb{R}^3 : \|\mathbf{x}\| = 1, x_3 \geq 0\}$ (Fig. 3.10).

3.30.2. In $\mathbb{R}$, let $A = \left\{\frac{1}{n} : n \in \mathbb{N}\right\}$. Then $\bar{A} = A \cup \{0\}$ and $A^\circ = \varnothing$, because for every $n \in \mathbb{N}$ and $\epsilon > 0$, the neighbourhood $B\left(\frac{1}{n}; \epsilon\right)$ contains points outside of A. Then $\partial A = \bar{A}$.

3.30.3. In $\mathbb{R}$, let $A = [0, 1] \cap \mathbb{Q}$ be the set of all rational points from $[0, 1]$. Then by density property of $\mathbb{Q}$, $\bar{A} = [0, 1]$, and by Example 3.5.2, $A^\circ = \varnothing$. Hence $\partial A = [0, 1]$.

Theorem 3.31 *A subset A of a metric space X is open iff $A \cap \partial A = \varnothing$.*

Proof Let $A \subset X$ be open. Then $A^\circ = A$. Hence

$$A \cap \partial A = A \cap (\bar{A} \setminus A^\circ) = A \cap (\bar{A} \setminus A) = \varnothing.$$

Conversely, let $A \cap \partial A = \varnothing$. Now $\partial A = \bar{A} \setminus A^\circ = \bar{A} \cap (A^\circ)^c$ and $A \subset \bar{A}$. Therefore,

$$\varnothing = A \cap \partial A = A \cap (\bar{A} \cap (A^\circ)^c) = A \cap (A^\circ)^c = A \setminus A^\circ.$$

Hence $A \subset A^\circ$, i.e. $A^\circ = A$, whence A is open. ■

Theorem 3.32 *In a metric space (X, d), a set $A \subset X$ is closed iff $\partial A \subset A$.*

Proof Let $A \subset X$ be closed. Then $\bar{A} = A$. Hence

$$\partial A = \bar{A} \setminus A^\circ = A \setminus A^\circ \subset A.$$

Conversely, let $\partial A \subset A$. Then $\partial A \setminus A = \varnothing$. Hence

$$\begin{aligned}\varnothing = \partial A \setminus A &= (\bar{A} \setminus A^\circ) \setminus A \\ &= (\bar{A} \cap (A^\circ)^c) \cap A^c \\ &= \bar{A} \cap (A^\circ \cup A)^c \\ &= \bar{A} \cap A^c = \bar{A} \setminus A.\end{aligned}$$

But $\bar{A} \setminus A = \varnothing$ implies that $\bar{A} \subset A$. Hence $\bar{A} = A$, i.e. A is closed. ∎

Another characterisation of closed sets is listed in Corollary 3.41.1.

Miscellaneous Results

3.1. Let (X, d_X) and (Y, d_Y) be two metric spaces and A, B be two open (or closed) sets in X and Y, respectively. Show that $A \times B$ is open (or closed) in $X \times Y$ w.r.t. the product metric d.

Proof. **Case 1.** *A and B are open*: If either of A or B is empty, then $A \times B$ is also empty, and we are done. Otherwise, choose $(a, b) \in A \times B$ arbitrarily. Since both A and B are open in X and Y, respectively, $\exists \delta_1, \delta_2 > 0$ such that $B_X(a; \delta_1) \subset A$ and $B_Y(b; \delta_2) \subset B$. Choose $\delta := \min\{\delta_1, \delta_2\} > 0$. Then by definition of product metric,

$$\begin{aligned}B_{X \times Y}((a, b); \delta) &= \{(x, y) \in X \times Y : d((x, y), (a, b)) < \delta\} \\ &= \{(x, y) \in X \times Y : \max\{d_X(x, a), d_Y(y, b)\} < \delta\} \\ &= \{(x, y) \in X \times Y : d_X(x, a) < \delta \text{ and } d_Y(y, b) < \delta\} \\ &\subset \{(x, y) \in X \times Y : d_X(x, a) < \delta_1 \text{ and } d_Y(y, b) < \delta_2\} \\ &= \{(x, y) \in X \times Y : x \in B_X(a; \delta_1) \text{ and } y \in B_Y(b; \delta_2)\} \\ &= B_X(a; \delta_1) \times B_Y(b; \delta_2) \\ &\subset A \times B.\end{aligned}$$

Hence (a, b) is an interior point of $A \times B$. Since (a, b) is arbitrary, $A \times B$ is open.

Case 2. *A and B are closed*: If $A \times B$ does not have a limit point, then we are done. Otherwise, let (a, b) be a limit point of $A \times B$. Then for every $\epsilon > 0$, $\hat{B}_{X \times Y}((a, b); \epsilon) \cap (A \times B) \neq \varnothing$. But as shown in the last case,

$$\hat{B}_{X \times Y}((a, b); \epsilon) = \hat{B}_X(a; \epsilon) \times \hat{B}_Y(b; \epsilon).$$

Hence

$$\begin{aligned}
&\forall \epsilon > 0,\ \hat{B}_{X\times Y}((a,b);\epsilon) \cap (A \times B) \neq \varnothing \\
\Longrightarrow\ &\forall \epsilon > 0,\ [\hat{B}_X(a;\epsilon) \times \hat{B}_Y(b;\epsilon)] \cap (A \times B) \neq \varnothing \\
\Longrightarrow\ &\forall \epsilon > 0,\ [\hat{B}_X(a;\epsilon) \cap A] \times [\hat{B}_Y(b;\epsilon) \cap B] \neq \varnothing \\
\Longrightarrow\ &\forall \epsilon > 0,\ \hat{B}_X(a;\epsilon) \cap A \neq \varnothing \text{ and } \hat{B}_Y(b;\epsilon) \cap B \neq \varnothing \\
\Longrightarrow\ &a \in A' \text{ and } b \in B' \\
\Longrightarrow\ &a \in A \text{ and } b \in B. \qquad \text{[as both } A \text{ and } B \text{ are closed sets]}
\end{aligned}$$

Therefore, $(a, b) \in A \times B$. Consequently, $A \times B$ is closed. □

Remark This result can be extended to a product of any finite number of spaces.

3.2. *A counter-intuitive result:* In a metric space, it is not always true that $\overline{B(x; r)} = B[x; r]$.

***Proof*.** In a discrete metric space (X, d) with more than one element, for $x \in X$, $B(x; 1) = \{x\} = \overline{B(x; 1)}$, but $B[x; 1] = X$. □

Remark A discrete metric space never stops to amaze us with counter-intuitive results. It is always a good practice to check with a discrete metric space before coining a proposition for metric spaces in general.

3.3. Let A be a non-empty subset of a metric space (X, d) and $x \in X$. Recall that the distance $d(x, A)$ of x from A defined as

$$d(x, A) = \inf\{d(x, a) : a \in A\}.$$

Show that $d(x, A) = 0$ if and only if $x \in \bar{A}$.

***Proof*.** Let $x \in \bar{A}$. Choose $\epsilon > 0$ arbitrarily small. Then by definition of $\bar{A}$, $\exists a \in A$ such that $d(x, a) < \epsilon$. Hence $0 \leq \inf\{d(x, a) : a \in A\} < \epsilon$, i.e. $0 \leq d(x, A) < \epsilon$. As $\epsilon > 0$ is arbitrary, $d(x, A) = 0$.

Conversely, let $d(x, A) := \inf\{d(x, a) : a \in A\} = 0$. Choose $\epsilon > 0$ arbitrarily. By definition of infimum, $\exists a \in A$ such that $0 \leq d(x, a) < \epsilon$. Therefore $x \in \bar{A}$.

Hence $\bar{A} = \{x \in X : d(x, A) = 0\}$. □

3.4. Let A and B be two non-empty disjoint closed sets in a metric space (X, d). Then there exist disjoint open sets U and V such that $A \subset U$ and $B \subset V$.

***Proof*.** For any point $a \in A$, the distance of a from B is positive. Because, by the above result, $d(a, B) = 0$ implies $a \in \bar{B} = B$, as B is closed. But this is a contradiction as $A \cap B = \varnothing$. Similarly, for every $b \in B$, $d(b, A) > 0$.

For $a \in A$, let $r_a := \frac{d(a,B)}{2} > 0$ and consider $U := \bigcup_{a \in A} B(a; r_a)$. Since each $B(a; r_a)$ is open and arbitrary union of open sets is open, U is open and $A \subset U$.

Similarly, for $b \in B$, let $r_b := \frac{d(b,A)}{2} > 0$ and consider $V := \bigcup_{b \in B} B(b; r_b)$. For the same reason as above, V is open and $B \subset V$.

We will now show $U \cap V = \emptyset$. Assume the contrary. Let $x \in U \cap V$. Then we can find $a_0 \in A$ and $b_0 \in B$ such that $d(x, a_0) < r_{a_0}$ and $d(x, b_0) < r_{b_0}$. WLOG, assume $r_{a_0} \leq r_{b_0}$. Then,

$$\begin{aligned} 2r_{b_0} &= d(b_0, A) \\ &\leq d(b_0, a_0) \qquad\qquad [\text{as } d(b_0, A) = \inf\{d(b_0, a) : a \in A\}] \\ &\leq d(b_0, x) + d(x, a_0) \\ &< r_{b_0} + r_{a_0} \leq 2r_{b_0}. \end{aligned}$$

But as $r_{b_0} > 0$, this implies $2 < 2$, a contradiction. Hence $U \cap V = \emptyset$. □

Remark A more concise proof of this fact is listed in Corollary 5.11.1.

3.5. *Bolzano-Weierstrass theorem in* $\mathbb{R}$: Every infinite bounded subset of $\mathbb{R}$ has a limit point in $\mathbb{R}$.

***Proof*.** Let S be an infinite bounded subset of $\mathbb{R}$. Then for some $M > 0$, $S \subset [-M, M]$. Construct a set H in the following way:

$$H := \{x \in \mathbb{R} : x \text{ is greater than infinitely many elements of } S\}.$$

Then clearly $M \in H$ and $-M \notin H$. Also, if $m < -M$, then $m \notin H$, showing that H is a non-empty bounded below subset of $\mathbb{R}$. Hence by completeness axiom of $\mathbb{R}$, $\inf H$ exists. Call $\xi = \inf H$. We will show ξ is a limit point of S (Fig. 3.11).

Since $\xi = \inf H$, for any arbitrarily chosen $\epsilon > 0$, $\exists h \in H$ such that $\xi \leq h < \xi + \epsilon$. But $h \in H$ implies that h is greater than infinitely many elements of S. Hence $\xi + \epsilon$ is also greater than infinitely many elements of S. Again, as $\xi = \inf H$, $\xi - \epsilon \notin H$. Hence $\xi - \epsilon$ is not greater than infinitely many elements of S, i.e. $\xi - \epsilon$ is greater than at most finitely many elements of S. Hence $B(\xi; \epsilon)$ contains infinitely many elements from S. Since $\epsilon > 0$ is arbitrary, ξ is a limit point of S. □

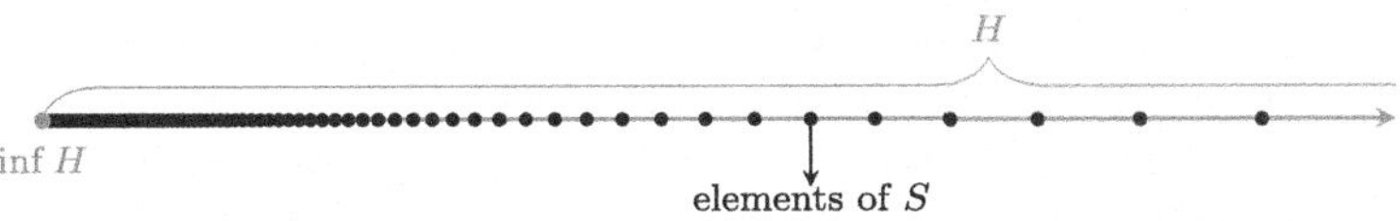

Fig. 3.11 Construction of H

Exercises: 3.2

3.8. Prove the following facts about derived sets:

(a) If $A \subset B$, then $A' \subset B'$.
(b) $(A')' \subset A'$, i.e. A' is a closed set.
(c) For two subsets $A, B \subset X$, $(A \cup B)' = A' \cup B'$.
(d) For two subsets $A, B \subset X$, $(A \cap B)' \subset A' \cap B'$. Provide a counter-example to show that the equality does not hold in general.

3.9. Prove the following facts about closures:

(a) If $A \subset B$, then $\bar{A} \subset \bar{B}$.
(b) For two subsets $A, B \subset X$, $\overline{A \cup B} = \bar{A} \cup \bar{B}$.
(c) For two subsets $A, B \subset X$, $\overline{A \cap B} \subset \bar{A} \cap \bar{B}$. Provide a counter-example to show that the equality does not hold in general.
(d) For a subset A, $\text{diam}(A) = \text{diam}(\bar{A})$.

3.10. (a) Show that the straight line $ax + by + c = 0$ is a closed subset of $\mathbb{R}^2$ with usual metric.
(b) Let $\mathbf{a} := (a_1, \ldots, a_n) \in \mathbb{R}^n \setminus \mathbf{0}$ be fixed. Show that the hyperplane

$$\mathbf{a} \cdot \mathbf{x} = \text{ constant}$$

is a closed subset of $\mathbb{R}^n$ with usual metric.

3.11. Assuming the definition of $\bar{A}$ as the smallest closed set containing A, show that $\bar{A} = A \cup A'$.

3.12. Show that if A is a non-empty closed and bounded subset of $\mathbb{R}$, then both $\sup A$ and $\inf A$ exist in A.

3.13. Let B denote a ball $B(a; r)$ or $B[a; r]$ (either open or closed) in a metric space X. Show that a point $x \in \partial B$ iff $d(x, a) = r$.

3.3 Subspaces of a Metric Space

In the algebraic systems like groups, vector spaces, etc., we have seen the concept of a *sub*-system. A similar concept is also there for metric spaces. However, unlike the algebraic counterparts, the definition of a metric subspace does not involve any condition on the concerned subset, rather it is defined via a very natural restriction on the metric. Still, it has some significant consequences on the topology of the subspace.

Definition 3.33 Let (X, d) be a metric space and Y be a non-empty subset of X. Then (Y, d_Y) is called a *subspace* of X where d_Y is the metric d restricted to Y defined as $d_Y = d\big|_{Y \times Y}$. d_Y is called the metric *induced* by d on Y.

The restriction of the usual metric to a subset of $\mathbb{R}$ results into some uncomfortable and counter-intuitive examples of both open and closed sets. Consider the following cases for instance:

- Refer to Example 3.1.2 $[0, 1]$ is acting as a subspace of $\mathbb{R}$ there. As shown in that example, $[0, 1)$ is the open unit ball in $[0, 1]$ and hence is open. But definitely it is not open in the entire space $\mathbb{R}$. The singleton set $\{1\}$ is the complement of the open set $[0, 1)$ in $[0, 1]$ and hence is closed. $\{1\}$ is also closed in $\mathbb{R}$ as well.
- If you consider another subspace $(0, 1)$ of $\mathbb{R}$, the set $\left(0, \frac{1}{2}\right]$ is closed in $(0, 1)$, which is not closed in $\mathbb{R}$.
- The singleton set $\{5\}$ is open in the subspace $\mathbb{Z}$ of $\mathbb{R}$, as the open ball $B\left(5; \frac{1}{2}\right) = \{5\}$ is in $\mathbb{Z}$.

The following description of the open (hence, closed) sets in a subspace of a metric space helps us to deal with these anomalies.

Theorem 3.34 *Let X be a metric space and Y be a subspace of X. Then $A \subset Y$ is open in Y iff there is $G \subset X$, open in X, such that $A = G \cap Y$.*

Proof Let for $a \in A$, $B_Y(a; \epsilon)$ denote the open ϵ-ball around a in Y, which is defined as

$$B_Y(a; \epsilon) = \{y \in Y : d(y, a) < \epsilon\}.$$

We start our proof by observing the fact that

$$B_Y(a; \epsilon) = \{x \in X : d(x, a) < \epsilon\} \cap Y = B(a; \epsilon) \cap Y, \tag{$*$}$$

where $B(a; \epsilon)$ is the corresponding open ϵ-ball around a in X.
Now, let $A \subset Y$ be open in Y. If $A = \varnothing$, then $A = \varnothing \cap Y$, and $\varnothing$ is open in X. Next, assume $A \neq \varnothing$. Choose $a \in A$ arbitrarily. Since A is open in Y, $\exists \epsilon_a > 0$ such that $B_Y(a; \epsilon_a) \subset A$. Since $a \in A$ is arbitrary, we have

$$\begin{aligned}
A &= \bigcup_{a \in A} B_Y(a; \epsilon_a) \\
&= \bigcup_{a \in A} [B(a; \epsilon_a) \cap Y] && \text{[by } (*)\text{]} \\
&= \left(\bigcup_{a \in A} B(a; \epsilon_a) \right) \cap Y \\
&= G \cap Y,
\end{aligned}$$

where $G := \bigcup_{a \in A} B(a; \epsilon_a)$. But G is open in X as all $B(a; \epsilon_a)$'s are open in X, and arbitrary union of open sets is open.

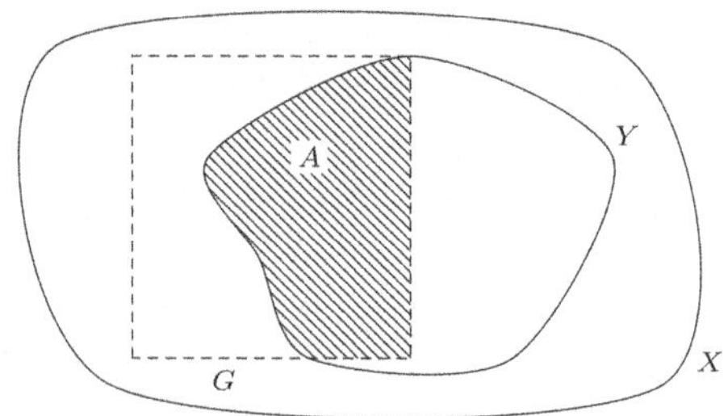

Fig. 3.12 An open (resp., closed) subset in a subspace is the intersection of the subspace with an open (resp., closed) subset from the parent space

Conversely, let $A \subset Y$ be such that $A = G \cap Y$, where G is open in X. Assuming $A \neq \varnothing$, choose $a \in A$ arbitrarily. Since $a \in G$, and G is open, $\exists \epsilon > 0$ such that $B(a; \epsilon) \subset G$. But this implies

$$B(a; \epsilon) \cap Y \subset G \cap Y \implies B_Y(a; \epsilon) \subset G \cap Y = A.$$

Hence a is an interior point of A in Y. Since $a \in A$ is arbitrary, A is open in Y. ■

See Fig. 3.12 for a visual illustration. The version describing the closed sets in a subspace comes as the following:

Corollary 3.34.1 *Let X be a metric space and Y be a subspace of X. Then $B \subset Y$ is closed in Y iff there is $F \subset X$, closed in X, such that $B = F \cap Y$.*

Proof

$$\begin{aligned}
& B \text{ is closed in } Y \\
\iff & Y \setminus B \text{ is open in } Y \\
\iff & Y \setminus B = G \cap Y, && \text{for some } G \text{ open in } X \\
\iff & (Y \cap B^c)^c = (G \cap Y)^c, && \text{[taking complements on both sides]} \\
\iff & Y^c \cup B = G^c \cup Y^c \\
\iff & Y \cap (Y^c \cup B) = Y \cap (G^c \cup Y^c), && \text{[taking intersection with } Y\text{]} \\
\iff & Y \cap B = Y \cap G^c \\
\iff & B = Y \cap F, && \text{where } F = G^c \text{ is closed in } X.
\end{aligned}$$

■

The following two corollaries follow immediately:

Corollary 3.34.2 *Let X be a metric space and Y be a subspace of X. Every $A \subset Y$ open in Y will be open in X iff Y is open in X.*

Corollary 3.34.3 *Let X be a metric space and Y be a subspace of X. Every $B \subset Y$ closed in Y will be closed in X iff Y is closed in X.*

Exercises: 3.3

3.14. Let A be the following subset of $\mathbb{R}^2$.

$$A := \left\{ \left(x, \sin \frac{1}{x} \right) : 0 < x \leq 1 \right\}.$$

Is A closed? Is A closed as a subset of the subspace $Y = (0, 1] \times [-1, 1]$?

3.15. Let Y be a subspace of the metric space X. Let $A \subset X$. If $\mathrm{Cl}_Y(A)$ denotes the closure of A in Y (i.e. the closure of the set $A \cap Y$ in Y), show that $\mathrm{Cl}_Y(A) = \bar{A} \cap Y$, where $\bar{A}$ is the usual closure of A in X.

3.4 Sequences in a Metric Space

We have already had our first encounter with sequences in real analysis and have seen how sequences make our lives easier. We will have the discussion of sequences along the similar line in metric spaces.

Definition 3.35 Let (X, d) be a metric space. A *sequence* in X is a map $f : \mathbb{N} \to X$.

Notation Borrowing the notation from real analysis, we will denote the nth term $f(n)$ of the sequence as x_n and the entire sequence by $\{x_n\}_{n \in \mathbb{N}}$, often abbreviated as $\{x_n\}$.

Remark The above notation for a sequence, though conventionally accepted, may become a bit ambiguous when we use the same to denote the range of a sequence. For example, writing the sequence $\{(-1)^n\}$ using this notation is $\{-1, 1 - 1, 1, \ldots\}$. Looking at this object from a set-theoretic point of view actually boils it down to the set $\{-1, 1\}$, which is not really a sequence. To avoid the confusions in these situations, we will specifically mention "the sequence $\{-1, 1 - 1, 1, \ldots\}$" to denote the sequence $\{(-1)^n\}$ and "the range $\{-1, 1\}$".

Definition 3.36 Let (X, d) be a metric space and $f : \mathbb{N} \to X$ be a sequence in X. Let $g : \mathbb{N} \to \mathbb{N}$ be a strictly increasing function. Then the function $f \circ g : \mathbb{N} \to X$ is a *subsequence* of the sequence $\{f(n)\}$.

Remark The definition of a subsequence may appear a bit jarring while reading, but it essentially describes a subsequence as we all know it. $\{x_{n_k}\}_{k \in \mathbb{N}}$ is a subsequence of the sequence $\{x_n\}$ if $\{n_k\}_{k \in \mathbb{N}}$ is a strictly increasing sequence of positive integers. In a subsequence, we select infinitely many terms from the original sequence in the same order as they appear there.

Definition 3.37 Let (X, d) be a metric space and $\{x_n\}$ be a sequence in X. $\{x_n\}$ is said to *converge* to a point $x \in X$ if

$$\text{for every } \epsilon > 0, \text{ there exists } k \in \mathbb{N}, \text{ such that } \forall n \geq k,\ d(x_n, x) < \epsilon.$$

In this case, we call $\{x_n\}$ a *convergent* sequence and the point x as the *limit* of the sequence $\{x_n\}$.

Notation If a sequence $\{x_n\}$ converges to a point x, we often denote it by writing $x_n \to x$ (Read: x_n "converges" (*or* "tends" *or* "approaches") to x).

Definition 3.38 (Equivalent Definitions) Let (X, d) be a metric space and $\{x_n\}$ be a sequence in X. Then:

- $\{x_n\}$ is said to converge to a point $x \in X$ iff the sequence $\{d(x_n, x)\}_n$ of real numbers converges to 0.
- $\{x_n\}$ is said to converge to a point $x \in X$ iff

$$\text{for every } \epsilon > 0, \text{ there exists } k \in \mathbb{N}, \text{ such that, } \forall n \geq k,\ x_n \in B(x; \epsilon).$$

The following result is carried forward from the real analysis to metric spaces, but it *does not* carry forward to general topological spaces.

Proposition 3.39 *In a metric space, limit of a convergent sequence is unique.*

Proof Let (X, d) be a metric space and let $\{x_n\} \subset X$ converge to the points x and x' in X. Assume, if possible, $x \neq x'$. Then by Hausdorff property, $\exists \epsilon > 0$ such that $B(x; \epsilon) \cap B(x'; \epsilon) = \varnothing$. But by definition, $\exists k_1, k_2 \in \mathbb{N}$ such that

$$\forall n \geq k_1,\ x_n \in B(x; \epsilon) \qquad \text{and} \qquad \forall n \geq k_2,\ x_n \in B(x'; \epsilon).$$

Choose $k = \max\{k_1, k_2\}$. Then

$$\forall n \geq k,\ x_n \in B(x; \epsilon) \text{ and } x_n \in B(x'; \epsilon).$$

But this implies $x_k \in B(x; \epsilon) \cap B(x'; \epsilon)$, a contradiction. Hence $x = x'$. ■

Examples

3.39.1. Let X be a metric space, and $x \in X$. A *constant sequence* $\{x_n\}$, where every $x_n = x$, converges to x.

3.39.2. The sequence $\left\{\frac{1}{n}\right\}_{n \in \mathbb{N}}$ converges to 0 in $\mathbb{R}$.

***Proof*.** Choose $\epsilon > 0$ arbitrarily. Then $\frac{1}{n} < \epsilon \iff n > \frac{1}{\epsilon}$. Choose $k = \left[\frac{1}{\epsilon}\right] + 1$, where $[\,\cdot\,]$ denotes the greatest integer function. Then $\forall n \geq k$, $|\frac{1}{n} - 0| < \epsilon$. Hence $\frac{1}{n} \to 0$. □

3.39.3. Let for each $k = 1, \ldots, n$, $\{x_k^{(m)}\}_m$ be a sequence in $\mathbb{R}$ converging to x_k. Then the sequence $\{\mathbf{x}^{(m)}\}_m$ converges to $\mathbf{x}$ in $\mathbb{R}^n$, where $\mathbf{x}^{(m)} = (x_1^{(m)}, \ldots, x_n^{(m)})$ and $\mathbf{x} = (x_1, \ldots, x_n)$ are points in $\mathbb{R}^n$.
Proof. Choose $\epsilon > 0$ arbitrarily. As $x_k^{(m)} \to x_k$ in $\mathbb{R}$, we have n positive integers m_k, $k = 1, \ldots, n$ such that $\forall m \geq m_k$, $|x_k^{(m)} - x_k| < \frac{\epsilon}{\sqrt{n}}$. Choose $m_0 := \max\{m_k : k = 1, \ldots, n\}$. Then $\forall m \geq m_0$, $|x_k^{(m)} - x_k| < \frac{\epsilon}{\sqrt{n}}$ holds for every coordinate $k \in \{1, \ldots, n\}$. Hence

$$\forall m \geq m_0, \quad \|\mathbf{x}^{(m)} - \mathbf{x}\| = \sqrt{\sum_{k=1}^{n} (x_k^{(m)} - x_k)^2} < \epsilon.$$

□

Remark The converse of this proposition is also true. If $\mathbf{x}^{(m)} = (x_1^{(m)}, \ldots, x_n^{(m)})$, and $\mathbf{x} = (x_1, \ldots, x_n)$ are points in $\mathbb{R}^n$ and $\mathbf{x}^{(m)} \to \mathbf{x}$, then for every coordinate k, $x_k^{(m)} \to x_k$ in $\mathbb{R}$. Proof of this fact is left as an exercise.

3.39.4. Let (X, d) be a discrete space and $\{x_n\}$ be a sequence in X. Now, in a discrete space, $d(x, y) = 0$ or 1 iff $x = y$ and $x \neq y$, respectively. So, for a point $x \in X$, the sequence $\{d(x_n, x)\}$ of real numbers has only elements 0 and 1 repeating (any one or both) infinitely often. Hence, in order to have $x_n \to x$, i.e. $d(x_n, x) \to 0$, $d(x_n, x)$ must be 0 after first finitely many terms, i.e. $x_n = x$ after first finitely many terms. A sequence that becomes constant after a finite number of terms is called as an *eventually constant* sequence. In a discrete space, the only sequences that converge are those that are eventually constant.

3.39.5. Let $X = \mathbb{R}_{m \times n}$ be the set of all real $m \times n$ matrices. Identify X with $\mathbb{R}^{mn}$ and impose the standard norm on it. Let $\{A^{(p)}\}_p \subset X$ be a sequence, where $A^{(p)} := [\![a_{ij}^{(p)}]\!]_{1 \leq i \leq m, 1 \leq j \leq n}$. Then by Example 3.39.3, $\{A^{(p)}\}$ converges to $A = [\![a_{ij}]\!] \in X$ iff each $a_{ij}^{(p)} \to a_{ij}$.

3.39.6. Let (X, d_X) and (Y, d_Y) be metric spaces. Consider $X \times Y$ with product metric. A sequence $\{(x_n, y_n)\} \subset X \times Y$ converges to a point $(x, y) \in X \times Y$ iff $x_n \to x$ in X and $y_n \to y$ in Y.
Proof.

$$\begin{aligned} & d((x_n, y_n), (x, y)) \to 0 \\ \iff \quad & \max\{d_X(x_n, x), d_Y(y_n, y)\} \to 0 \\ \iff \quad & d_X(x_n, x) \to 0 \text{ and } d_Y(y_n, y) \to 0, \end{aligned}$$

i.e. $\{(x_n, y_n)\}$ converges to (x, y) in $X \times Y$ iff $\{x_n\}$ converges to x and $\{y_n\}$ converges to y in X and Y, respectively. □

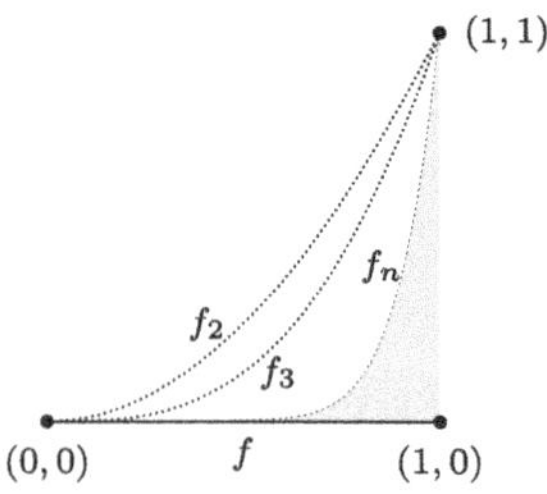

Fig. 3.13 Shaded area is $\int_0^1 |f_n - f|$

3.39.7. Let $X = \mathcal{C}_{\sup}[a, b]$ and $\{f_n\} \subset X$ be a sequence converging to a function $f \in \mathcal{C}[a, b]$. Then for every $\epsilon > 0$, $\exists k \in \mathbb{N}$ such hat $\forall n \geq k$, $\|f_n - f\|_\infty < \epsilon$, i.e. $\sup\{|f_n(x) - f(x)| : x \in [a, b]\} < \epsilon$, i.e. $\forall x \in [a, b]$, $|f_n(x) - f(x)| < \epsilon$. In terminologies of real analysis, this implies the sequence of functions $\{f_n\}$ converges uniformly to the function f.
Conversely, let the sequence of functions $\{f_n\}$ converge uniformly to the function f. Then for any chosen $\epsilon > 0$, $\exists k \in \mathbb{N}$ such that $\forall n \geq k$, $|f_n(x) - f(x)| < \epsilon$ holds for every $x \in [a, b]$. Whence $\forall n \geq k$, $\sup\{|f_n(x) - f(x)| : x \in [a, b]\} < \epsilon$, i.e. the sequence $\{f_n\}$ converges to f in $\mathcal{C}_{\sup}[a, b]$.

3.39.8. Let $f_n(x) := x^n$, $x \in [0, 1]$, $n \in \mathbb{N}$. Let $f(x) = 0$, $x \in [0, 1]$. Then all f_n's and f are continuous functions on $[0, 1]$. Consider $f_n, f \in \mathcal{C}_{\text{int}}[0, 1]$. Then

$$\|f_n - f\|_1 = \int_0^1 |f_n - f| = \int_0^1 x^n \, dx = \frac{1}{n+1} \to 0 \text{ as } n \to \infty.$$

Hence $\{f_n\}$ converges to $f \equiv 0$ in $\mathcal{C}_{\text{int}}[0, 1]$ (see Fig. 3.13).

But as we know the sequence of functions $\{f_n\}$ does not converge uniformly in $[0, 1]$, $\{f_n\}$ is not convergent in $\mathcal{C}_{\sup}[0, 1]$.

3.39.9. One can cook up metrics to disrupt the typical behaviour of sequence convergence. The metric mentioned in Exercise 2.14 supplies the main ingredient for this recipe. For example, define an injective map $f: \mathbb{R} \to \mathbb{R}$ by

$$f(x) = \begin{cases} n, \text{ when } x = \frac{1}{n}, \\ \frac{1}{n}, \text{ when } x = n, \\ x, \text{ otherwise} \end{cases};$$

where $n \in \mathbb{N}$ is a positive integer. Consider the metric $d(x, y) = |f(x) - f(y)|$ and the sequences $\{n\}$ and $\{\frac{1}{n}\}$. In this metric, the sequence $\{n\}$ converges to 0, because

$$d(n, 0) = |f(n) - f(0)| = |\tfrac{1}{n} - 0| \to 0.$$

However, the sequence $\{\frac{1}{n}\}$ diverges, because for any fixed real number α, the quantity

$$d(\tfrac{1}{n}, \alpha) = |f(\tfrac{1}{n}) - f(\alpha)| = |n - f(\alpha)|$$

cannot be made arbitrarily small. Hence $\{\frac{1}{n}\}$ cannot converge to α.

3.4.1 Sequential Equivalence of Topological Entities

Many of the definitions in the earlier sections have equivalent forms in terms of sequences. These equivalent definitions make those entities much easier to work with. We will illustrate here two major sequential equivalent definitions. The first one is of the equivalent metrics.

Theorem 3.40 *Let X be a non-empty set, and d, d' be two metrics defined on X. A necessary and sufficient condition (NASC) for the metrics d and d' to be* equivalent *is that for any sequence $\{x_n\} \subset X$ and $x \in X$,*

$$d(x_n, x) \to 0 \iff d'(x_n, x) \to 0.$$

Proof *Necessary part*: Let d and d' be equivalent metrics on X. Let $\{x_n\} \subset X$ be a sequence which converges to $x \in X$ in the metric d, i.e. $d(x_n, x) \to 0$. We will show $d'(x_n, x) \to 0$ as well. Choose $\epsilon' > 0$ arbitrarily. As d and d' are equivalent metrics, Theorem 3.18 guarantees that there exists $\epsilon > 0$ such that $B_d(x; \epsilon) \subset B_{d'}(x; \epsilon')$. Now as $\{x_n\}$ converges to x in the metric d, $\exists k \in \mathbb{N}$ such that $\forall n \geq k$, $x_n \in B_d(x; \epsilon)$. Consequently, $\forall n \geq k$, $x_n \in B_{d'}(x; \epsilon')$ as well, and as $\epsilon' > 0$ is arbitrary, $d'(x_n, x) \to 0$. By similar arguments, we can show that $d'(x_n, x) \to 0 \implies d(x_n, x) \to 0$.

Sufficient part: Assume the hypothesis. We will show that the conditions in Theorem 3.18 are satisfied. Choose $\epsilon > 0$ arbitrarily and consider $B_d(x; \epsilon)$. If for no $\epsilon' > 0$, does $B_{d'}(x; \epsilon')$ contain in $B_d(x; \epsilon)$, then the following statement is true:

$$\forall r' > 0, \quad B_{d'}(x; r') \setminus B_d(x; \epsilon) \neq \varnothing. \tag{$*$}$$

Now, for each positive integer n, choose $r = \frac{1}{n}$ inductively, and choose

$$x_n \in B_{d'}\left(x; \frac{1}{n}\right) \setminus B_d(x; \epsilon).$$

Consider the sequence $\{x_n\}$. For each n, $0 \leq d'(x_n, x) < \frac{1}{n}$. Hence $d'(x_n, x) \to 0$. But on the other hand, for each n, $d(x_n, x) \geq \epsilon$, showing that $d(x_n, x) \not\to 0$, a contradiction to the hypothesis.

Therefore, for the chosen $\epsilon > 0$, $\exists \epsilon' > 0$ such that $B_{d'}(x; \epsilon') \subset B_d(x; \epsilon)$. By similar arguments, we can show $\forall \delta' > 0$, $\exists \delta > 0$ such that $B_d(x; \delta) \subset B_{d'}(x; \delta')$. Hence by Theorem 3.18, d and d' are equivalent metrics. ■

Following are more examples of equivalent (or non-equivalent) metrics. First two are redoing of the manipulations done earlier to show how this theorem makes them easier to execute. More examples can be found in the Exercises.

Examples

3.40.1. Consider Example 3.18.1. Let $\{x_n\} \subset X$ converge to $x \in X$ in the metric d. By algebra of limits of sequences of real numbers, we get

$$d(x_n, x) \to 0 \implies \frac{d(x_n, x)}{1 + d(x_n, x)} \to 0 \implies d'(x_n, x) \to 0.$$

On the other hand, let $\{x_n\}$ converge to x in the metric d'. If $\{x_n\}$ has a constant subsequence each of whose term is x, then that subsequence converges to x in the metric d as well. We are interested in the other subsequences of $\{x_n\}$. So WLOG, assume that $x_n \neq x$, i.e. $d(x_n, x) \neq 0$.

Now $d(x_n, x) \to 0$ is implied by saying

$$\forall \epsilon > 0,\ \exists k \in \mathbb{N},\ \text{such that } \forall n \geq k,\ d(x_n, x) < \epsilon.$$

But this is same as saying that $\forall \epsilon > 0$, $\exists k \in \mathbb{N}$, such that $\forall n \geq k$,

$$\frac{1}{d(x_n, x)} > \frac{1}{\epsilon} \impliedby \frac{1}{1 + \frac{1}{d(x_n, x)}} < \frac{1}{1 + \frac{1}{\epsilon}} \impliedby d'(x_n, x) < \frac{\epsilon}{1 + \epsilon}. \tag{$*$}$$

As $\frac{\epsilon}{1+\epsilon}$ is a positive real number and $d'(x_n, x) \to 0$, $(*)$ is satisfied.

Hence $d(x_n, x) \to 0 \iff d'(x_n, x) \to 0$, and therefore, d and d' are equivalent.

3.40.2. Consider Example 3.18.2. Let $\{x_n\} \subset X$ be an arbitrary sequence. Then for a point $x \in X$

$$d(x_n, x) \to 0 \iff \min\{1, d(x, x_n)\} \to 0 \iff d''(x_n, x) \to 0;$$

i.e. d and d'' are equivalent metrics.

3.40.3. Theorem 3.40, along with Example 3.39.8, proves that the metrics d_1 and d_∞ on $\mathcal{C}[a, b]$ are not equivalent.

3.40.4. On the set X of all bounded sequences of real and complex numbers, define the metrics

$$d(\mathbf{x}, \mathbf{y}) := \sum_{n=1}^{\infty} \left(a_n \cdot \frac{|x_n - y_n|}{1 + |x_n - y_n|} \right), \qquad \text{[refer Exercise 2.7];}$$

and

$$d_\infty(\mathbf{x}, \mathbf{y}) := \sup\{|x_n - y_n| : n \in \mathbb{N}\}, \qquad \text{[refer Example 2.2.7],}$$

where $\mathbf{x} = \{x_n\}$, $\mathbf{y} = \{y_n\} \in X$, and $\sum a_n$ is a convergent series of positive reals. Consider the sequence $\{\mathbf{e}^{(n)}\}_n \subset X$ where $\mathbf{e}^{(n)} = \{e_k^{(n)}\}_k := \begin{cases} 1, \text{ if } k = n \\ 0, \text{ if } k \neq n \end{cases}$, i.e.

$$\mathbf{e}^{(n)} = \{e_k^{(n)}\}_k := \{0, \dots, 0, \underset{\substack{\uparrow \\ n^{\text{th}} \text{ term}}}{1}, 0, \dots\}.$$

If $\mathbf{0} \in X$ denotes the constant sequence $\{0, 0, \dots\}$, then

$$d(\mathbf{e}^{(n)}, \mathbf{0}) = \sum_{k=1}^{\infty} \left(a_k \cdot \frac{e_k^{(n)}}{1 + e_k^{(n)}} \right) = a_n \to 0, \text{ as } \sum a_n \text{ is convergent,}$$

but

$$d_\infty(\mathbf{e}^{(n)}, \mathbf{0}) = \sup\{e_k^{(n)} : k \in \mathbb{N}\} = 1 \not\to 0 \text{ as } n \to \infty.$$

Hence d and d_∞ are not equivalent metrics on X.

The second sequential equivalent definition is of the adherent points.

Theorem 3.41 *Let (X, d) be a metric space and $A \subset X$. A point $x \in X$ is an adherent point of A iff there is a sequence $\{x_n\}$ of points from A converging to x.*

Proof *Necessary part*: Let $x \in \bar{A}$. Then by definition, $\forall \epsilon > 0$, $B(x; \epsilon) \cap A \neq \varnothing$. For each positive integer n, choose $\epsilon = \frac{1}{n}$ inductively. Therefore, for each index n, we find points $x_n \in B\left(x; \frac{1}{n}\right) \cap A$. Consider the sequence $\{x_n\}$. Then $\{x_n\} \subset A$, and for each n, $0 \leq d(x_n, x) < \frac{1}{n}$. Hence $d(x_n, x) \to 0$, i.e. $x_n \to x$.

Sufficient part: Let $A \subset X$ and $\exists\{x_n\} \subset A$ which converges to x. Choose $\epsilon > 0$ arbitrarily. As $x_n \to x$, for the chosen $\epsilon > 0$, $\exists k \in \mathbb{N}$ such that $d(x_k, x) < \epsilon$. Hence $x_k \in B(x; \epsilon) \cap A$, showing $B(x; \epsilon) \cap A \neq \varnothing$. Since $\epsilon > 0$ is arbitrary, $x \in \bar{A}$. ■

Corollary 3.41.1 *Let (X, d) be a metric space and $A \subset X$. Then A is closed iff every convergent sequence $\{x_n\} \subset A$ converges to a point $x \in A$.*

Remark A point x is called a *subsequential limit* of a sequence $\{x_n\} \subset X$ if a subsequence of $\{x_n\}$ converges to x. Let $A = \{x_n : n \in \mathbb{N}\}$ be the range set of the sequence $\{x_n\}$. Then x is a subsequential limit of $\{x_n\}$ iff it is an adherent point of A.

Example 3.41.1 Consider the metric spaces in Example 2.2.9 and Exercise 2.8. Then $\mathcal{C}[a, b]$ is a closed subset of $\mathcal{B}[a, b]$.

Proof Let f be a limit point of $\mathcal{C}[a, b]$. Then there is a sequence of continuous functions $\{f_n\}$ converging to f in the metric d_∞. Then by Example 3.39.7, the sequence $\{f_n\}$ converges uniformly to f on $[a, b]$. Borrowing results from real analysis, we conclude that f is continuous on $[a, b]$ as well. Hence $f \in \mathcal{C}[a, b]$, i.e. $\mathcal{C}[a, b]$ is closed. □

Miscellaneous Results

3.6. Let (X, d) be a metric space and $A \subset X$. If a sequence $\{x_n\} \subset X$ converges to a point $a \in A$, then $\exists k \in \mathbb{N}$ such that $\forall n \geq k$, $x_n \in A$. Show that A is open.

***Proof*.** If $A = \varnothing$, we are done. Otherwise, choose $a \in A$ arbitrarily. We will show a is an interior point of A.

Assume the contrary. Then for every $\epsilon > 0$, $B(a; \epsilon) \not\subset A$, i.e.

$$\forall \epsilon > 0, \quad B(a; \epsilon) \setminus A \neq \varnothing.$$

For each $n \in \mathbb{N}$, inductively choose $x_n \in B\left(a; \frac{1}{n}\right) \setminus A$, and consider the sequence $\{x_n\}$. Clearly, $d(x_n, a) < \frac{1}{n}$, and hence $x_n \to a$. But each $x_n \in X \setminus A$, a contradiction to the hypothesis. □

3.7. Another version of Bolzano-Weierstrass theorem in $\mathbb{R}$ is "Every bounded sequence of real numbers has a convergent subsequence".

***Proof*.** We will show that this version bi-implies the version mentioned in Miscellaneous result 3.5. Assume every bounded infinite subset of $\mathbb{R}$ has a limit point in $\mathbb{R}$. Let $\{x_n\} \subset \mathbb{R}$ be a bounded sequence. Consider the range $X = \{x_n : n \in \mathbb{N}\}$. Since the sequence $\{x_n\}$ is bounded, X is also so. Now, if X is a finite set, then some element x in the sequence $\{x_n\}$ has repeated infinitely often. Hence the constant subsequence $\{x_{n_k} = x\}_k$ is convergent. Otherwise, if X is an infinite set, by hypothesis, X has a limit point, say x. Since x is also an adherent point of X, by Theorem 3.41, there exists a convergent sequence in X converging to x, i.e. to say, there exists a convergent subsequence of $\{x_n\}$ converging to x.

Conversely, assume every bounded sequence of real numbers has a convergent subsequence. Let $X \subset \mathbb{R}$ be a bounded infinite set. Since X is infinite, we can always choose a sequence of distinct elements $\{x_n\} \subset X$. But by hypothesis, $\{x_n\}$ has a convergent subsequence, say $\{x_{n_k}\}_k$ converging to a point x, say. Then for any chosen $\epsilon > 0$, $\exists k \in \mathbb{N}$ such that $x_{n_k} \in B(x; \epsilon)$. Since x_{n_k}'s are all distinct, the ϵ-neighbourhood of x contains infinitely many elements of X. As $\epsilon > 0$ is arbitrary, x is a limit point of X. □

3.8. Let (X, d) be a metric space. The diagonal $D := \{(x, x) : x \in X\} \subset X \times X$ is closed in $X \times X$ w.r.t. the product metric.

***Proof*.** Let $(x, x') \in X \times X$ be a limit point of D. Then by Theorem 3.41, $\exists\{(x_n, x_n)\} \subset D$ converging to (x, x'). But then by Example 3.39.6, $\{x_n\}$ converges to both x and x'. Hence by Proposition 3.39, $x = x'$, i.e. $(x, x') \in D$. Therefore, D is closed. □

Exercises: 3.4

3.16. A sequence $\{x_n\}$ in a metric space X is called *bounded* if there is a point $a \in X$ and some $M > 0$ so that the entire sequence $\{x_n\} \subset B(a; M)$. Show that a convergent sequence is always bounded.

3.17. Show that x is a limit point of a set A in a metric space X if and only if there is a sequence of distinct points from A converging to x.

3.18. In the metric space $\mathbb{R}$ with usual metric d, call $A := \mathbb{N}$ and $B := \left\{n + \frac{1}{2n} : n \in \mathbb{N}\right\}$. Show that both A and B are closed sets, $d(A, B) = 0$, and $A \cap B = \varnothing$. If this result is not counter-intuitive enough, consider this fact in the light of Miscellaneous result 3.4.

3.19. Let $\{x_n\}$ be a sequence in a metric space X so that every subsequence of $\{x_n\}$ has a subsequence converging to a fixed point $x \in X$. Show that $x_n \to x$.

3.20. If $\mathbf{x}^{(m)} = (x_1^{(m)}, \ldots, x_n^{(m)})$ and $\mathbf{x} = (x_1, \ldots, x_n)$ are points in $\mathbb{R}^n$ and $\mathbf{x}^{(m)} \to \mathbf{x}$, then show that for every k, $x_k^{(m)} \to x_k$ in $\mathbb{R}$.

3.21. Prove *Bolzano-Weierstrass theorem in* $\mathbb{R}^n$: Every bounded sequence in $\mathbb{R}^n$ has a convergent subsequence.

Is this result true in any arbitrary metric space? Justify.

3.22. In a metric space (X, d), let $\{x_n\}$ be a convergent sequence. Show that for every point $x' \in X$, the set $\{d(x_n, x') : n \in \mathbb{N}\}$ is a bounded subset of real numbers.

3.23. In a metric space (X, d), let the sequences $\{x_n\}$ and $\{y_n\}$ converge to the points x and y, respectively. Show that the sequence $\{d(x_n, y_n)\}_n \subset \mathbb{R}$ converges to $d(x, y)$ w.r.t. the usual metric.

3.24. Use Exercise 2.14 to produce a metric d on $\mathbb{R}$ so that both the sequences $\{x_n = \frac{1}{n}\}$ and $\{y_n = -\frac{1}{n}\}$ converge in $(\mathbb{R}, d)$, but $\lim(x_n + y_n) \neq \lim x_n + \lim y_n$.

Remark The usual metric on $\mathbb{R}$ is induced by the usual norm on $\mathbb{R}$, which is again generated by considering $\mathbb{R}$ as vector space over itself w.r.t. the usual addition and multiplication of real numbers. In that circumstance, the result $\lim(x_n + y_n) = \lim x_n + \lim y_n$ holds for two convergent sequences $\{x_n\}$ and $\{y_n\}$. This interplay of algebra and analysis on $\mathbb{R}$ does not happen for an arbitrary metric.

3.25. Let $\{(X_k, d_k) : k = 1, \ldots, n\}$ be n metric spaces and $X = X_1 \times \cdots \times X_n$. Let $\mathbf{x} = (x_1, \ldots, x_n)$ and $\mathbf{y} = (y_1, \ldots, y_n) \in X$ be arbitrary. For $p \geq 1$, define the metric d_p on X by

$$d_p(\mathbf{x}, \mathbf{y}) = \left(\sum_{k=1}^{n} [d_k(x_k, y_k)]^p\right)^{1/p},$$

and the metric d_∞ by

$$d_\infty(\mathbf{x}, \mathbf{y}) = \max\{d_k(x_k, y_k) : k = 1, \ldots, n\}.$$

(See Exercise 2.12.) Show that each d_p metric is equivalent to the d_∞ metric on X.

3.26. Let (X, d) be a metric space. Let $f: [0, \infty) \to [0, \infty)$ be a continuous map satisfying:

i) $f(t) = 0$ iff $t = 0$.
ii) For any $t_1 < t_2$, $f(t_1) \leq f(t_2)$.
iii) For every t_1, t_2, $f(t_1 + t_2) \leq f(t_1) + f(t_2)$.

Show that $f \circ d$ is a metric on X equivalent to d.

3.27. Refer to Example 2.2.9 and Exercise 2.9. For functions $f, g \in \mathcal{C}^1[a, b]$, call

$$d(f, g) := \sup\{|f(x) - g(x)| : x \in [a, b]\}$$

and

$$\rho(f, g) := |f(a) - g(a)| + \sup\{|f'(x) - g'(x)| : x \in [a, b]\}.$$

Show that d and ρ are not equivalent metrics on $\mathcal{C}^1[a, b]$. [*Hint*: Consider the sequence of functions $\left\{\frac{x^n}{n}\right\}_n$.]

3.5 Allied Terminologies and Results

In this section, we will discuss some related concepts and allied results. We start with the following subsection.

3.5.1 Exterior and Boundary

Definition 3.42 Let X be a metric space and $A \subset X$. A point $x \in X$ is called an *exterior point* of A if x is an interior point of $X \setminus A$.
The set of all exterior points of A is called the *exterior* of A and is denoted by $\operatorname{ext} A$, i.e. $\operatorname{ext} A = (A^c)^\circ$.

Remark Exterior of a set A is the largest open set outside of A.

Example 3.42.1 Consider the metric space $\mathbb{R}^2$ with usual metric. Let

$$A := \{(x, y) \in \mathbb{R}^2 : x^2 + y^2 < 1\}$$

be the open unit disc at (0, 0). Then $A^c = \{(x, y) \in \mathbb{R}^2 : x^2 + y^2 \geq 1\}$. Hence

$$\operatorname{ext} A = \operatorname{int}(A^c) = \{(x, y) \in \mathbb{R}^2 : x^2 + y^2 > 1\}. \quad \text{(See Fig. 3.14.)}$$

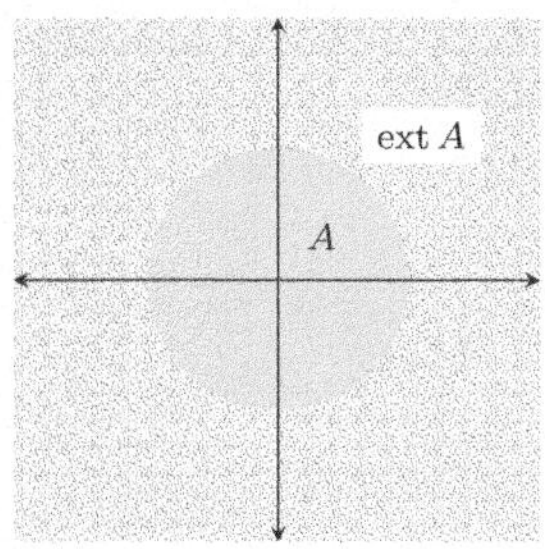

Fig. 3.14 Exterior of open unit disc at origin in $\mathbb{R}^2$

The following proposition lists a couple of properties of exterior of a subset. Proofs of them are left for the reader as exercises.

Proposition 3.43 *Let (X, d) be a metric space and $A, B \subset X$. Then:*

i) $\operatorname{ext} A = \left(\bar{A}\right)^c = \{x \in X : d(x, A) > 0\}$.
ii) $\operatorname{ext}(A \cup B) = \operatorname{ext} A \cap \operatorname{ext} B$.

Miscellaneous Results

3.9. Let (X, d) be a metric space and $A \subset X$. Show that $\operatorname{int} A$, ∂A, and $\operatorname{ext} A$ form a partition of X.

Proof Recall that $\partial A = \bar{A} \setminus \operatorname{int} A$. We need to show that:

- $\operatorname{int} A \cap \partial A = \varnothing$.
- $\partial A \cap \operatorname{ext} A = \varnothing$.
- $\operatorname{ext} A \cap \operatorname{int} A = \varnothing$.
- $X = \operatorname{int} A \cup \partial A \cup \operatorname{ext} A$.

Now, by definition of ∂A, $\operatorname{int} A \cap \partial A = \varnothing$. To show $\partial A \cap \operatorname{ext} A = \varnothing$, choose $x \in \operatorname{ext} A$. Then $\exists \delta > 0$ such that $B(x; \delta) \subset X \setminus A$. Hence $x \notin \bar{A}$, i.e. $\operatorname{ext} A \subset (X \setminus \bar{A})$. Therefore, $\partial A \cap \operatorname{ext} A = \varnothing$. Lastly, $\operatorname{ext} A \cap \operatorname{int} A = \varnothing$ because $\operatorname{int} A \subset A$ and $\operatorname{ext} A \subset X \setminus A$.
By definition of ∂A, $\bar{A} = \operatorname{int} A \cup \partial A$, and by Proposition 3.43, $\operatorname{ext} A = X \setminus \bar{A}$. Hence $X = \operatorname{int} A \cup \partial A \cup \operatorname{ext} A$. This completes the proof. □

Definition 3.44 Let X be a metric space and $A \subset X$. A point $x \in X$ is said to be a *boundary point* of A if x is neither an interior point nor an exterior point of A.

3.5.2 Dense Subsets and Separable Spaces

Definition 3.45 Let X be a metric space and $A \subset X$. A is called *dense* in X if $\bar{A} = X$.

A subset is considered dense if it can approximate any point in the space. Typically, it is a well-distributed portion of the space, spread throughout its entirety.

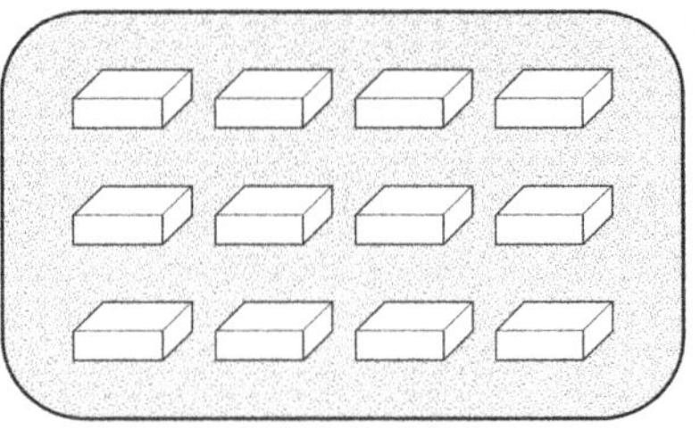

Fig. 3.15 Brick-laid muddy ground

Here is a real-life analogy. Suppose you are walking on a muddy and slippery ground. To move "normally" (without the constant fear of slipping)—you would lay bricks evenly and densely across the field (Fig. 3.15). The bricks should be placed so closely that you can reach arbitrarily near any point on the field while stepping only on the bricks. In this case, the set of bricks is dense in the combined set of the field and bricks. However, the bricks should not cover the entire field, as that would not be economical.

Thus, a dense set is a convenient subset of the entire space, allowing one to approximate any point in the space through these convenient points. Formally, this means that for any point in the space, there exists a sequence of points from the dense set that converges to it.

For example, consider the sequence

$$\left\{\frac{[10^n\sqrt{2}]}{10^n}\right\}_n,$$

which produces successive terms $\{1.4, 1.41, 1.414, \dots\}$. Although each term is a rational number, the sequence ultimately converges to the irrational number $\sqrt{2}$. In practice, rational numbers are often more convenient to work with than irrationals, and indeed, $\mathbb{Q}$ is dense in $\mathbb{R}$ (with respect to the usual metric).

Of course, the set of all irrational numbers is also dense in $\mathbb{R}$, but they are usually not as convenient to manipulate. This is why we are generally less concerned with their density[1].

Similarly, polynomials are often easier to handle than arbitrary continuous functions, and they are dense in the space of continuous functions with respect to the uniform metric—this is the famous Weierstrass' approximation theorem. We will explore proofs of both of these facts.

With this understanding, we arrive at the following immediate consequence.

[1] There is a joke about the irrational numbers: Irrational numbers were frustrated about being called "irrationals", so they went to God to complain. Looking at them, God exclaimed, "Oh my God! You are a crowd. I cannot listen to all of you together. Come one by one!"

Proposition 3.46 *Let X be a metric space and $D \subset X$. Then D is dense in X if and only if for every point $x \in X$, there is a sequence $\{x_n\}$ of points from D converging to x.*

Use Theorem 3.41 to prove this proposition.

Examples

3.46.1. For $n \geq 1$, $\mathbb{Q}^n$ is dense in $\mathbb{R}^n$.

Proof. We will first prove the density property of $\mathbb{Q}$ in $\mathbb{R}$.

Density property of $\mathbb{Q}$: Let x, y be two real numbers with $x < y$. Then there exists a rational number r such that $x < r < y$.

Proof: $y - x > 0$. Therefore, by Archimedean property of $\mathbb{R}$, $\exists n \in \mathbb{N}$ such that $n(y-x) > 1$, i.e. $nx + 1 < ny$. Again by Archimedean property, it can be shown that $\exists m \in \mathbb{Z}$ such that $m - 1 \leq nx < m$. Combining everything, we get

$$nx < m \leq nx + 1 < ny.$$

So, finally, we have $nx < m < ny$, and whence, $x < \frac{m}{n} < y$. $\frac{m}{n}$ is the required rational number r. This completes the proof of density property of $\mathbb{Q}$.

Now let $x \in \mathbb{R}$ be arbitrary. Then for each $n \in \mathbb{N}$, by density property of $\mathbb{Q}$, $\exists r_n \in \mathbb{Q}$ such that $|x - r_n| < \frac{1}{n}$. But this implies $\{r_n\}$ converges to x. So for each $x \in \mathbb{R}$, $\exists \{r_n\} \subset \mathbb{Q}$ such that $r_n \to x$. Now let $\mathbf{x} = (x_1, \dots, x_n) \in \mathbb{R}^n$ be arbitrary. Then for every $k = 1, \dots, n$, there exist sequences $\{r_k^{(m)}\}_m \subset \mathbb{Q}$ converging to x_k. Hence by Example 3.39.3, the sequence $\{\mathbf{r}^{(m)}\}_m \subset \mathbb{Q}^n$ converges to $\mathbf{x}$, where $\mathbf{r}^{(m)} = (r_1^{(m)}, \dots, r_n^{(m)})$. Therefore, $\mathbb{Q}^n$ is dense in $\mathbb{R}^n$. □

3.46.2. Let $q > 0$ be an irrational number. Denote by $\mathbb{Z}[q]$ the set

$$\mathbb{Z}[q] := \{a + bq : a, b \in \mathbb{Z}\}.$$

Then $\mathbb{Z}[q]$ is dense in $\mathbb{R}$.

Proof. We will show that for any two real numbers x and y with $x < y$, $\exists z \in \mathbb{Z}[q]$ such that $x < z < y$.

For $b \in \mathbb{Z}$, consider the number $s_b := bq - [bq]$, where $[\cdot]$ is the greatest integer function. Then $0 \leq s_b < 1$ and $s_b \in \mathbb{Z}[q]$. Also for $b, b' \in \mathbb{Z}$,

$$s_b = s_{b'} \implies bq - [bq] = b'q - [b'q] \implies q(b - b') = [bq] - [b'q].$$

Since q is an irrational, the last equality is true only if $b = b'$. Hence for every $b \neq b'$, $s_b \neq s_{b'}$, and therefore the map $b \mapsto s_b$ is one to one. As a result, the set $S := \{s_b : b \in \mathbb{Z}\}$ is an infinite subset of $[0, 1) \cap \mathbb{Z}[q]$.

Choose $\epsilon > 0$ arbitrarily. Divide the interval $[0, 1)$ into k subintervals such that each subinterval is of length $< \epsilon$. Then at least one subinterval must contain at least two points from S, say s_b and $s_{b'}$. WLOG, let $s_b < s_{b'}$. Then $0 < s_{b'} - s_b < \epsilon$. But $s_{b'} - s_b = ([bq] - [b'q]) + (b' - b)q$ is again a point in $\mathbb{Z}[q]$. Calling $s := s_{b'} - s_b$, we conclude that for each $\epsilon > 0$, $\exists s \in \mathbb{Z}[q]$ such that $0 < s < \epsilon$.

Now choose $x, y \in \mathbb{R}$ arbitrarily with $x < y$ and fix $0 < \epsilon < y - x$. For any arbitrary $0 < \delta < \epsilon$, let n be the smallest integer such that $n\delta > x$, i.e. $(n-1)\delta \le x < n\delta$. Then $n\delta < y$, because otherwise, $y \le n\delta$ produces

$$y - x \le n\delta - (n-1)\delta = \delta > \epsilon,$$

a contradiction. Hence $x < n\delta < y$. Now, for the chosen $\epsilon > 0$, $\exists s \in \mathbb{Z}[q]$ such that $0 < s < \epsilon$. Hence choosing $\delta = s$, we can find $n \in \mathbb{N}$ such that $x < ns < y$. But $s \in \mathbb{Z}[q]$ implies $ns \in \mathbb{Z}[q]$ as well. Call $ns = z$. Therefore, for every $x, y \in \mathbb{R}$, with $x < y$, $\exists z \in \mathbb{Z}[q]$ such that $x < z < y$. □

3.46.3. $\mathbb{R} \setminus \mathbb{Q}$ is dense in $\mathbb{R}$.

Proof. We will show that for any two real numbers x, y with $x < y$, there exists an irrational number q such that $x < q < y$. This is true because $x < y \implies x\sqrt{2} < y\sqrt{2}$, and by density property of $\mathbb{Q}$, $\exists r \in \mathbb{Q}$ such that $x\sqrt{2} < r < y\sqrt{2}$, which is to say $\exists r \in \mathbb{Q}$ such that $x < \frac{r}{\sqrt{2}} < y$. But as r is rational, $\frac{r}{\sqrt{2}}$ is irrational. □

3.46.4. *Weierstrass approximation theorem:* Denote by $\mathcal{P}[a, b]$, the set of all polynomials on $[a, b]$ with real coefficients. Then $\mathcal{P}[a, b]$ is dense in $\mathcal{C}[a, b]$ w.r.t. the uniform metric.

Remark Proof of this very important fact, deserving its own section, can be found in Appendix C.

To deliberately channelise our efforts to "convenient" dense sets, we have the following:

Definition 3.47 Let (X, d) be a metric space. X is called *separable* if it contains a countable dense subset.

X is *inseparable* if it is not separable.

Examples

3.47.1. For $n \ge 1$, $\mathbb{R}^n$ is separable.

Proof. Since $\mathbb{Q}$ is countable, product of n copies of $\mathbb{Q}$ is also countable. Since by Example 3.46.1, $\mathbb{Q}^n$ is dense in $\mathbb{R}^n$, $\mathbb{R}^n$ is separable. □

Remark As seen in Example 3.46.2, $\mathbb{Z}[q]$ is also a countable dense subset of $\mathbb{R}$.

3.47.2. $\mathcal{C}_{\sup}[a, b]$ is separable.

Proof. Fix $\epsilon > 0$ arbitrarily. By Weierstrass' approximation theorem, for every continuous function $f \in \mathcal{C}[a, b]$, there will exist a polynomial $p \in \mathcal{P}[a, b]$ such that

$$\forall x \in [a, b], \quad |f(x) - p(x)| < \epsilon. \qquad (*)$$

Let $p(x) = a_0 + a_1 x + \cdots + a_n x^n$, where $a_0, \ldots, a_n \in \mathbb{R}$ and $x \in [a, b]$. Now for every $k \in \{0, \ldots, n\}$, let $\{a_k^{(m)}\}_m$ be a sequence of rational numbers converging to a_k. Then by Example 3.39.3, $\exists m_0 \in \mathbb{N}$ such that for every index $k = 0, \ldots, n$, $|a_k^{(m)} - a_k| < \frac{\epsilon}{n}$. Call

$$p^{(m_0)}(x) := a_0^{(m_0)} + a_1^{(m_0)} x + \cdots + a_n^{(m_0)} x^n.$$

Then

$$\begin{aligned}
\|p - p^{(m_0)}\|_\infty &= \sup_{x\in[a,b]} |p(x) - p_{m_0}(x)| \\
&= \sup_{x\in[a,b]} \left[\sum_{k=0}^{n} \left| a_k^{(m_0)} x^k - a_k x^k \right| \right] \\
&\le \sum_{k=0}^{n} \sup_{x\in[a,b]} \left| a_k^{(m_0)} x^k - a_k x^k \right| \\
&= \sum_{k=0}^{n} \left| a_k^{(m_0)} - a_k \right| \sup_{x\in[a,b]} x^k \\
&< \sum_{k=0}^{n} \frac{\epsilon}{n}.B = B\epsilon,
\end{aligned}$$

where $B := \sup\{x^k : x \in [a, b], k = 0, \ldots, n\}$. Then by $(*)$,

$$\|f - p^{(m_0)}\|_\infty \le \|f - p\|_\infty + \|p - p^{(m_0)}\|_\infty < \epsilon + B\epsilon = (B + 1)\epsilon.$$

Now, let D_n be the set of all polynomials of degree $< n$ with rational coefficients. Then cardinality of each D_n is same as the cardinality of product of n copies of $\mathbb{Q}$, i.e. D_n is countable. Call

$$D := \bigcup_{n\in\mathbb{N}} D_n.$$

Then D is countable union of countable sets and hence is countable. Our above analysis shows that D is dense in $\mathcal{C}_{\text{sup}}[a, b]$. Hence $\mathcal{C}_{\text{sup}}[a, b]$ is separable. □

3.47.3. $\mathcal{C}_{\text{int}}[a, b]$ is separable.

Proof. Consider the same countable set D constructed in the previous example. As D is dense in $\mathcal{C}[a, b]$ in the sup metric, for a given $\epsilon > 0$, and for any continuous function $f \in \mathcal{C}[a, b]$, there exists a polynomial p from

D such that for every point $x \in [a, b]$, $|f(x) - p(x)| < \frac{\epsilon}{b-a}$. Consequently,

$$d_1(f, p) = \int_a^b |f - g| < \int_a^b \frac{\epsilon}{b-a} = \epsilon.$$

This completes the proof. □

3.47.4. ℓ_p (Example 2.2.6) is separable.

Remark With the technique of the previous example fresh in mind, one might be tempted to consider the set of all rational sequences which are there in ℓ_p. But observe that the product $\mathbb{Q}^\infty$ of countably infinitely many copies of $\mathbb{Q}$ is not countable. So there is a possibility that $\mathbb{Q}^\infty \cap \ell_p$ may not be countable as well. Indeed, the set $\prod_{n\in\mathbb{N}}\left(\left[-\frac{1}{2^n}, \frac{1}{2^n}\right] \cap \mathbb{Q}\right)$ is an uncountable subset of ℓ_p. To overcome this obstacle, let us carefully re-examine the previous proof. There, we constructed each D_n in bijection with a product of finitely many copies of $\mathbb{Q}$, thus keeping each of them countable. We replicate that technique here in the following way:

Proof. Consider the eventually constant sequences of rational numbers converging to 0. For every $n \in \mathbb{N}$, let

$$D_n := \left\{\{x_k\}_k \subset \mathbb{Q} \text{ such that for every } k > n,\ x_k = 0\right\}.$$

Then each $D_n \subset \ell_p$ and has the cardinality of $\mathbb{Q}^n$, hence countable. Therefore

$$D = \bigcup_{n\in\mathbb{N}} D_n$$

is also contained in ℓ_p and is countable. We will show that D is dense in ℓ_p.

Let $\mathbf{x} = \{x_k\} \in \ell_p$. Fix $\epsilon > 0$ arbitrarily. Then $\exists k_0 \in \mathbb{N}$ such that $\sum_{k>k_0}|x_k|^p < \frac{\epsilon^p}{2}$. Consider the point $\mathbf{x}' = (x_1, \dots, x_{k_0}) \in \mathbb{R}^{k_0}$. Since $\mathbb{Q}^{k_0}$ is dense in $\mathbb{R}^{k_0}$, $\exists \mathbf{r}' = (r_1, \dots, r_{k_0}) \in \mathbb{Q}^{k_0}$ such that $(\|\mathbf{x}' - \mathbf{r}'\|_p)^p < \frac{\epsilon^p}{2}$. Therefore for the point $\mathbf{r} = (r_1, \dots, r_{k_0}, 0, 0, \dots)$,

$$(\|\mathbf{x}' - \mathbf{r}'\|_p)^p = \sum_{k\in\mathbb{N}} |x_k - r_k|^p = \sum_{k=1}^{k_0} |x_k - r_k|^p + \sum_{k>k_0} |x_k|^p < \frac{\epsilon^p}{2} + \frac{\epsilon^p}{2} = \epsilon^p,$$

i.e. $\|\mathbf{x} - \mathbf{r}\|_p < \epsilon$. But $\mathbf{r} \in D$. Hence the proof. □

3.47.5. If X is an uncountable set with discrete metric, then it is inseparable. Proof of this fact is left for the reader as an exercise.

3.47.6. ℓ_∞ (Example 2.2.7) is inseparable.

***Proof*.** We will see the proof in two steps:

Step 1: Let $A := \{\mathbf{a} = \{a_k\}_k \in \ell_\infty : a_k = 0 \text{ or } 1\}$. Then $A \subset \ell_\infty$, and cardinality of A is $2^{\aleph_0}$, which is also the cardinality of power set of $\mathbb{N}$. By Cantor's theorem, there cannot exist a bijection between a set and its power set. Hence A is uncountable.

Step 2: If possible, let $D \subset \ell_\infty$ be a countable dense set. Let $D := \{\mathbf{x}^{(1)}, \mathbf{x}^{(2)}, \dots\}$. As D is dense, for any $\epsilon > 0$, $\bigcup_{n\in\mathbb{N}} B(\mathbf{x}^{(n)}; \epsilon) = \ell_\infty$. Choose, in particular, $\epsilon = \frac{1}{3}$. Then $A \subset \bigcup_{n\in\mathbb{N}} B\left(\mathbf{x}^{(n)}; \frac{1}{3}\right)$ as well. Call $\mathcal{B} := \left\{B\left(\mathbf{x}^{(n)}; \frac{1}{3}\right) : n \in \mathbb{N}\right\}$. Then $\mathcal{B}$ is countable. Since by step 1, A is uncountable, there is at least one $B\left(\mathbf{x}^{(n_0)}; \frac{1}{3}\right)$ in $\mathcal{B}$ containing two distinct points $\mathbf{a}, \mathbf{b} \in A$. But by construction, $\mathbf{a} \neq \mathbf{b}$ implies $d_\infty(\mathbf{a}, \mathbf{b}) = 1$. Hence

$$1 = d_\infty(\mathbf{a}, \mathbf{b}) \leq d_\infty(\mathbf{a}, \mathbf{x}^{(n_0)}) + d_\infty(\mathbf{x}^{(n_0)}, \mathbf{b}) < \frac{1}{3} + \frac{1}{3} = \frac{2}{3},$$

a contradiction. Hence our assumption that $D \subset \ell_\infty$ is a countable dense set is not tenable. Therefore, ℓ_∞ is inseparable. □

3.5.3 Countability Axioms

The concepts in this subsection are mostly designed for topological spaces. Restriction of our discussion to metric spaces fetches some trivial facts. Nevertheless, some concepts do matter in bringing out metric spaces with pleasant properties.

Definition 3.48 Let (X, d) be a metric space and $x \in X$. A collection $\mathcal{U}_x := \{U_\lambda : \lambda \in \Lambda\}$ of open sets in X, each containing the point x, is called a *local base* at x if for every open subset G of X containing x, there exists $U_{\lambda_0} \in \mathcal{U}_x$ such that $x \in U_{\lambda_0} \subset G$.

Definition 3.49 Let (X, d) be a metric space. X is called *first countable* if at every point $x \in X$, there is a countable local base $\mathcal{U}_x$.

Examples

3.49.1 Consider the metric space $\mathbb{R}$ with usual metric. Let $x \in \mathbb{R}$ be arbitrary. Define

$$\mathcal{U}_x := \{(r, s) : r < x < s \text{ and } r, s \in \mathbb{Q}\}.$$

Let $G \subset \mathbb{R}$ be an open set containing x. Then $\exists\delta > 0$ such that $B(x; \delta) \subset G$. Choose rational numbers p and q so that $x - \delta < p < x < q < x + \delta$. Then

$x \in (p, q) \subset G$ and $(p, q) \in \mathcal{U}_x$. Hence $\mathcal{U}_x$ is a local base at x. Moreover, $\mathcal{U}_x$ is countable (cardinality of $\mathcal{U}_x$ is that of $\mathbb{Q} \times \mathbb{Q}$). Since $x \in \mathbb{R}$ is arbitrary, $\mathbb{R}$ is first countable.

Remark The above example actually generalises to every metric space as proved in the following:

Proposition 3.50 *Every metric space is first countable.*

Proof Let (X, d) be a metric space and $x \in X$ be arbitrary. Consider

$$\mathcal{U}_x := \{B(x; r) : r \in \mathbb{Q}\}.$$

Then $\mathcal{U}_x$ is countable. We will show $\mathcal{U}_x$ is a local base at x.
Let $G \subset X$ be any open set containing x. Then $\exists \delta > 0$ such that $B(x; \delta) \subset G$. Using density property of rational numbers, choose $r \in \mathbb{Q}$ such that $0 < r < \delta$. Then $x \in B(x; r) \subset B(x; \delta) \subset G$. Hence $\mathcal{U}_x$ is a local base at x. Since $x \in X$ is arbitrary, X is first countable. ■

Examples

3.50.1. Consider the metric space $\mathbb{R}^2$ with usual metric. Let $\mathbf{x} = (x_1, x_2) \in \mathbb{R}^2$ be arbitrary. Denote by

$$U_\lambda^{(1)}(\mathbf{x}) = \{\mathbf{y} = (y_1, y_2) \in \mathbb{R}^2 : (x_1 - y_1)^2 + (x_2 - y_2)^2 < \lambda^2\}$$

and

$$U_\lambda^{(2)}(\mathbf{x}) = \{\mathbf{y} = (y_1, y_2) \in \mathbb{R}^2 : (x_1 - y_1)^2 + 2(x_2 - y_2)^2 < \lambda^2\}.$$

Then both the collections $\mathcal{U}_{\mathbf{x}}^{(1)} := \{U_\lambda^{(1)}(\mathbf{x}) : \lambda > 0\}$ and $\mathcal{U}_{\mathbf{x}}^{(2)} := \{U_\lambda^{(2)}(\mathbf{x}) : \lambda > 0\}$ are local bases at $\mathbf{x}$. Proof of this fact is left for the reader as an exercise.

Definition 3.51 Let (X, d) be a metric space. A collection $\mathcal{U} := \{U_\lambda : \lambda \in \Lambda\}$ of open subsets of X is called a *base* for X if $\mathcal{U}$ is a local base for every point $x \in X$, i.e. for any open set G of X and for every point $x \in G$, there exists $U_{\lambda_x} \in \mathcal{U}$ such that $x \in U_{\lambda_x} \subset G$.

Example 3.51.1 Let (X, d) be metric space. The collection $\mathcal{B} := \{B(x; r) : x \in X, r > 0\}$ of all open balls is a base for X.

Proof Follows trivially from the definition of an open set. Let $G \subset X$ be open and $x \in G$. Then $\exists \delta > 0$ such that $x \in B(x; \delta) \subset G$. ■

Proposition 3.52 *Let (X, d) be a metric space. A collection $\mathcal{U}$ of open sets from X is a base for X iff every open subset of X is union of a subcollection of $\mathcal{U}$.*

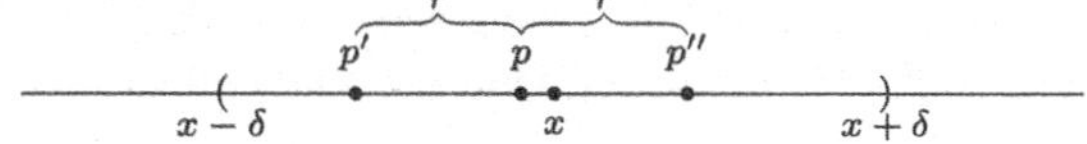

Fig. 3.16 $\mathbb{R}$ is second countable

Proof Let $\mathcal{U}$ be a base for X and G be a non-empty open subset of X. Then for every $x \in G$, we can find $U_x \in \mathcal{U}$ such that $x \in U_x \subset G$. Hence

$$G = \{x : x \in G\} \subset \bigcup_{x \in G} U_x \subset G.$$

Hence $G = \bigcup_{x \in G} U_x$. If $G = \varnothing$, it is union of an empty subcollection of $\mathcal{U}$. Conversely, let $\mathcal{U}$ be a collection of open subsets of X and every open subset of X is union of a subcollection of $\mathcal{U}$. Let G be an arbitrary non-empty open subset of X and $x \in G$. Then there is a subcollection $\mathcal{U}_G \subset \mathcal{U}$ such that $G = \bigcup_{U \in \mathcal{U}_G} U$. Hence for each $x \in G$, we can find $U_x \in \mathcal{U}_G$ such that $x \in U_x$. As $U_x \subset G$, $\mathcal{U}$ is a base for X. ■

Definition 3.53 Let (X, d) be a metric space. X is called *second countable* if X admits a countable base.

Example 3.53.1 $\mathbb{R}$ with usual metric is second countable.

Proof Consider the collection

$$\mathcal{B} := \{B(p; r) : p, r \in \mathbb{Q}\}.$$

Then $\mathcal{B}$ is a countable collection of open sets in $\mathbb{R}$. Now, for any open set $G \subset \mathbb{R}$ and a point $x \in G$, $\exists \delta > 0$ so that $(x - \delta, x + \delta) \subset G$. By density property of rational numbers, choose $p', p'' \in \mathbb{Q}$ such that $x - \delta < p' < x < p'' < x + \delta$. Call $p := \frac{p'+p''}{2}$ and $r := p - p'$. Then $B(p; r) \in \mathcal{B}$ and $x \in B(p; r) \subset B(x; \delta) \subset G$. See Fig. 3.16. Hence $\mathcal{B}$ is a base for $\mathbb{R}$. ■

We can imitate the same arguments to prove the much general

Theorem 3.54 *A metric space is second countable if and only if it is separable.*

Proof *Necessary part*: Let X be second countable and $\mathcal{B} := \{B_n : n \in \mathbb{N}\}$ be a countable base for X. For $n \in \mathbb{N}$, fix points $x_n \in B_n$. Call $D := \{x_n : n \in \mathbb{N}\}$. We will show that D is dense in X.
Let G be any non-empty open set in X and $x \in G$. Since $\mathcal{B}$ is a base for X, there is some $B_{n_0} \in \mathcal{B}$ such that $x \in B_{n_0} \subset G$. Therefore the point $x_{n_0} \in B_{n_0}$ is also an element of G. Hence $x_{n_0} \in D \cap G$, showing that $D \cap G \neq \varnothing$. Since G is any arbitrary non-empty open set in X, by Exercise 3.28, D is dense in X. Since D is countable, X is separable.

Sufficient part: Let X be a separable metric space. Let $D := \{x_n : n \in \mathbb{N}\} \subset X$ be a countable dense set. Consider the collection

$$\mathcal{B} := \{B(x_n; r) : x_n \in D, r \in \mathbb{Q}\}.$$

Clearly $\mathcal{B}$ is countable as both D and $\mathbb{Q}$ are countable. We will show that $\mathcal{B}$ is a base for X.
Let $G \subset X$ be open and $x \in G$. Since G is open, $\exists \delta > 0$ such that $B(x; \delta) \subset G$. Again, as D is dense in X, $\exists x_k \in D$ such that $d(x, x_k) < \frac{\delta}{2}$. Clearly, $x \in B\left(x_k; \frac{\delta}{2}\right) \subset B(x; \delta)$. Because,

$$y \in B\left(x_k; \tfrac{\delta}{2}\right) \implies d(x, y) \leq d(x, x_k) + d(x_k, y) < \tfrac{\delta}{2} + \tfrac{\delta}{2} = \delta \implies y \in B(x; \delta).$$

Now, by density property of $\mathbb{Q}$, we get $r_0 \in \mathbb{Q}$ such that $d(x_k, x) < r_0 < \frac{\delta}{2}$.
Hence $x \in B(x_k; r_0) \subset B\left(x_k; \frac{\delta}{2}\right) \subset B(x; \delta) \subset G$. Hence $\mathcal{B}$ is a base. ■

Exercises: 3.5

3.28. Show that a subset A of a metric space X is dense if and only if A intersects every ball (open or closed) in X. In other words, A is dense iff A intersects every non-empty open set in X.

3.29. Prove Proposition 3.43.

3.30. Prove the following facts about boundary of a set:

(a) $\partial A = (\text{int}\, A \cup \text{ext}\, A)^c$
$= \{x \in X : \forall \delta > 0, B(x; \delta) \cap A \neq \varnothing \text{ and } B(x; \delta) \cap A^c \neq \varnothing\}$
$= \bar{A} \cap \overline{A^c}$;
i.e. boundary of A is the set of all boundary points of A.

(b) $\bar{A} = A \cup \partial A$.

(c) $\partial A = \partial A^c$.

(d) A is open iff $\partial A \cap A = \varnothing$.

(e) A set A is called *clopen* if A is both closed and open (see Section 7.1 for more details). Show that A is clopen iff $\partial A = \varnothing$.

3.31. A function $f : [0, 1] \to \mathbb{R}$ is called *piecewise linear* if there is a partition $0 = x_0 < x_1 < \cdots < x_n = 1$ of the interval $[0, 1]$ so that f is linear on each of the subintervals (x_{r-1}, x_{r+1}), i.e. there are constants a_r, b_r so that

$$f(x) = a_r x + b_r, \qquad x \in (x_{r-1}, x_r),$$

$r = 1, \ldots, n$. (See Definition 4.13 for an example.) Prove the following:

(a) Show that a piecewise linear function is continuous iff it is linear on each of the closed subintervals $[x_{r-1}, x_r]$, $r = 1, \ldots, n$.

(b) The set of piecewise linear continuous functions on $[0, 1]$ is dense in the metric space $\mathcal{C}[0, 1]$ with uniform metric, i.e. given a continuous function $g \in \mathcal{C}[0, 1]$ and $\epsilon > 0$ however small, there always exists a piecewise linear function f on $[0, 1]$ so that for each point $x \in [0, 1]$, $|g(x) - f(x)| < \epsilon$. [*Hint:* For given $\epsilon > 0$, find $\delta > 0$ by uniform continuity of g. Obtain $k \in \mathbb{N}$ so that $\frac{1}{k} < \delta$. Divide $[0, 1]$ into partition with length of subintervals $\frac{1}{k}$. At each point x_r of subdivision, define $f(x_r) = g(x_r)$. Extend f linearly to all of $[0, 1]$.]

3.32. Show that if X is an uncountable set with discrete metric, then it is *inseparable*.

3.33. Consider the metric space $\mathbb{R}^2$ with usual metric. Let $\mathbf{x} = (x_1, x_2) \in \mathbb{R}^2$ be arbitrary. For fixed $a, b > 0$, define

$$U_\lambda(\mathbf{x}) = \{\mathbf{y} = (y_1, y_2) \in \mathbb{R}^2 : a(x_1 - y_1)^2 + b(x_2 - y_2)^2 < \lambda^2\}.$$

Show that $\mathcal{U}_{\mathbf{x}} := \{U_\lambda(\mathbf{x}) : \lambda > 0\}$ is a local base at $\mathbf{x}$. Hence conclude that in Example 3.50.1, both $\mathcal{U}_{\mathbf{x}}^{(1)}$ and $\mathcal{U}_{\mathbf{x}}^{(2)}$ are local bases at $\mathbf{x}$.

3.34. Show that a subspace of a second countable metric space is also second countable.

Chapter 4
Completeness

After the first course on real analysis, you might have realised that much of the richness of the set of all real numbers and its associated contexts are attributed to the completeness axiom of real numbers. This fundamental property underpins key results like the Archimedean property, the Bolzano-Weierstrass theorem, the Cantor's intersection theorem, and the convergence of Cauchy sequences, amongst others. This property also distinguished the real numbers from the rationals in the axiomatic development.

However, as we extend our study of analysis beyond the real line to arbitrary metric spaces, we no longer have an immediate way to incorporate this concept. The completeness axiom for $\mathbb{R}$ relied on the order axioms, which are not generally available in an arbitrary metric space. This chapter is dedicated to overcoming this limitation.

In the first two sections, we will explore the completeness of a metric space, establish a necessary and sufficient condition for a space to be complete, and introduce the Baire category theorem. The chapter concludes with a discussion on a method for completing a given space.

4.1 Definition and Examples

To introduce completeness in a general metric space, we seek a property that not only implies the completeness axiom in $\mathbb{R}$ but can also be naturally defined in a metric space. Here, the notion of Cauchy sequence proves to be a valuable tool. A Cauchy sequence in a metric space is defined as follows:

Definition 4.1 Let (X, d) be a metric space. A sequence $\{x_n\} \subset X$ is called a *Cauchy sequence* if for every positive ϵ, there exists a positive integer k such that for every natural number $m, n \geq k$, $d(x_m, x_n) < \epsilon$. Or in other words, for every natural number p, and $n \geq k$, $d(x_{n+p}, x_n) < \epsilon$.

S. Paul, *Metric Spaces*, University Texts in the Mathematical Sciences,
https://doi.org/10.1007/978-981-96-9259-0_4

We already know that every Cauchy sequence of real numbers is convergent. Here, we formally record a proof of this fact. If you are already familiar with the arguments from your earlier course, you may choose to skip the proof.

Theorem 4.2 *A Cauchy sequence in* $\mathbb{R}$ *is convergent.*

Proof Let $\{x_n\} \subset \mathbb{R}$ be a Cauchy sequence. We follow the following steps to show $\{x_n\}$ is convergent.

Step 1: Choose $\epsilon = 1$. Then we can find a positive integer k_1 such that whenever $m, n \geq k_1$, we have $|x_m - x_n| < 1$, i.e. in particular, for each $n \geq k_1$, $|x_n - x_{k_1}| < 1$, i.e. $x_{k_1} - 1 < x_n < x_{k_1} + 1$. Call $b = \min\{x_1, \ldots, x_{k_1-1}, x_{k_1} - 1\}$, and $B = \max\{x_1, \ldots, x_{k_1-1}, x_{k_1} + 1\}$. Then for every $n \in \mathbb{N}$, $b \leq x_n \leq B$, i.e. $\{x_n\}$ is bounded.

Step 2: Since $\{x_n\}$ is bounded, by Bolzano-Weierstrass theorem (miscellaneous result 3.7), there exists a convergent subsequence $\{x_{n_k}\}$ of $\{x_n\}$, converging to x, say. Therefore, for $\epsilon > 0$ chosen arbitrarily, we can find $k_2 \in \mathbb{N}$ such that for each index $k \geq k_2$, $|x_{n_k} - x| < \frac{\epsilon}{2}$. $\cdots (*)$

Step 3: Since $\{x_n\}$ is Cauchy, for the chosen $\epsilon > 0$, we have another positive integer k_3 such that whenever $m, n \geq k_3$, we have $|x_m - x_n| < \frac{\epsilon}{2}$. Call $k_0 = \max\{k_2, k_3\}$. Then $n_{k_0} \geq k_0 \geq k_3$, and hence for each $n \geq k_0$, choosing $m = n_{k_0}$, we see that $|x_n - x_{n_{k_0}}| < \frac{\epsilon}{2}$. Also $k_0 \geq k_2$, and hence by $(*)$, $|x_{n_{k_0}} - x| < \frac{\epsilon}{2}$. Therefore, for each $n \geq k_0$,

$$|x_n - x| \leq |x_n - x_{n_{k_0}}| + |x_{n_{k_0}} - x| < \frac{\epsilon}{2} + \frac{\epsilon}{2} = \epsilon.$$

Since $\epsilon > 0$ is arbitrary, $\{x_n\}$ is convergent.

■

Observe that the proof relies on the Bolzano–Weierstrass theorem, which, in turn, follows from the completeness axiom. What is particularly intriguing in this context is that the converse also holds: The convergence of all Cauchy sequences in $\mathbb{R}$ implies the completeness axiom. We now proceed to prove this assertion.

Theorem 4.3 *Assume that every Cauchy sequence in* $\mathbb{R}$ *is convergent. Then every non-empty bounded above subset of* $\mathbb{R}$ *has a least upper bound in* $\mathbb{R}$.

Proof Recall the definition of least upper bound from Definition 1.39. Let A be a non-empty bounded above subset of $\mathbb{R}$. Since $A \neq \varnothing$, choose $a_0 \in A$. Also, as A is bounded above, let b_0 $(> a_0)$ be an upper bound of A. Call $c_1 = \frac{a_0+b_0}{2}$. If c_1 is an upper bound of A, call $b_1 = c_1$ and $a_1 = a_0$. Otherwise, call $a_1 = c_1$ and $b_1 = b_0$. Similarly let $c_2 = \frac{a_1+b_1}{2}$. If c_2 is an upper bound of A, call $b_2 = c_2$ and $a_2 = a_1$. Otherwise, call $a_2 = c_2$ and $b_2 = b_1$ (see Fig. 4.1). Continuing inductively in this fashion, we construct two sequences (with possibilities of infinite repetition) $\{a_n\}$ and $\{b_n\}$, where for every $n \in \mathbb{N}$, a_n is not an upper bound of A and b_n is an upper

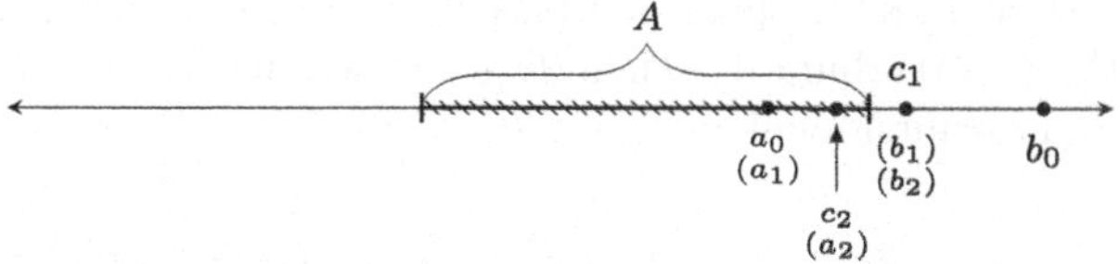

Fig. 4.1 Construction of *lub*

bound of A. Observe that for every $n \in \mathbb{N}$,

$$a_0 \leq a_1 \leq a_2 \leq \cdots \leq a_{n-1} \leq a_n < b_n \leq b_{n-1} \cdots \leq b_2 \leq b_1 \leq b_0,$$

and $b_n - a_n = \frac{b_0 - a_0}{2^n}$, i.e. $b_n - a_n \to 0$ as $n \to \infty$. Choose $\epsilon > 0$ arbitrarily. Then we can find $k \in \mathbb{N}$ such that $b_k - a_k < \epsilon$. Now for $m, n \geq k$, $a_m, a_n, b_m, b_n \in [a_k, b_k]$, and this implies $|a_m - a_n| < \epsilon$ and $|b_m - b_n| < \epsilon$, i.e. both the sequences $\{a_n\}$ and $\{b_n\}$ are Cauchy sequences. By hypothesis, both $\{a_n\}$ and $\{b_n\}$ are convergent. Call $\alpha = \lim a_n$ and $\beta = \lim b_n$. But $\lim(b_n - a_n) = 0$ implies that $\alpha = \beta = M$, say. We will show that $M = \sup A$.

First we claim that M is an upper bound of A. To prove this claim, assume the contrary. Then some element $a \in A$ will exceed M. Choose $\epsilon > 0$ so that $M+\epsilon < a$. As $b_n \to M$, we will have some term b_k from the sequence $\{b_n\}$ so that $b_k < M + \epsilon < a$. But this is a contradiction, as b_k is an upper bound of A.

Next we show that M is the smallest upper bound of A. Assuming the contrary again, we get an upper bound $b < M$ of A. Choose $\epsilon > 0$ so that $b < M - \epsilon$. As $a_n \to M$, we will have some term a_p from the sequence $\{a_n\}$ so that $b < M - \epsilon < a_p$. This is again a contradiction, as by construction, a_p is not an upper bound of A.

Consequently, M is the least upper bound of A and the proof is complete. ■

Thus, the convergence of Cauchy sequences allows us to extend the concept of the completeness axiom of $\mathbb{R}$ to an arbitrary metric space. However, we observe that not every space possesses the same richness as the real numbers. For instance, consider the space $X = (0, 1]$, where the Cauchy sequence $\{x_n = \frac{1}{n}\}_n$ fails to find its limit within X. This highlights an important distinction: The convergence of a sequence in an arbitrary space depends not only on the sequence itself but also on the presence of its limit within the space.

To formally distinguish spaces that guarantee a limit for every "eligible" (Cauchy) sequence, we introduce the following definition:

Definition 4.4 A metric space (X, d) is called *complete* if every Cauchy sequence in X is convergent.

A metric space is *incomplete* if it is not complete.

On the other hand, whether a metric space is complete or incomplete, every convergent sequence is necessarily a Cauchy sequence. More precisely, we assert the following:

Proposition 4.5 *In a metric space, a convergent sequence is always Cauchy.*

Proof Let (X, d) be a metric space and $\{x_n\} \subset X$ be a sequence converging to a point $x \in X$. Fix $\epsilon > 0$ arbitrarily. Then $\exists k \in \mathbb{N}$ such that $\forall n \geq k$, $d(x_n, x) < \frac{\epsilon}{2}$. Therefore for every natural number $m, n \geq k$,

$$d(x_m, x_n) \leq d(x_m, x) + d(x_n, x) < \frac{\epsilon}{2} + \frac{\epsilon}{2} = \epsilon.$$

Consequently, $\{x_n\}$ is Cauchy. ∎

Examples

4.5.1. $\mathbb{R}$ is complete with respect to the usual metric.
Proof. Precisely what we proved in Theorem 4.2. □

4.5.2. $\mathbb{Q}$ is incomplete with respect to the usual induced metric.
Proof. The sequence $\left\{\left(1+\frac{1}{n}\right)^n\right\}_n$ is a sequence of rational numbers converging to e in $\mathbb{R}$. Since it is convergent in $\mathbb{R}$, it is Cauchy in $\mathbb{R}$ and hence is so in $\mathbb{Q}$ in the induced metric as well. But as e is not rational, this sequence is not convergent in $\mathbb{Q}$. Consequently, $\mathbb{Q}$ is incomplete. □

4.5.3. Refer Example 2.2.4. For $p \geq 1$, $\mathbb{R}^n$ is complete w.r.t. the metric d_p.
Proof. Let $\{\mathbf{x}^{(m)}\}_m$ be a Cauchy sequence in $\mathbb{R}^n$, where $\mathbf{x}^{(m)} = (x_1^{(m)}, \ldots, x_n^{(m)})$. We will show $\{\mathbf{x}^{(m)}\}$ converges to a point in $\mathbb{R}^n$. Now $\{\mathbf{x}^{(m)}\}$ is Cauchy implies that for any $\epsilon > 0$ arbitrary, $\exists N \in \mathbb{N}$ such that $\forall m, m' \geq N, d_p(\mathbf{x}^{(m)}, \mathbf{x}^{(m')}) < \epsilon$, i.e. for every $m, m' \geq N$,

$$\sum_{k=1}^{n} |x_k^{(m)} - x_k^{(m')}|^p < \epsilon^p \implies \forall k = 1, \ldots, n,\ |x_k^{(m)} - x_k^{(m')}| < \epsilon.$$

But this shows that for every coordinate $k = 1, \ldots, n$, $\{x_k^{(m)}\}_m$ is a Cauchy sequence of real numbers, hence convergent. Let for kth coordinate, $x_k := \lim_{m\to\infty} x_k^{(m)}$, and call $\mathbf{x} := (x_1, \ldots, x_n) \in \mathbb{R}^n$. Let for $N_k \in \mathbb{N}$, $|x_k^{(m)} - x_k| < \frac{\epsilon}{n^{1/p}}$. Choose $N_0 := \max\{N_k : k = 1, \ldots, n\}$. Then for every $m \geq N_0$,

$$\|\mathbf{x}^{(m)} - \mathbf{x}\|_p^p = \sum_{k=1}^{n} |\mathbf{x}_k^{(m)} - x_k|^p < \epsilon^p \implies \|\mathbf{x}^{(m)} - \mathbf{x}\|_p < \epsilon.$$

Hence $\mathbf{x}^{(m)} \to \mathbf{x}$. This completes the proof. □

4.5.4. Refer Example 2.2.5. $\mathbb{R}^n$ is complete w.r.t. the metric d_∞. Proof of this fact is very similar to the previous example and is left as an exercise.

4.5.5. Refer Example 2.2.6. ℓ_p is a complete metric space.
Proof. We start with the claim that for two points $\mathbf{x} := \{x_k\}_k$, and $\mathbf{y} := \{y_k\}_k$ from ℓ_p, $\mathbf{x} + \mathbf{y} \in \ell_p$ as well. $(*)$

Proof of this is evident from the triangle inequality of the norm metric d_p as

$$\left[\sum_{k=1}^{\infty}|x_k + y_k|^p\right]^{1/p} = \|\mathbf{x}+\mathbf{y}\|_p \leq \|\mathbf{x}\|_p + \|\mathbf{y}\|_p$$

$$= \left[\sum_{k=1}^{\infty}|x_k|^p\right]^{1/p} + \left[\sum_{k=1}^{\infty}|y_k|^p\right]^{1/p} < \infty.$$

Now consider a Cauchy sequence $\{\mathbf{x}^{(n)}\} \subset \ell_p$, where $\mathbf{x}^{(n)} := \{x_1^{(n)}, x_2^{(n)}, \dots\}$. We need to show that $\{\mathbf{x}^{(n)}\}$ converges to a point $\mathbf{x} \in \ell_p$.

As $\{\mathbf{x}^{(n)}\}$ is Cauchy, for arbitrarily fixed $\epsilon > 0, \exists\, N \in \mathbb{N}$ such that $\forall m, n \geq N, d_p(\mathbf{x}^{(m)}, \mathbf{x}^{(n)}) < \epsilon$, i.e.

$$\forall m, n \geq N, \quad \sum_{k=1}^{\infty}|x_k^{(m)} - x_k^{(n)}|^p < \epsilon^p. \tag{4.1}$$

From here, we conclude the following:

$$\text{for each } k \in \mathbb{N}, \text{ and } m, n \geq N, \quad |x_k^{(m)} - x_k^{(n)}| < \epsilon \tag{4.2}$$

and

$$\text{for each } K \in \mathbb{N}, \text{ and } m, n \geq N, \quad \sum_{k=1}^{K}|x_k^{(m)} - x_k^{(n)}|^p < \epsilon^p. \tag{4.3}$$

Since $\epsilon > 0$ is arbitrary, from (4.2) we conclude that for each coordinate $k \in \mathbb{N}$, the sequence $\{x_k^{(n)}\}_n$ is a Cauchy sequence in $\mathbb{R}$, hence convergent (see Fig. 4.2).

For $k \in \mathbb{N}$, let $x_k = \lim_{n\to\infty} x_k^{(n)}$ and call $\mathbf{x} = (x_1, x_2, \dots)$.

Now in (4.3) letting $m \to \infty$ while keeping n fixed, we get

$$\forall K \in \mathbb{N} \text{ and } \forall n \geq N, \quad \sum_{k=1}^{K}|x_k - x_k^{(n)}|^p \leq \epsilon^p. \tag{4.4}$$

Again in (4.4), letting $K \to \infty$, we get

$$\forall n \geq N, \quad \sum_{k=1}^{\infty}|x_k - x_k^{(n)}|^p \leq \epsilon^p. \tag{4.5}$$

Choosing $n = N$ in (4.5), we see that $\{x_k - x_k^{(N)}\}_k \in \ell_p$, i.e. $\mathbf{x} - \mathbf{x}^{(N)} \in \ell_p$. Finally, using the claim $(*)$, we conclude $\mathbf{x} = (\mathbf{x} - \mathbf{x}^{(N)}) + \mathbf{x}^{(N)} \in \ell_p$.

$\{\mathbf{x}^{(n)}\}_n$	$=$	$\{x_1^{(n)}\}_n$	$\{x_2^{(n)}\}_n$	$\{x_3^{(n)}\}_n$	$\cdots$	$\{x_k^{(n)}\}_n$	$\cdots$
$\mathbf{x}^{(1)}$	$=$	$x_1^{(1)}$	$x_2^{(1)}$	$x_3^{(1)}$	$\cdots$	$x_k^{(1)}$	$\cdots$
$\mathbf{x}^{(2)}$	$=$	$x_1^{(2)}$	$x_2^{(2)}$	$x_3^{(2)}$	$\cdots$	$x_k^{(2)}$	$\cdots$
$\vdots$		$\vdots$	$\vdots$	$\vdots$		$\vdots$	
$\mathbf{x}^{(n)}$	$=$	$x_1^{(n)}$	$x_2^{(n)}$	$x_3^{(n)}$	$\cdots$	$x_k^{(n)}$	$\cdots$
$\vdots$		$\vdots$	$\vdots$	$\vdots$		$\vdots$	
$\downarrow$		$\downarrow$	$\downarrow$	$\downarrow$		$\downarrow$	
$\mathbf{x}$	$=$	x_1	x_2	x_3	$\cdots$	x_k	$\cdots$

Fig. 4.2 Convergence in ℓ_p

Also, (4.5) translates to $\forall n \geq N, d_p(\mathbf{x}^{(n)}, \mathbf{x}) \leq \epsilon$. Since $\epsilon > 0$ is arbitrary,

$$\lim_{n\to\infty} \mathbf{x}^{(n)} = \mathbf{x}.$$

Hence the proof. □

4.5.6. Refer Example 2.2.7. ℓ_∞ is also complete. Proof of this fact is very similar to the previous example and is left for the reader as an exercise.

4.5.7. Refer Example 2.2.9. $\mathcal{C}_{\sup}[a, b]$ is complete.

Proof. Let $\{f_n\}$ be a Cauchy sequence in $\mathcal{C}[a, b]$. Then for any chosen $\epsilon > 0$, $\exists k \in \mathbb{N}$ such that

$$\forall m, n \geq k, \quad \sup_{x\in[a,b]} |f_m(x) - f_n(x)| < \frac{\epsilon}{2}. \tag{$*$}$$

But this implies that for every natural number $m, n \geq k$, and for each $x \in [a, b]$, $|f_m(x) - f_n(x)| < \frac{\epsilon}{2}$. This amounts to saying that for each point $x \in [a, b]$, $\{f_n(x)\}_n$ is a Cauchy sequence of real numbers in usual metric, hence convergent.

Let $f(x) = \lim_{n\to\infty} f_n(x)$, $x \in [a, b]$. We will show that f is continuous on $[a, b]$ (i.e. $f \in \mathcal{C}[a, b]$) and $f_n \to f$ w.r.t. the metric d_∞.

Now for $x \in [a, b]$, and $n \geq k$, by $(*)$ we have

$$\begin{aligned}|f_n(x) - f(x)| &= \lim_{m\to\infty} |f_n(x) - f_m(x)| \\ &\leq \lim_{m\to\infty} \sup_{x\in[a,b]} |f_n(x) - f_m(x)| \leq \frac{\epsilon}{2}.\end{aligned}$$

Hence, $\forall n \geq k, d_\infty(f_n, f) = \sup_{x\in[a,b]} |f_n(x) - f(x)| \leq \frac{\epsilon}{2} < \epsilon$. $\cdots (**)$

At this point, let us recall a result from real analysis relevant in this regard:

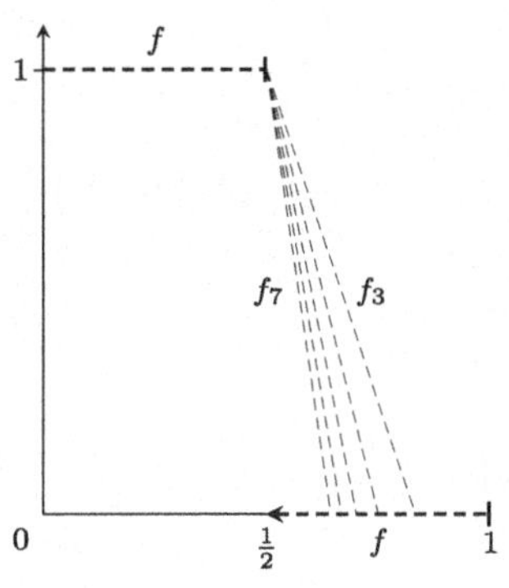

(a): Functions f_3 to f_7 and the limit function f

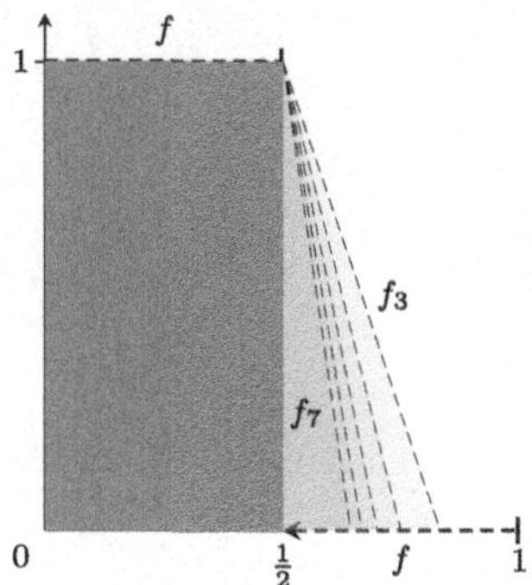

(b): Convergence in the d_1 metric

Fig. 4.3 $\mathcal{C}_{\text{int}}[0, 1]$ is incomplete

> Let D be a non-empty subset of $\mathbb{R}$, and let $\{f_n\}$ be a sequence of functions converging pointwise to a function f on D. Let $M_n := \sup_{x\in D}|f_n(x)-f(x)|$. Then $\{f_n\}$ converges uniformly to f on D iff $\lim M_n = 0$.

But $(**)$ implies that the required condition is satisfied in this case. So the sequence $\{f_n\}$ converges uniformly to f on $[a, b]$ w.r.t. usual metric on $\mathbb{R}$. Therefore, as f_n's are all continuous on $[a, b]$, f is also so. Hence the proof. □

4.5.8. Refer Example 2.2.8. $\mathcal{C}_{\text{int}}[0, 1]$ is incomplete.

Proof. The idea here is to approximate a discontinuous function by a sequence of continuous functions in terms of the area.

Let $\{f_n(x)\}_{n\geq 3}$ on $[0, 1]$ be defined by

$$f_n(x) = \begin{cases} 1, & x \in \left[0, \frac{1}{2}\right] \\ 1 - n\left(x - \frac{1}{2}\right), & x \in \left(\frac{1}{2}, \frac{1}{2} + \frac{1}{n}\right] \\ 0, & x \in \left(\frac{1}{2} + \frac{1}{n}, 1\right] \end{cases}.$$

We will first show that $\{f_n\}$ is a Cauchy sequence of continuous functions on $[0, 1]$ (Fig. 4.3). Next we will show $\{f_n\}$ does not converge to any continuous function f on $[0, 1]$.

To see that $\{f_n\}$ is a Cauchy sequence, observe that for arbitrary natural numbers n and p,

$$\begin{aligned} d_1(f_{n+p}, f_n) &= \int_0^1 |f_{n+p} - f_n| \\ &= \int_0^1 (f_n - f_{n+p}), \qquad [\because f_{n+p} \leq f_n \text{ on } [0, 1]] \end{aligned}$$

$$= \int_{\frac{1}{2}}^{\frac{1}{2}+\frac{1}{n}} (f_n - f_{n+p}),$$

$$\left[\because f_n = f_{n+p} \text{ outside of } \left(\tfrac{1}{2}, \tfrac{1}{2} + \tfrac{1}{n}\right]\right]$$

$$\leq \int_{\frac{1}{2}}^{\frac{1}{2}+\frac{1}{n}} f_n, \qquad [\because f_{n+p} \geq 0]$$

$$= \int_{\frac{1}{2}}^{\frac{1}{2}+\frac{1}{n}} 1 - n\left(x - \frac{1}{2}\right) dx = \frac{1}{2n}.$$

For any arbitrary fixed $\epsilon > 0$, choose $k \in \mathbb{N}$ such that $\frac{1}{2k} < \epsilon$. Hence, for every $n \geq k$ and all natural numbers p,

$$d_1(f_{n+p}, f_n) \leq \frac{1}{2n} \leq \frac{1}{2k} < \epsilon.$$

Therefore, $\{f_n\}$ is a Cauchy sequence.
Now we show that $\{f_n\}$ does not converge to any continuous function on $[0, 1]$. Assume the contrary. Let $\{f_n\}$ has a limit $f \in \mathcal{C}[0, 1]$. Then $f_n \to f$ implies $\int_0^1 |f_n - f| \to 0$. Since the integrand $|f_n - f|$ is always non-negative, using the linearity property of Riemann integration, we can say that for every pair of real numbers $a < b$ in $[0, 1]$,

$$\int_a^b |f_n - f| \to 0.$$

Choose $a = 0$ and $b = \frac{1}{2}$. As each $f_n = 1$ on $\left[0, \frac{1}{2}\right]$, this particular choice of a, b yields

$$\int_0^{\frac{1}{2}} |f_n - f| \to 0 \implies \int_0^{\frac{1}{2}} |1 - f| \to 0 \text{ as } n \to \infty.$$

Since the integrand is free from n, this implies $\int_0^{\frac{1}{2}} |1 - f| = 0$; and as f is continuous on $\left[0, \frac{1}{2}\right]$, $f = 1$ on $\left[0, \frac{1}{2}\right]$. (*)
For another choice of a and b, let $\frac{1}{2} < a < 1$ arbitrary, and $b = 1$. Choose $k \in \mathbb{N}$ big enough so that $\frac{1}{2} + \frac{1}{k} < a$. Then $\forall n \geq k$, $f_n = 0$ on $[a, 1]$. Hence

$$\lim_{n\to\infty} \int_a^1 |f_n - f| = 0 \implies \int_a^1 |f| = 0 \implies f = 0 \text{ on } [a, 1],$$

as f is continuous on $[a, 1]$. Since $\frac{1}{2} < a < 1$ is arbitrary, $f = 0$ on $\left(\frac{1}{2}, 1\right]$. This fact, combined with $(*)$, is a contradiction to the assumption f is continuous. Hence the proof. □

Miscellaneous Results

4.1. Give a counter-example to establish the following fact: Let X be a non-empty set and d and d' be two equivalent metrics on X. X is complete w.r.t. d does not necessarily imply that X is also complete w.r.t. d'.

Proof. On the set of natural numbers $\mathbb{N}$, let d denote the discrete metric. Then by Exercise 4.3, $\mathbb{N}$ is complete w.r.t. the metric d. Now define the metric d' on $\mathbb{N}$ by

$$d'(m, n) = \left|\frac{1}{m} - \frac{1}{n}\right|, \quad m, n \in \mathbb{N}.$$

By Example 3.16.1, d' is equivalent to d. Now consider the sequence

$$\{x_n\} := \{n\}_{n \in \mathbb{N}} = \{1, 2, 3, \ldots\}$$

in $(\mathbb{N}, d')$. For any $\epsilon > 0$ arbitrary,

$$d'(x_m, x_n) < \epsilon \Longleftarrow \left|\frac{1}{m} - \frac{1}{n}\right| < \epsilon \Longleftarrow \frac{1}{m} + \frac{1}{n} < \epsilon \Longleftarrow m, n \geq k',$$

where $k' := \left[\frac{2}{\epsilon}\right] + 1$. Hence $\{x_n\}$ is a Cauchy sequence in $(\mathbb{N}, d')$. We claim that $\{x_n\}$ is not convergent in $(\mathbb{N}, d')$. To see this claim, assume otherwise. Let for $p \in \mathbb{N}$, $x_n \to p$. Then for any arbitrary $\epsilon > 0$, there will exist a positive integer $k \in \mathbb{N}$ such that $\forall n \geq k$, $d'(x_n, p) < \frac{\epsilon}{2}$. WLOG, choose k big enough so that $\frac{1}{k} < \frac{\epsilon}{2}$. Then

$$d'(x_k, p) < \epsilon \Longrightarrow \left|\frac{1}{k} - \frac{1}{p}\right| < \frac{\epsilon}{2} \Longrightarrow \frac{1}{p} < \frac{1}{k} + \frac{\epsilon}{2} < \epsilon.$$

Since $\epsilon > 0$ is arbitrary, $\frac{1}{p} < \epsilon$ implies that $\frac{1}{p} = 0$. But there is no positive integer p such that $\frac{1}{p} = 0$. Hence the claim.

Therefore the Cauchy sequence $\{x_n\}$ in $(\mathbb{N}, d')$ is not convergent. Consequently, $(\mathbb{N}, d')$ is not complete. So d and d' are equivalent metrics on $\mathbb{N}$, but $\mathbb{N}$ is complete w.r.t. d but incomplete w.r.t. d'. □

4.2. Let $\mathbb{R}^\omega$ denote the set of all sequences of real numbers. For points $\mathbf{x} := \{x_k\}_k$ and $\mathbf{y} := \{y_k\}_k$ from $\mathbb{R}^\omega$, define

$$d(\mathbf{x}, \mathbf{y}) = \sum_{k=1}^{\infty} \frac{1}{k!} \cdot \frac{|x_k - y_k|}{1 + |x_k - y_k|}.$$

Show that $(\mathbb{R}^\omega, d)$ is a complete metric space.

Proof. We know that $\sum_k \frac{1}{k!}$ is a convergent series of positive reals. So by Exercise 2.7, d is a metric on $\mathbb{R}^\omega$. To show $(\mathbb{R}^\omega, d)$ is complete, consider a Cauchy sequence $\{\mathbf{x}^{(n)}\}_n \subset \mathbb{R}^\omega$, where $\mathbf{x}^{(n)} := \{x_1^{(n)}, x_2^{(n)}, \ldots\}$. We will show that $\{\mathbf{x}^{(n)}\}$ converges in $\mathbb{R}^\omega$.

For each kth coordinate of any two terms $\mathbf{x}^{(m)}$ and $\mathbf{x}^{(n)}$ of the sequence $\{\mathbf{x}^{(n)}\}$, we have

$$\begin{aligned}
\frac{|x_k^{(m)} - x_k^{(n)}|}{1 + |x_k^{(m)} - x_k^{(n)}|} &= k! \left(\frac{1}{k!} \cdot \frac{|x_k^{(m)} - x_k^{(n)}|}{1 + |x_k^{(m)} - x_k^{(n)}|} \right) \\
&\le k! \left(\sum_k \frac{1}{k!} \cdot \frac{|x_k^{(m)} - x_k^{(n)}|}{1 + |x_k^{(m)} - x_k^{(n)}|} \right) \\
&= k! \cdot d(\mathbf{x}^{(m)}, \mathbf{x}^{(n)}).
\end{aligned}$$

Since $\{\mathbf{x}^{(n)}\}$ is Cauchy, for an arbitrarily chosen $\epsilon > 0$, $\exists N_k \in \mathbb{N}$ such that

$$\forall m, n \ge N_k, \quad k! \cdot d(\mathbf{x}^{(m)}, \mathbf{x}^{(n)}) < \frac{\epsilon}{1 + \epsilon}.$$

But this implies,

$$\forall m, n \ge N_k, \quad \frac{|x_k^{(m)} - x_k^{(n)}|}{1 + |x_k^{(m)} - x_k^{(n)}|} < \frac{\epsilon}{1 + \epsilon}, \quad \text{i.e. } |x_k^{(m)} - x_k^{(n)}| < \epsilon.$$

Whence we conclude that for each coordinate k, $\{x_k^{(n)}\}_n$ (see Fig. 4.2) is a Cauchy sequence of real numbers and hence convergent. Call $x_k := \lim_{n\to\infty} x_k^{(n)}$ and $\mathbf{x} := \{x_1, x_2, \ldots\}$.

Clearly, $\mathbf{x} \in \mathbb{R}^\omega$. We will show that $\lim_{n\to\infty} \mathbf{x}^{(n)} = \mathbf{x}$.

Fix $\epsilon > 0$ arbitrarily. As $\sum \frac{1}{k!}$ is convergent, $\exists k_0 \in \mathbb{N}$ such that

$$\sum_{k > k_0} \frac{1}{k!} < \frac{\epsilon}{2}. \qquad (*)$$

Also, for each kth coordinate, $x_k^{(n)} \to x_k$, i.e. $\frac{|x_k^{(n)}-x_k|}{1+|x_k^{(n)}-x_k|} \to 0$ as $n \to \infty$. So for first k_0 coordinates, we can find $N_0 \in \mathbb{N}$ such that

$$\forall n \geq N_0, \quad \frac{|x_k^{(n)} - x_k|}{1 + |x_k^{(n)} - x_k|} \leq k! \cdot \frac{\epsilon}{2k_0} \text{ holds for every } k \leq k_0. \qquad (**)$$

Therefore, for every $n \geq N_0$, we have

$$\begin{aligned} d(\mathbf{x}^{(n)}, \mathbf{x}) &= \sum_{k=1}^{\infty} \frac{1}{k!} \cdot \frac{|x_k^{(n)} - x_k|}{1 + |x_k^{(n)} - x_k|} \\ &= \sum_{k \leq k_0} \frac{1}{k!} \cdot \frac{|x_k^{(n)} - x_k|}{1 + |x_k^{(n)} - x_k|} + \sum_{k > k_0} \frac{1}{k!} \cdot \frac{|x_k^{(n)} - x_k|}{1 + |x_k^{(n)} - x_k|} \\ &< \sum_{k \leq k_0} \frac{\epsilon}{2k_0} + \sum_{k > k_0} \frac{1}{k!}, && \text{(by } (**)) \\ &< \epsilon. && \text{(by } (*)) \end{aligned}$$

As $\epsilon > 0$ is arbitrary, $\mathbf{x}^{(n)} \to \mathbf{x}$. Consequently, $(\mathbb{R}^\omega, d)$ is complete. □

Exercises: 4.1

4.1. Show that a Cauchy sequence having a convergent subsequence is convergent.
4.2. Prove or disprove: A closed subspace of *any* metric space is complete.
4.3. Show that a discrete metric space is always complete.
4.4. Refer Example 2.2.5. Show that $\mathbb{R}^n$ is a complete metric space w.r.t. the metric d_∞.
4.5. Show that ℓ_∞ is a complete metric space.
4.6. Let X be the set of all convergent real sequences. For elements $\mathbf{x} = \{x_n\}$ and $\mathbf{y} = \{y_n\}$ from X, define

$$d(\mathbf{x}, \mathbf{y}) := \sup\{|x_n - y_n| : n \in \mathbb{N}\}.$$

Show that (X, d) is a complete metric space.
4.7. Let (X, d_X) and (Y, d_Y) be two metric spaces. Show that $X \times Y$ is complete w.r.t. the product metric if and only if both (X, d_X) and (Y, d_Y) are complete.

4.8. Let for every $n \in \mathbb{N}$, (X_n, d_n) denotes a complete metric space. The product space

$$X := \prod_{n\in\mathbb{N}} X_n$$

is the set of all sequences $\mathbf{x} = \{x_n\}$ where $x_n \in X_n$. For two points $\mathbf{x} = \{x_n\}$ and $\mathbf{y} = \{y_n\}$ from X, define

$$d(\mathbf{x}, \mathbf{y}) := \sum_{n\in\mathbb{N}} \frac{1}{n!} \cdot \frac{d_n(x_n, y_n)}{1 + d_n(x_n, y_n)}.$$

Show that (X, d) is a complete metric space.

4.9. Denote by $\mathcal{C}(\mathbb{R})$, the set of all real-valued continuous functions defined over $\mathbb{R}$. For each $n \in \mathbb{N}$, let

$$d_n(f, g) := \sup_{x\in[-n,n]} |f(x) - g(x)|, \qquad f, g \in \mathcal{C}[a, b].$$

For functions $f, g \in \mathcal{C}[a, b]$, show that the map d defined by

$$d(f, g) := \sum_{n\in\mathbb{N}} \frac{1}{n!} \cdot \frac{d_n(f, g)}{1 + d_n(f, g)}$$

is a complete metric on X.

4.10. Let X be a metric space. Let us introduce the following notations:

- $\mathcal{B}(X)$ denotes the set of all real-valued bounded functions on X.
- $\mathcal{C}(X)$ denotes the set of all real-valued continuous functions on X.
- $\mathcal{C}_b(X)$ denotes the set of all real-valued bounded continuous functions on X.

Show that all of the above spaces are complete w.r.t. the metric

$$d_\infty(f, g) := \sup_{x\in X} |f(x) - g(x)|.$$

4.2 Results and Consequences

Theorem 4.6 *A subspace Y of a complete metric space X is complete iff Y is closed in X.*

Proof *Necessary Part:* Let Y be complete as a subspace of X. We will show Y is closed in X. Let $y \in X$ be a limit point of Y. Then for each natural number n, the ball $B\left(y; \frac{1}{n}\right)$ contains at least a point $y_n \in Y$. Consider the sequence $\{y_n\} \subset Y$. Clearly, $\{y_n\}$ converges to y in X. Hence, by Proposition 4.5, $\{y_n\}$ is Cauchy as well. And since Y is complete, $\{y_n\}$ is convergent in Y. Therefore, $y \in Y$, and hence Y is closed.

Sufficient Part: Let Y be a closed subset of a complete metric space X and $\{y_n\} \subset Y$ be a Cauchy sequence. Then $\{y_n\}$ is a Cauchy sequence in X as well, and since X is complete, it converges to a point $y \in X$, say. Then by Theorem 3.41, $y \in \overline{Y}$. Since Y is closed in X, $y \in Y$, i.e. $\{y_n\}$ is convergent in Y. Consequently, Y is complete. ■

The following theorem provides an NASC for a metric space to be complete.

Theorem 4.7 (Cantor's Intersection Theorem) *A metric space (X, d) is complete if and only if intersection of every nested sequence of non-empty closed sets $\{F_n\}$ with $\lim_{n\to\infty} \operatorname{diam}(F_n) = 0$ has precisely one point, i.e. $\bigcap_n F_n$ is a singleton set.*

Proof *Necessary Part:* Let (X, d) be a complete metric space and $\{F_n\}$ be a sequence of non-empty closed sets such that:

i) For every $n \in \mathbb{N}$, $F_{n+1} \subset F_n$.
ii) $\lim_{n\to\infty} \operatorname{diam}(F_n) = 0$.

Choose a point x_n from each F_n and consider the sequence $\{x_n\}$. Since diam $F_n \to 0$, for any chosen $\epsilon > 0$, $\exists k \in \mathbb{N}$ such that diam $F_k < \epsilon$. But $F_{n+1} \subset F_n$ implies that for each $n \geq k$, $x_n \in F_k$. Then

$$\forall m, n \geq k,\ d(x_m, x_n) \leq \operatorname{diam} F_k < \epsilon,$$

showing that the sequence $\{x_n\}$ is Cauchy. Hence by hypothesis, $\{x_n\}$ is convergent. Let $x_n \to x \in X$. We will show $\bigcap_n F_n = \{x\}$.

$x_n \to x$ implies for the chosen $\epsilon > 0$, $\exists p \in \mathbb{N}$ such that $\forall n \geq p$, $x_n \in B(x; \epsilon)$, i.e. $\forall n \geq p$, $B(x; \epsilon) \cap F_n \neq \varnothing$. Also, as F_n's are nested, $x_p \in F_n$ for every $1 \leq n < p$. Hence for every $n \in \mathbb{N}$, $B(x; \epsilon) \cap F_n \neq \varnothing$, showing that $x \in \overline{F_n}$ for every $n \in \mathbb{N}$. But as F_n's are closed sets, $\overline{F_n} = F_n$, and hence, for every n, $x \in F_n$, i.e. $x \in \bigcap_n F_n$.

To show $\bigcap_n F_n$ does not have any other element than x, assume the contrary. Let $x' \in \bigcap_n F_n$. Then $x, x' \in F_k$. But diam $F_k < \epsilon$ implies $d(x, x') < \epsilon$. Since $\epsilon > 0$ is arbitrary, $d(x, x') = 0$, i.e. $x = x'$. Hence $\bigcap_n F_n = \{x\}$.

Sufficient Part: Assume that in the metric space (X, d), intersection of every nested sequence of non-empty closed sets $\{F_n\}$ with $\lim_{n\to\infty} \operatorname{diam}(F_n) = 0$ has precisely one point. We will show that X is complete.

Let $\{x_n\} \subset X$ be a Cauchy sequence. For $n \in \mathbb{N}$, construct

$$H_n := \{x_n, x_{n+1}, \dots\}.$$

Then for every n, $H_{n+1} \subset H_n$. As $\{x_n\}$ is Cauchy, for any arbitrary $\epsilon > 0$, $\exists k \in \mathbb{N}$ such that $\forall n \geq k$, $d(x_{n+p}, x_n) < \frac{\epsilon}{2}$ holds for every $p \in \mathbb{N}$. But this implies $\forall n \geq k$, diam $H_n \leq \frac{\epsilon}{2} < \epsilon$, i.e. diam $H_n \to 0$ as $n \to \infty$. Call $F_n := \overline{H_n}$. Then F_n's are closed and $x_n \in F_n$. Also, by Exercise 3.9, $F_{n+1} \subset F_n$, and diam $F_n =$ diam $H_n \to 0$. Hence by hypothesis, $\bigcap_n F_n$ is a singleton set, $\{x\}$, say. Now diam $F_n \to 0$ implies that for the chosen $\epsilon > 0$, $\exists p \in \mathbb{N}$ such that $\forall n \geq p$, diam $F_n < \epsilon$. Again $x \in \bigcap_n F_n$ implies that $\forall n \geq p$, $d(x_n, x) < \epsilon$. Therefore $x_n \to x$. Since $\{x_n\} \subset X$ is any arbitrary Cauchy sequence, X is complete. ■

We use Cantor's intersection theorem to prove the mighty Baire category theorem demanding its own couple of subsections.

4.2.1 Baire Category Theorem

We often explore the notion of "smallness" in an abstract setting, where its meaning varies depending on the context. In our present discussion, we attempt to characterise a set as "small" based on its cardinality. From this perspective, the smallest non-trivial set is a singleton, and we investigate spaces generated by different collections of singletons.

A finite collection of singletons, meaning a space with only finitely many points, always forms a discrete space (Exercise 3.7). Next, we consider an enumerable collection of singletons—an enumerable space—and ask how rich such a space can be. Specifically, can it be a complete space? Here, the discrete metric again poses a challenge, as every discrete space, regardless of its cardinality, is complete. To avoid this pathological case, we refine our notion of "smallness" by identifying it with nowhere dense sets, formally defined as follows:

Definition 4.8 Let X be a metric space. A subset $A \subset X$ is called *nowhere dense* in X if the interior of its closure does not have any non-empty open set, i.e. int $\overline{A} = \varnothing$.

In a discrete space, the only nowhere dense set is the empty set. Now, we continue our investigation: Does there exist a countable collection of nowhere dense sets that still yields a complete metric space? The Baire category theorem ensures that such a possibility does not exist. This result is fundamental in analysis and has far-reaching consequences. In this subsection, we examine the theorem itself, followed by some of its applications in the next.

To motivate the above definition, we recall our discussion on dense subsets in the previous chapter (Sect. 3.5.2). By Definition 3.45, a "dense subset" is dense everywhere in the usual sense of the word. However, if we relax this "everywhere" condition, we obtain sets like $A = (0, 1) \cap \mathbb{Q}$. Clearly, A is not dense throughout $\mathbb{R}$, but it is dense somewhere—specifically, in $[0, 1]$. Continuing in this direction, we arrive at the other extreme: finite sets, which are merely scattered collections of isolated points. (A point a is called an *isolated point* of a set A if, for some $r > 0$, the neighbourhood $B(a; r) \cap A$ contains only the point a).

Thus, we define a nowhere dense set as a collection of isolated points while ensuring that it remains hollow in its interior. One way to enforce this hollowness is to require that its closure does not contain any non-empty open set. This condition is natural: If a subset A contains an open ball $B(a; r)$, then A will necessarily be dense at least in $B[a; \frac{r}{2}]$.

We now proceed to a more formal discussion. An equivalent formulation of the above definition is as follows:

Definition 4.9 Let X be a metric space. A non-empty subset $A \subset X$ is called *nowhere dense* in X if for every non-empty open set $U \subset X$, there is a non-empty open subset $V \subset U$ such that $A \cap V = \varnothing$.

Remarks

(a) The proof of equivalence of these two definitions and a few more versions is left for the reader in Exercise 4.11.
(b) Subset of a nowhere dense set is again nowhere dense.

Examples

4.9.1. In the metric space $\mathbb{R}$ with usual metric, $\mathbb{Z}$ is nowhere dense.
4.9.2. In the metric space $\mathbb{R}^2$ with usual metric, the subset $\mathbb{R}$ identified as $\mathbb{R} \equiv \{(x, 0) : x \in \mathbb{R}\}$ is nowhere dense. In fact, any proper subspace of the real or complex vector spaces $\mathbb{R}^n$ or $\mathbb{C}^n$ is nowhere dense. A proof of this fact is asked in Exercise 4.15.
4.9.3. In a discrete metric space, $\varnothing$ is the only nowhere dense subset.

Definition 4.10 Let X be a metric space and $A \subset X$. A is said to be a subset of *category I* if A is a countable union of nowhere dense subsets. Such sets are also often called as *meagre sets.*

A is said to be a subset of *category II* if it is not of category I. A set of category II is also known as *non-meagre set.*

Examples

4.10.1. In a metric space, the empty set is of category I.
4.10.2. The set of all rational numbers is of category I in the metric space of real numbers with usual metric. Each rational number r, considered as the set $\{r\}$, is nowhere dense. Hence, the countability of $\mathbb{Q}$ guarantees that it is of category I.

Remarks

(a) Every subset of a metric space is either of category I or of category II.
(b) A subset of a category I subset in a metric space is again of category I.
(c) A countable union of category I subsets is again of category I.

Theorem 4.11 (Baire Category Theorem) *Every complete metric space is of category II.*

Proof Assume the contrary. Let X be a complete metric space and $\{A_n\}_{n\in\mathbb{N}}$ be a sequence of nowhere dense sets such that $X = \bigcup_n A_n$. As each A_n is nowhere dense, by Exercise 4.11(b), $(\overline{A_n})^c$ is everywhere dense and hence is non-empty.

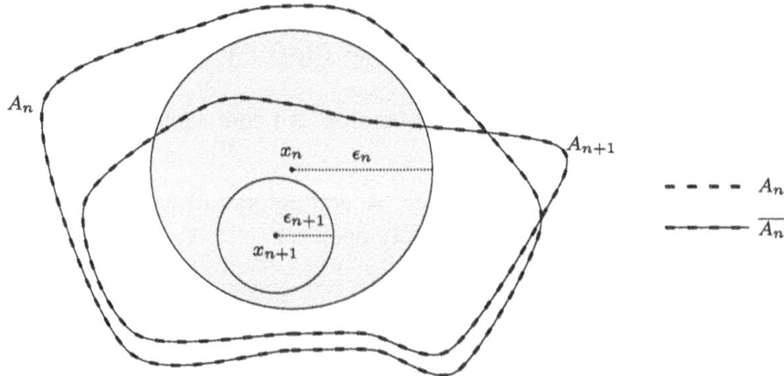

Fig. 4.4 Construction of $B(x_n; \epsilon_n)$'s. A_n consists solely of points on its boundary, with no inside points included

Choose $x_n \in (\overline{A_n})^c$. As $(\overline{A_n})^c$ is open, there exists some $r > 0$ such that $B(x_n; r) \subset (\overline{A_n})^c$. Then for any $0 < \epsilon_n < r$,

$$\overline{B(x_n; \epsilon_n)} \subset B(x_n; r) \subset (\overline{A_n})^c \subset (A_n)^c, \text{ i.e. } \overline{B(x_n; \epsilon_n)} \cap A_n = \varnothing.$$

Inductively, we claim to obtain $x_{n+1} \in (\overline{A_{n+1}})^c$ and $\epsilon_{n+1} > 0$ with the following properties (Fig. 4.4):

i) $\overline{B(x_{n+1}; \epsilon_{n+1})} \cap A_{n+1} = \varnothing$.
ii) $\overline{B(x_{n+1}; \epsilon_{n+1})} \subset \overline{B(x_n; \epsilon_n)}$.
iii) $\epsilon_{n+1} \leq \frac{1}{2}\epsilon_n$.

Proof of Claim Look at $B(x_n; \epsilon_n) \cap (\overline{A_{n+1}})^c$. If $B(x_n; \epsilon_n) \cap (\overline{A_{n+1}})^c = \varnothing$, then $B(x_n; \epsilon_n) \subset \overline{A_{n+1}}$. This contradicts that A_{n+1} is nowhere dense. Hence $B(x_n; \epsilon_n) \cap (\overline{A_{n+1}})^c \neq \varnothing$. Choose $x_{n+1} \in B(x_n; \epsilon_n) \cap (\overline{A_{n+1}})^c$. Now, as $B(x_n; \epsilon_n) \cap (\overline{A_{n+1}})^c$ is open, there is some $r > 0$ such that

$$B(x_{n+1}; r) \subset B(x_n; \epsilon_n) \cap (\overline{A_{n+1}})^c.$$

Choosing $0 < \epsilon_{n+1} \leq \frac{1}{2} \min\{r, \epsilon_n\}$, we establish the claim.

Observe that

$$\operatorname{diam} \overline{B(x_n; \epsilon_n)} = 2\epsilon_n \leq \frac{\epsilon_1}{2^{n-2}}.$$

Hence, $\epsilon_n \to 0$ as $n \to \infty$.

Thus we have a nested sequence $\{\overline{B(x_n; \epsilon_n)}\}_n$ of closed sets with diameters approaching to zero in the complete metric space X. Hence by Cantor's intersection theorem (Theorem 4.7), we have a point

$$\xi \in \bigcap_n \overline{B(x_n; \epsilon_n)}.$$

But for every n, $\overline{B(x_n; \epsilon_n)} \cap A_n = \varnothing$, implying $\xi \notin A_n$. Hence $\xi \notin \bigcup_n A_n = X$, a contradiction to our assumption.

Therefore, every complete metric space must be of category II. ■

The following statements are equivalent forms of Baire category theorem.

Corollary 4.11.1 (Baire Category Theorem) *Let X be a complete metric space and $\{A_n\}_n$ be a sequence of subsets of X such that $\bigcup_n A_n = X$. Then the closure of at least one A_n must have non-empty interior.*

Corollary 4.11.2 (Baire Category Theorem) *Let X be a complete metric space and $\{U_n\}_n$ be a sequence of open dense subsets of X. Then $\bigcap_n U_n$ is non-empty.*

Remark George F Simmons in [12] says:

> This theorem is admittedly rather technical in nature, and the reader can hardly be expected to appreciate its significance at the present stage of our work. He will find, however, that a need for it crops up from time to time, and when this need arises, Baire's theorem is an indispensable tool.

4.2.2 Applications of Baire Category Theorem

This subsection will present three significant results obtained using the Baire category theorem, along with a few more applications. The sheer variety of topics in Mathematics these results dwell on will provide an idea about the might of this theorem.

Uniform Boundedness Principle

Theorem 4.12 *Let X be a complete space and $\mathcal{C}(X, \mathbb{R})$ denote the set of all real-valued continuous functions defined on X. Let $\mathcal{F}$ be a non-empty collection of $\mathcal{C}(X, \mathbb{R})$. Assume that for every point $x \in X$, there is some $M_x > 0$ such that for every function $f \in \mathcal{F}$,*

$$|f(x)| \le M_x.$$

Then there exist a non-empty open set $U \subset X$ and a constant $M > 0$ such that

$$\text{for every function } f \in \mathcal{F}, \quad |f(x)| \leq M \text{ holds for every } x \in U.$$

Proof For $n \in \mathbb{N}$ and $f \in \mathcal{F}$, define

$$A_{n,f} := \{x \in \mathbb{R} : |f(x)| \leq n\} = f^{-1}([-n, n]).$$

Then each $A_{n,f}$ is closed, by Theorem iv). (It is an easy result from the next chapter of *Continuity*. Essentially, it is no different from its simpler version in real analysis other than the notations. See footnote[1] for the simpler version.)

Now, call

$$A_n := \bigcap_{f \in \mathcal{F}} A_{n,f}.$$

Then A_n is closed (Theorem 3.21). We further claim that $\bigcup_n A_n = X$. Because for each $x \in X$, there will exist $M_x > 0$ such that for each function $f \in \mathcal{F}$, $|f(x)| \leq M_x$. Hence for some positive integer $n_x > M_x$, $x \in A_{n_x}$. Therefore, $x \in A_{n_x} \subset \bigcup_n A_n$, and hence the claim. Since X is complete, by Corollary 4.11.1 of Baire category theorem, there is a positive integer M, such that A_M has a non-empty interior, say U. Therefore by construction, for each $f \in \mathcal{F}$, $|f(x)| \leq M$ holds for every $x \in U$. ■

Remark The above theorem can be restated in other words as: *Any pointwise bounded family of real-valued continuous functions defined over a complete space X must be uniformly bounded on a non-empty open subset U of X.*

Everywhere Continuous, Nowhere Differentiable Functions

When asked to provide an example of a real number, a layperson will almost always name an integer. This common response reflects an intuition that fails to grasp the overwhelming abundance of irrational numbers compared to integers. A similar phenomenon occurs in when a mathematics student is asked to sketch the graph of a real-valued continuous function over an interval. In most cases, the resulting graph will be smooth, with at most a few isolated points of non-differentiability. However, mathematically, the reality is quite the opposite.

[1] Let $n \in \mathbb{N}$, $f \in \mathcal{F}$ be arbitrary, and x_0 be a limit point of $A_{n,f}$, i.e. $\exists\{x_k\} \subset A_{n,f}$ converging to x_0. Then by definition, for every k, $|f(x_k)| \leq n$. Hence by continuity of f and $|\cdot|$,

$$|f(x_0)| = |f(\lim x_k)| = |\lim f(x_k)| = \lim|f(x_k)| \leq n.$$

Whence, $x_0 \in f^{-1}([-n, n]) = A_{n,f}$, showing that $A_{n,f}$ is closed.

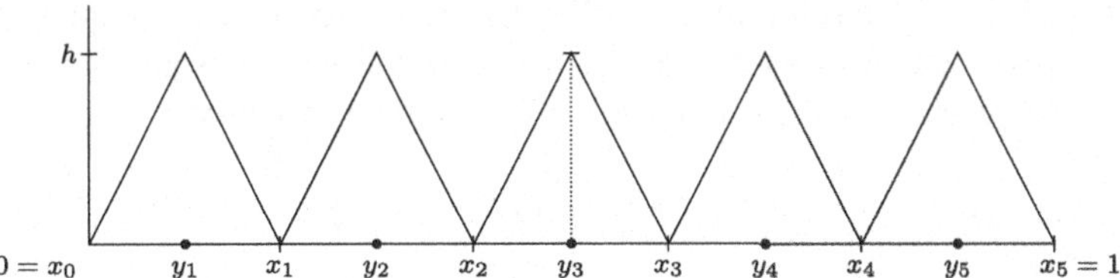

Fig. 4.5 Saw-tooth function

If a computer, devoid of any programmed aesthetic bias, were tasked with selecting random examples of continuous functions, it would almost always produce functions that are nowhere differentiable. This striking fact highlights how differentiable functions, despite their prominence in a typical calculus course, form a relatively small subset of all continuous functions.

The proof presented below establishes this theorem without explicitly constructing any nowhere differentiable functions. A concrete example of such a pathological function will follow the proof.

Before proceeding, we introduce the construction of a special class of functions that will be essential to our argument.

Definition 4.13 A function $f: [0, 1] \to \mathbb{R}$ is called a *saw-tooth function* if there is a positive real number h and a partition $0 = x_0 < x_1 < \cdots < x_n = 1$ of the interval $[0, 1]$ such that for each index r, $f(x_r) = 0$, $f(y_r) = h$, where $y_r = \frac{1}{2}(x_{r-1} + x_r)$ is the midpoint of the subinterval $[x_{r-1}, x_r]$, and extended linearly to all of $[0, 1]$. The number h is called the *height* of the function.

Remarks

(a) The domain $[0, 1]$ can be scaled to any closed and bounded interval $[a, b]$.
(b) For any $h > 0$ fixed, we can have the lines of any wished steepness by considering n big enough. (More points of subdivision will guarantee a higher (absolute value of) slope of the lines. See Fig. 4.5.)
(c) Saw-tooth function is an example of a continuous piecewise linear function. (See Exercise 3.31.)

Theorem 4.14 *The set of all real-valued continuous functions defined on* $[0, 1]$ *having no finite derivative at any point is a set of category II.*

Proof Call $X = \mathcal{C}[0, 1]$. Recall that X is a complete metric space w.r.t. the uniform metric (Example 4.5.7). We will show that the set $\mathcal{D}_R$ of all functions in X having finite right-handed derivative at some point in $[0, 1]$ is a set of category I. The set $\mathcal{D}$ of all functions in X having finite derivative at some point in $[0, 1]$, being a subset of $\mathcal{D}_R$, will eventually become a set of category I. Therefore, Baire category theorem will imply that the set $X \setminus \mathcal{D}$ of all functions in X having no finite derivative at any point in $[0, 1]$ is a set of category II.

A function $f \in X$ has a finite right-handed derivative at some point $x \in [0, 1]$ if there is some number ρ so that for each $\epsilon > 0$, there exists some $\delta > 0$ so that whenever we have $x + h \in [0, 1]$ with $0 < h < \delta$, the inequality

$$\left| \frac{f(x+h) - f(x)}{h} - \rho \right| < \epsilon$$

holds. Emphasising on the "finite" part of the derivative ρ, we construct the following sets: Let $\mathcal{D}_n$ denote the collection of functions $f \in X$ so that for some point $x \in [0, 1 - \frac{1}{n}]$, the inequality

$$\left| \frac{f(x+h) - f(x)}{h} \right| \leq n$$

holds for each $0 < h < \frac{1}{n}$. In other words, $\mathcal{D}_n$ consists of those functions $f \in X$ whose right-hand derivative, wherever exists, is bounded by n in its absolute value. This way, if a function f has finite right-handed derivative at some point, then $f \in \mathcal{D}_n$ for some positive integer n. Thus $\mathcal{D}_R \subset \bigcup_{n=1}^{\infty} \mathcal{D}_n$. We will show each $\mathcal{D}_n$ is nowhere dense. Then $\bigcup_{n=1}^{\infty} \mathcal{D}_n$ is a set of category I and thus $\mathcal{D}_R$ is also so as well.

To show each $\mathcal{D}_n$ is nowhere dense, we start by showing $\mathcal{D}_n$ is closed. Let $g \in \overline{\mathcal{D}_n}$ be arbitrary. Then we have a sequence $\{f_k\}$ from $\mathcal{D}_n$ converging to g. Fix $h \in (0, \frac{1}{n})$ arbitrarily. From the fact that each $f_k \in \mathcal{D}_n$, we obtain a sequence $\{x_k\}$ of points from $[0, 1 - \frac{1}{n}]$ so that for each index k, the inequality

$$\left| \frac{f_k(x_k + h) - f_k(x_k)}{h} \right| \leq n$$

holds. As $\{x_k\}$ is a bounded sequence, by Bolzano-Weierstrass theorem, we obtain a convergent subsequence $\{x_p\}$ of $\{x_k\}$ converging to some point $\xi \in [0, 1 - \frac{1}{n}]$. We want to show

$$\left| \frac{g(\xi + h) - g(\xi)}{h} \right| \leq n,$$

thus concluding $g \in \mathcal{D}_n$. We estimate the numerator in the LHS as follows:

$$\begin{aligned} |g(\xi + h) - g(\xi)| &\leq |g(\xi + h) - g(x_p + h)| \\ &\quad + |g(x_p + h) - f_p(x_p + h)| \\ &\quad + |f_p(x_p + h) - f_p(x_p)| \\ &\quad + |f_p(x_p) - g(x_p)| \\ &\quad + |g(x_p) - g(\xi)| \\ &= \text{(A)} + \text{(B)} + \text{(C)} + \text{(D)} + \text{(E)}, \text{ (resp.) say.} \end{aligned} \tag{$*$}$$

We now estimate each of the summands in the RHS of $(*)$ one by one. As $g \in \overline{\mathcal{D}}_n \subset X$ is continuous and $x_p \to \xi$, we have some $p_1 \in \mathbb{N}$ so that for each $p \geq p_1$,

$$|g(\xi + h) - g(x_p + h)| \leq \frac{nh}{5}, \qquad \text{i.e., (A)} \leq \frac{nh}{5}.$$

As $f_k \to g$, $f_p \to g$ as well, and hence we obtain $p_2 \geq p_1$ so that for each $p \geq p_2$,

$$|g(x_p + h) - f_p(x_p + h)| \leq \frac{nh}{5}, \qquad \text{i.e., (B)} \leq \frac{nh}{5}.$$

As each $f_p \in D_n$, by construction of D_n's, we obtain for each p that,

$$|f_p(x_p + h) - f_p(x_p)| \leq \frac{nh}{5}, \qquad \text{i.e., (C)} \leq \frac{nh}{5}.$$

As $f_p \to g$ (uniform convergence), we obtain for each $p \geq p_2$ that,

$$|f_p(x_p) - g(x_p)| \leq \frac{nh}{5}, \qquad \text{i.e., (D)} \leq \frac{nh}{5}.$$

Finally, as g is (uniformly) continuous and $x_p \to \xi$, we obtain that for each $p \geq p_1$,

$$|g(x_p) - g(\xi)| \leq \frac{nh}{5}, \qquad \text{i.e., (E)} \leq \frac{nh}{5}.$$

Combining all the estimations above, we obtain from $(*)$ that for each $p \geq p_2$,

$$|g(\xi + h) - g(\xi)| \leq \frac{nh}{5} + \frac{nh}{5} + \frac{nh}{5} + \frac{nh}{5} + \frac{nh}{5} = nh.$$

Thus $\left|\frac{g(\xi+h)-g(\xi)}{h}\right| \leq n$, showing that $g \in \mathcal{D}_n$. Consequently, $\mathcal{D}_n$ is closed.

We will now show $X \setminus \mathcal{D}_n$ is everywhere dense in X and use Exercise 4.11(b) to conclude that $\mathcal{D}_n$ is nowhere dense.

Fix $f \in X$ arbitrarily. Then f is uniformly continuous over $[0, 1]$. Thus for any $\epsilon > 0$, there exists $\delta > 0$ so that for any two points $x, x' \in [0, 1]$ with $|x - x'| < \delta$, we will have $|f(x) - f(x')| < \frac{\epsilon}{2}$. Choose a positive integer $m > n$ so that $\frac{1}{m} < \delta$. Divide $[0, 1]$ into m subintervals of equal length at $0 = x_0 < x_1 < \cdots < x_m = 1$. In each kth subinterval $I_k = [x_{k-1}, x_k]$, consider the parallelogram $PQRS$ where

$$P(k) = \left(x_{k-1}, f(x_{k-1}) - \frac{\epsilon}{2}\right), \qquad Q(k) = \left(x_k, f(x_k) - \frac{\epsilon}{2}\right),$$
$$R(k) = \left(x_k, f(x_k) + \frac{\epsilon}{2}\right), \qquad S(k) = \left(x_{k-1}, f(x_{k-1}) + \frac{\epsilon}{2}\right).$$

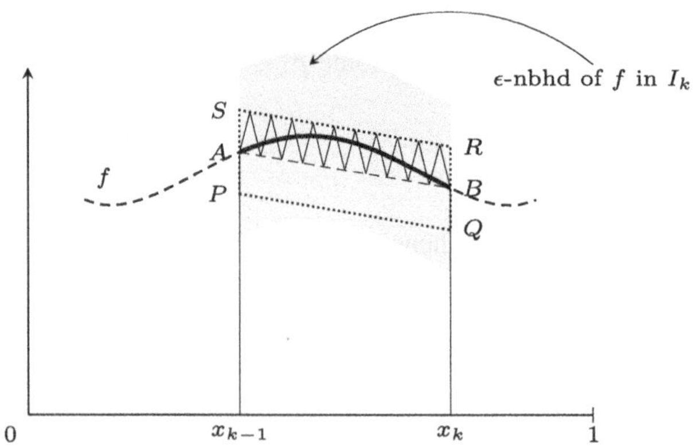

Fig. 4.6 Approximation by saw-tooth function

Also, for brevity, call the points

$$A(k) = (x_{k-1}, f(x_{k-1})), \qquad B(k) = (x_k, f(x_k)).$$

See Fig. 4.6. Observe that $PQRS$ stays inside the ϵ-neighbourhood of f in I_k. (Prove this!) Let g_k be the function on I_k obtained by joining the points A and B by a straight line. Then g_k stays inside the parallelogram $PQRS$. Let s_k be a saw-tooth function defined on I_k with height $\frac{\epsilon}{2}$ with slope of the lines greater than M in absolute value, where M is a positive integer to be determined later. Then the map $h_k := g_k + s_k$ is a piecewise linear continuous function with minimum slope of lines

$$M - \text{absolute value of slope of } g_k = M - \left| \frac{f(x_k) - f(x_{k-1})}{x_k - x_{k-1}} \right|.$$

Choose M so large that $M - \left| \frac{f(x_k)-f(x_{k-1})}{x_k-x_{k-1}} \right| > n$. Thus the map h_k satisfies the following properties in I_k:

i) h_k is a continuous function that stays inside the parallelogram $PQRS$, i.e. inside the ϵ-neighbourhood of f in I_k.
ii) Derivative of h_k, wherever exists, has absolute value greater than n.

Finally, extend h_k to all of $[0, 1]$ by defining $h_k(x) = 0$ for points x outside of I_k. Now construct the function

$$h := h_1 + \cdots + h_m.$$

Then h satisfies both the above properties mentioned above, but this time, over all of $[0, 1]$. Thus $h \in X \setminus \mathcal{D}_n$ and $\|h - f\| < \epsilon$. Hence $X \setminus \mathcal{D}_n$ is dense in X.

This completes the proof. ■

We now present a concrete example of a function that is continuous everywhere but differentiable nowhere. Finding such a member requires effort, akin to spotting a transcendental number among their uncountable abundance in the set of all real numbers. The following construction, originally attributed to van der Waerden, offers a striking illustration of this surprising phenomenon.

Example 4.14.1 The idea here is to construct a sequence of saw-tooth functions with decreasing heights and increasing frequencies. The sum of such a sequence will eventually converge to a desired function. With this context, let us define the primary saw-tooth function $h\colon \mathbb{R} \to \mathbb{R}$ by

$$h(x) = \begin{cases} x, \text{ if } 0 \le x \le \frac{1}{2} \\ 1 - x, \text{ if } \frac{1}{2} < x \le 1 \end{cases}$$

and extend to all of $\mathbb{R}$ periodically under the condition $h(x+1) = h(x)$. ($h(x)$ is denoted as $h_0(x)$ in Fig. 4.7.) Then h is a continuous function over $\mathbb{R}$ with period 1. For $n \ge 0$, call

$$h_n(x) = \frac{h(2^n x)}{2^n}.$$

The multiplier 2^n in the argument of h in the numerator will increase the frequency of tooth in the saw-tooth function h; in contrast, its appearance in the denominator will decrease its amplitude. Observe that for any real x, $|h_n(x)| \le \frac{1}{2}.\frac{1}{2^n}$. Thus the series $\sum_{n\ge 0} h_n(x)$ converges uniformly over $\mathbb{R}$. Therefore the sum function

$$H(x) = \sum_{n=0}^{\infty} h_n(x) = \sum_{n=0}^{\infty} \frac{h(2^n x)}{2^n}$$

is continuous over $\mathbb{R}$. (In Fig. 4.7, we have used the notation $H_n(x) = \sum_{k=0}^{n} h_k(x)$ for the partial sum functions.) We will show that H is nowhere differentiable on $\mathbb{R}$. As h is a periodic function with period 1, so is H. Therefore, it suffices to show H is not differentiable at any point $a \in [0, 1)$.

Fix $a \in [0, 1)$ arbitrarily. Observing the presence of the multiplier 2^n, we switch to the binary representation for the remaining calculations. Let

$$a = 0.a_1 a_2 \ldots$$

be the non-terminating binary representation of a, where each a_k is either 0 or 1. Construct a sequence $\{a^{(m)}\}$ from $[0, 1)$ in the following way: $a^{(m)} = 0.a_1^{(m)} a_2^{(m)} \ldots$, where

$$a_k^{(m)} = \begin{cases} a_k, \text{ when } k \ne m \\ 1 - a_k, \text{ when } k = m \end{cases},$$

i.e. $a^{(m)}$ differs from a only in its mth digit after the binary point. Thus $a^{(m)} \to a$. Now, observe that for any $n \ge m$, the quantities $2^n a^{(m)}$ and $2^n a$ differ only by an

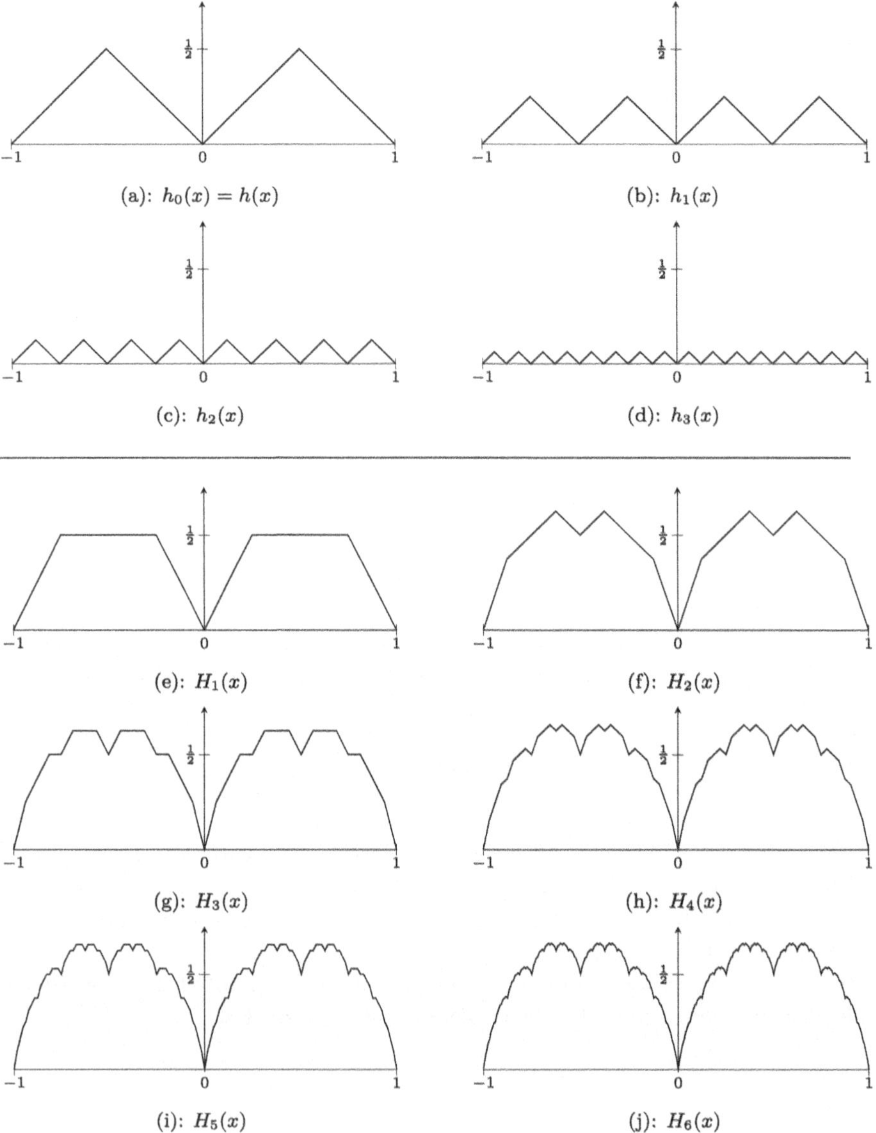

Fig. 4.7 Construction of H_n's illustrated on $[-1, 1]$

integer. Thus, for such values of n, $h(2^n a^{(m)}) - h(2^n a) = 0$. On the other hand, when $n < m$, $2^n a^{(m)}$ and $2^n a$ differ only at the $(m-n)$th digit after binary point. Combining these two observations, we obtain

$$h(2^n a^{(m)}) - h(2^n a) = \begin{cases} \pm\frac{1}{2^{m-n}}, \text{ when } n < m \\ 0, \text{ when } n \geq m \end{cases}.$$

Therefore,

$$\begin{aligned}\frac{H(a^{(m)})-H(a)}{a^{(m)}-a} &= \sum_{n=0}^{\infty}\frac{h(2^n a^{(m)})-h(2^n a)}{2^n(a^{(m)}-a)}\\ &= \sum_{n=0}^{m-1}\frac{h(2^n a^{(m)})-h(2^n a)}{2^n(a^{(m)}-a)}\\ &= \sum_{n=0}^{m-1}\frac{\pm 1/2^{m-n}}{2^n(\pm 1/2^m)}\\ &= \sum_{n=0}^{m-1}\pm 1. \qquad (*)\end{aligned}$$

Thus for each m, the quantity $\frac{H(a^{(m)})-H(a)}{a^{(m)}-a}$ is sum of m terms each of which is either 1 or -1. Therefore, $\frac{H(a^{(m)})-H(a)}{a^{(m)}-a}$ is an odd integer when m is odd and is an even integer when m is even. It follows that

$$\lim_{m\to\infty}\frac{H(a^{(m)})-H(a)}{a^{(m)}-a}$$

cannot exist. This proves that the limit

$$\lim_{x\to a}\frac{H(x)-H(a)}{x-a}$$

does not exist. Consequently, H is not differentiable at a. This completes the proof.

Set of Points of Discontinuity of a Real-Valued Function Defined over $\mathbb{R}$

In the chapter on continuity in real analysis, we encountered the pathological example of the *popcorn function* or Thomæ's function (Fig. 4.8). Defined on the interval [0, 1], this function exhibits the intriguing property of being continuous at every irrational point while being discontinuous at every rational point. The formal definition of this function goes as $f\colon [0,1] \to \mathbb{R}$, where

$$f(x) = \begin{cases} 0, & \text{when } x = 0 \text{ or an irrational}\\ \frac{1}{n}, & \text{when } x = \frac{m}{n}, \text{ where } m \text{ and } n \text{ are positive integers with } \gcd(m,n)=1. \end{cases}$$

This naturally raises the following question: "Does there exist a function that is continuous at every rational point and discontinuous at every irrational point?"

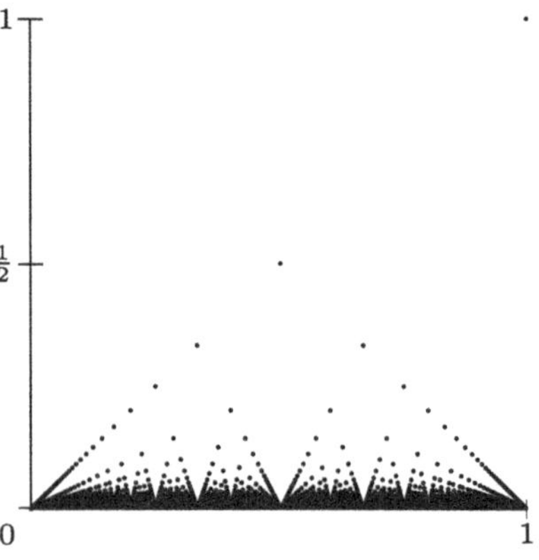

Fig. 4.8 Popcorn function

We now seek an answer to this question. To avoid suspense, we reveal upfront that the answer is "No". We will establish this claim systematically, beginning with the definition of a special class of subsets of $\mathbb{R}$.

Definition 4.15 A subset A of $\mathbb{R}$ is called of *type* $\boldsymbol{F_\sigma}$ if A is the union of enumerable closed subsets of $\mathbb{R}$, i.e.

$$A = \bigcup_{n=1}^{\infty} A_n,$$

where each A_n is a closed subset of $\mathbb{R}$.

We wish to eventually prove the following facts:

i) The set of points of discontinuity of a function $f : \mathbb{R} \to \mathbb{R}$ is of type F_σ
ii) The set of all irrationals is *not* a set of type F_σ

thus proving our claim. Before proceeding further, let us see some examples of F_σ sets.

Examples

4.15.1. Every closed subset is a F_σ set. If A is closed, then A can be written as $A = \bigcup_{n\geq 1} A_n$, where $A_1 = A$ and $A_n = \varnothing$ for every $n > 1$.
4.15.2. The set $\mathbb{Q}$ of all rationals is a F_σ set. As $\mathbb{Q}$ is enumerable, let $\{r_n\}_{n=1}^{\infty}$ be an enumeration of $\mathbb{Q}$. Therefore, $\mathbb{Q} = \bigcup_{n\geq 1}\{r_n\}$. Since each singleton is closed in $\mathbb{R}$, it follows that $\mathbb{Q}$ is a F_σ set.
4.15.3. A bounded open interval (a, b) is of type F_σ. We can write

$$(a, b) = \bigcup_{n=m}^{\infty} \left[a + \frac{1}{n}, b - \frac{1}{n}\right],$$

where m is a fixed positive integer so that $a + \frac{1}{m} < b - \frac{1}{m}$, i.e. $m > \frac{2}{b-a}$. As each $[a + \frac{1}{n}, b - \frac{1}{n}]$ is a closed set, it proves our claim.

We now introduce a concept as an NASC to check continuity of a function $f:\mathbb{R} \to \mathbb{R}$ at a point.

Definition 4.16 Let $f:\mathbb{R} \to \mathbb{R}$ and I be any bounded open interval. The *oscillation* of f *over* $\boldsymbol{I}$ is defined as the quantity

$$\omega(f, I) = \sup_{x\in I} f(x) - \inf_{x\in I} f(x).$$

If a is any point, the *oscillation* of f *at* $\boldsymbol{a}$ is defined as the quantity

$$\omega(f, a) = \inf_{I} \omega(f, I),$$

where the infimum is taken over all bounded open intervals I containing a.

It is an easy exercise in real analysis to verify the following:

Proposition 4.17 *Let $f:\mathbb{R} \to \mathbb{R}$ and $x \in \mathbb{R}$. f is continuous at x if and only if the oscillation of f at x, $\omega(f, x) = 0$.*

With these frameworks, we arrive at the following:

Theorem 4.18 *Let $f:\mathbb{R} \to \mathbb{R}$. For $n \geq 1$, denote by E_n, the set*

$$E_n = \left\{x \in \mathbb{R} : \omega(f, x) \geq \frac{1}{n}\right\}.$$

Then each E_n is closed. Moreover, if E is the set of all points of discontinuity of f, then

$$E = \bigcup_{n=1}^{\infty} E_n.$$

Thus E is a set of type F_σ.

Proof For any fixed n, let y be a limit point of E_n. Consider any bounded interval I containing y. Then I must contain a point $x \in E_n$. Therefore,

$$\omega(f, I) \geq \omega(f, x) \geq \frac{1}{n}. \qquad (*)$$

As $(*)$ is true for any I containing y,

$$\omega(f, y) = \inf_I \omega(f, I) \geq \frac{1}{n}.$$

Thus $y \in E_n$. Consequently, E_n is closed.

For the second part, we use Proposition 4.17. For any $n \geq 1$, $x \in E_n$ implies $\omega(f, x) > 0$. Therefore x is a point of discontinuity of f, i.e. $x \in E$. Consequently, $\bigcup_{n\geq 1} E_n \subset E$. To prove the converse, let x be a point of discontinuity of f. Then $\omega(f, x) > 0$. By Archimedean property of $\mathbb{R}$, find $n_0 \in \mathbb{N}$ so that $\frac{1}{n_0} < \omega(f, x)$. Hence $x \in E_{n_0}$. Thus $E \subset \bigcup_{n\geq 1} E_n$ as well. This completes the proof. ■

Lastly, we prove:

Proposition 4.19 *The set of irrationals is not of type F_σ.*

Proof Assume the contrary. Let for closed sets $\{A_n\}$, $\mathbb{R} \setminus \mathbb{Q} = \bigcup_n A_n$. However, as each A_n is closed and contains only irrational points, the interior of $\overline{A}_n$ is empty. Thus, each A_n is nowhere dense. Therefore, $R \setminus \mathbb{Q}$ is of category I, which contradicts Exercise 4.14. Hence the proof. ■

We finish this section with a couple more applications of Baire category theorem.

Miscellaneous results:

4.3. Write $X = [0, \infty)$. Let a continuous function $f: X \to \mathbb{R}$ be such that for each point $x \in X$, $\lim_{n\to\infty} f(nx) = 0$. Show that $\lim_{x\to\infty} f(x) = 0$.
Proof. Fix $\epsilon > 0$. For $n \in \mathbb{N}$, let

$$A_n := \{x \in X : |f(nx)| \leq \epsilon\}.$$

As f is continuous, the map $x \mapsto f(nx)$ is also continuous. Hence A_n, which is the pulled-back image of the closed set $[-\epsilon, \epsilon]$ under the map $f(nx)$, is also closed (Theorem 5.4.(*iii*)). Now, set $F_k = \bigcap_{n\geq k} A_n$. As any intersection of closed sets is closed, each F_k is closed as well. Observe that any point $x \in F_k$ satisfies that for every $n \geq k$, $|f(nx)| \leq \epsilon$. We claim that

$$\bigcup_{k\in\mathbb{N}} F_k = X.$$

Because by the hypothesis, for any point $x_0 \in X$, $\lim_{n\to\infty} f(nx_0) = 0$, i.e. for some $k_0 \in \mathbb{N}$, $x_0 \in F_{k_0}$.
As X is complete, by the version of Baire category theorem mentioned in Corollary 4.11.1, at least one of the F_k must have a non-empty interior. Assume

the set F_p contains an interval (a, b). Therefore,

$$\text{for each point } y \in (a, b) \text{ and for each } n \geq p, \quad |f(ny)| \leq \epsilon.$$
$$\Longrightarrow \forall n \geq p, \text{ and } \forall t \in (na, nb), \quad |f(t)| \leq \epsilon.$$
$$\Longrightarrow \forall t \in \bigcup_{n \geq p}(na, nb), \quad |f(t)| \leq \epsilon.$$

Choose n big enough so that the set $\bigcup_{n \geq p}(na, nb)$ contains the entire interval $[M, \infty)$ for some $M > 0$. Precisely, choose n big enough, so that

$$(n+1)a < nb, \text{ i.e. } n > \frac{a}{b-a}.$$

Therefore, for any point $t \geq M$, $|f(t)| \leq \epsilon$, i.e. $f(t) \to 0$ as $t \to \infty$. This completes the proof. □

4.4. Let X be a complete metric space. Let $f_n \colon X \to \mathbb{R}$ be a sequence of continuous functions converging pointwise to a function $f \colon X \to \mathbb{R}$. Show that there exists a dense subset D of X so that f is continuous at every point from D.
Proof. Fix $\epsilon > 0$. For any $m, n \in \mathbb{N}$, set

$$A_{m,n}(\epsilon) := \{x \in X : |f_m(x) - f_n(x)| \leq \epsilon\}.$$

As each function in the sequence $\{f_n\}$ is continuous, so is $|f_m - f_n|$. Thus by Theorem 5.4.(*iii*), each $A_{m,n}(\epsilon)$, which is the pulled-back image of the closed set $[-\epsilon, \epsilon]$, is closed. Now, for any fixed $k \in \mathbb{N}$, construct

$$F_k(\epsilon) = \bigcap_{m,n \geq k} A_{m,n}.$$

$F_k(\epsilon)$ being an intersection of closed sets is closed again. Also, observe that for any point $x \in F_k(\epsilon)$, it is satisfied that for every $m, n \geq k$, $|f_m(x) - f_n(x)| \leq \epsilon$. We claim that

$$\bigcup_{k \in \mathbb{N}} F_k(\epsilon) = X. \tag{4.6}$$

Because for any point $x_0 \in X$, the sequence $f_n(x_0) \to f(x_0)$, i.e. the sequence $\{f_n\}$, is Cauchy at the point x_0, i.e. for some $k_0 \in \mathbb{N}$, $x_0 \in F_{k_0}(\epsilon)$.
Since X is given to be complete and each $F_k(\epsilon)$ is closed, by the version of Baire category theorem mentioned in Corollary 4.11.1, at least one of the $F_k(\epsilon)$ must have a non-empty interior. Construct

$$U(\epsilon) := \bigcup_{k \in \mathbb{N}} \operatorname{int} F_k(\epsilon). \tag{4.7}$$

Then $U(\epsilon)$ is a non-empty open subset of X for every $\epsilon > 0$. Therefore, we can construct the sequence $\{U_n\}$ of open subsets by $U_n := U(\frac{1}{n})$. We will show that each U_n is dense in X.

We will use Exercise 3.28. Let $B[x; r]$ be any closed ball in X. Write B for $B[x; r]$. Since (4.6) is true for any $\epsilon > 0$, we have

$$B = \bigcup_k \left(B \cap F_k \left(\frac{1}{n} \right) \right).$$

Now, B, being a closed subset of a complete space, is itself complete. Therefore, by Baire category theorem again, at least one of $B \cap F_k(\frac{1}{n})$ has non-empty interior, i.e. for some $k_0 \in \mathbb{N}$, $\text{int}[B \cap F_{k_0}(\frac{1}{n})] \neq \varnothing$. However, by Exercise 3.2,

$$\text{int} \left[B \cap F_{k_0} \left(\frac{1}{n} \right) \right] = \text{int}\, B \cap \text{int}\, F_{k_0} \left(\frac{1}{n} \right) \subset B \cap \text{int}\, F_{k_0} \left(\frac{1}{n} \right) \subset B \cap U_n.$$

Hence, it follows that $B \cap U_n \neq \varnothing$, i.e. each U_n is dense in X. Thus $\{U_n\}$ is a sequence of open dense subsets in X. Therefore, by the stronger version of the Baire category theorem recorded in Exercise 4.17,

$$U := \bigcap_n U_n$$

is dense in X. We claim that every point of U is a point of continuity for f. Fix $a \in U$ and $\epsilon > 0$ arbitrarily. Choose $p \in \mathbb{N}$ so that $\frac{1}{p} < \frac{\epsilon}{3}$. Now, by construction,

$$a \in U \implies a \in U_p = U \left(\frac{1}{p} \right) = \bigcup_{k \in \mathbb{N}} \text{int}\, F_k \left(\frac{1}{p} \right). \qquad \text{[by (4.7)]}$$

Therefore, for some $k_0 \in \mathbb{N}$, $a \in \text{int}\, F_{k_0}(\frac{1}{p})$. Thus there exists some $\delta_1 > 0$ so that the neighbourhood $(a - \delta_1, a + \delta_1) \subset F_{k_0}(\frac{1}{p})$. This paraphrases to

$$\text{for any point } x \in (a - \delta_1, a + \delta_1), \text{ and for any } m, n \geq k_0,$$
$$|f_m(x) - f_n(x)| \leq \frac{1}{p} < \frac{\epsilon}{3}.$$

Letting $m \to \infty$, we obtain that

$$\text{for any point } x \in (a - \delta_1, a + \delta_1), \text{ and for any } n \geq k_0,$$
$$|f(x) - f_n(x)| \leq \frac{1}{p} < \frac{\epsilon}{3}. \qquad (4.8)$$

As f_{k_0} is continuous, for the chosen $\epsilon > 0$ and for the concerned point a, there exists some $\delta_0 > 0$ so that for any point $x \in (a-\delta_0, a+\delta_0)$, $|f_{k_0}(x) - f_{k_0}(a)| < \frac{\epsilon}{3}$. Call $\delta = \min\{\delta_0, \delta_1\} > 0$. Therefore, for any point $x \in (a-\delta, a+\delta)$, from (4.8), we have

$$\begin{aligned} |f(x) - f(a)| &\le |f(x) - f_{k_0}(x)| + |f_{k_0}(x) - f_{k_0}(a)| + |f_{k_0}(a) - f(a)| \\ &< \frac{\epsilon}{3} + \frac{\epsilon}{3} + \frac{\epsilon}{3} = \epsilon. \end{aligned}$$

Thus, f is continuous at a. As $a \in U$ is arbitrary, f is continuous on the dense subset U of X. □

Exercises: 4.2

4.11. Let X be a metric space and A be a non-empty subset of X. Show that the following are equivalent:

(a) A is nowhere dense.
(b) $\overline{A}^c$ is everywhere dense.
(c) Each non-empty open set has a non-empty open subset disjoint from $\overline{A}$.
(d) Each non-empty open set has a non-empty open subset disjoint from A.
(e) Each non-empty open set contains an open ball disjoint from A.

4.12. Show that a non-empty open interval is of category II in $\mathbb{R}$.

4.13. Show that the statements in Theorem 4.11 and Corollaries 4.11.1 and 4.11.2 are equivalent.

4.14. Show that the set of irrationals is of category II in $\mathbb{R}$.

4.15. Show that any proper subspace of the real or complex vector spaces $\mathbb{R}^n$ or $\mathbb{C}^n$ is nowhere dense. Hence conclude that these spaces cannot be expressed as a countable union of its proper subspaces.

4.16. Show that any non-empty complete metric space without isolated points is uncountable. Hence conclude that $\mathbb{R}$ is uncountable.

4.17. Prove the stronger version of Corollary 4.11.2: Let $\{U_n\}$ be a sequence of open, dense subsets in a complete metric space X. Show that $\bigcap_n U_n$ is also dense in X.

4.18. Show that $\mathbb{R}$ is a set of type F_σ.

4.19. Prove Proposition 4.17.

4.3 Completion of a Metric Space

Incomplete spaces present an unfortunate limitation: A Cauchy sequence in such a space has terms that get arbitrarily close to each other but may fail to converge to a limit within the space. Our goal in this section is to repair this vulnerability and construct a completion of an incomplete space.

An organic approach to achieving this is to augment the existing space with new points that serve as the limits of non-convergent Cauchy sequences. Let (X, d) be the incomplete parent space, and let $\widetilde{X}$ be the set obtained by adjoining these new limit points to X. However, $\widetilde{X}$ does not yet possess a metric structure. Our next task is to define a metric $\widetilde{d}$ on $\widetilde{X}$ such that it agrees with the original metric on X, i.e. for the pre-existing points $x_1, x_2 \in X$, we require $\widetilde{d}(x_1, x_2) = d(x_1, x_2)$. If $\widetilde{d}$ successfully transforms $\widetilde{X}$ into a complete metric space, we might consider our construction complete—but there is an important additional requirement.

As a proper completion of X, the space $\widetilde{X}$ should be augmented with *only* the necessary new points—those that arise as limits of Cauchy sequences in X. In other words, X should be dense in $\widetilde{X}$, ensuring that every point in $\widetilde{X}$ is the limit of a sequence of points from X. Only then can we consider our construction fully satisfactory.

This process of introducing new points requires careful mathematical formulation. To achieve it rigorously, we need to embed X within the extended space $\widetilde{X}$ in a way that preserves its original structure. With this in mind, we now introduce the necessary formal definitions.

Definition 4.20 Let (X, d_X) and (Y, d_Y) be two metric spaces. A map $f: X \to Y$ is called an *isometry* if f preserves the metric, i.e. for any two points x_1, x_2 from X,

$$d_Y(f(x_1), f(x_2)) = d_X(x_1, x_2).$$

Furthermore, if f is a bijection between X and Y, then the spaces X and Y are called *isometric*.

Remarks

(a) Isometric spaces are essentially identical. Analogically, a bijective isometry between metric spaces is what an isomorphism is between algebraic structures.
(b) An isometry is always injective. If $f: X \to Y$ is an isometry, then for points $x_1, x_2 \in X$,

$$f(x_1) = f(x_2) \implies 0 = d_Y(f(x_1), f(x_2)) = d_X(x_1, x_2) \implies x_1 = x_2.$$

Therefore in case of an isometry $f: X \to \widetilde{X}$, $f(X)$ is an isometric (hence identical) image of X embedded in $\widetilde{X}$. So the metric $\widetilde{d}$ on the extended space $\widetilde{X}$ must necessarily satisfy the property that for points $y_1, y_2 \in f(X) \subset \widetilde{X}$,

$$\widetilde{d}(y_1, y_2) = d(x_1, x_2),$$

where $y_1 = f(x_1)$ and $y_2 = f(x_2)$.
Finally, we also want $\widetilde{d}$ to make $f(X)$ dense in $\widetilde{X}$. Summarising all the discussions we have had so far, let us agree to the following:

Definition 4.21 Let X be a metric space. A complete metric space $\widetilde{X}$ is said to be a *completion* of X if there exists a map $f: X \to \widetilde{X}$ such that:

a) f is an isometry from X into $\widetilde{X}$.
b) The image $f(X)$ is dense in $\widetilde{X}$.

Our main objective in this section is to prove that for any metric space, there exists a unique completion of the space (for a complete space, the completion is the space itself). But before that, let us see a few examples of isometry.

Examples

4.21.1. The map from $\mathbb{C}$ to $\mathbb{R}^2$ given by $z = x + iy \mapsto (x, y)$ is an isometry.
Proof. Proof of this fact is straightforward and left to the reader. □

4.21.2. If A is an $n \times n$ real orthogonal matrix, then for $\mathbf{x} \in \mathbb{R}^n$, the map $\mathbf{x} \mapsto A\mathbf{x}$ is an isometry.
Proof. For a real orthogonal matrix A (i.e. $A^T A = I$), we know that for arbitrary $\mathbf{x} \in \mathbb{R}^n$,

$$\|A\mathbf{x}\| := \sqrt{\langle A\mathbf{x}, A\mathbf{x}\rangle} = \sqrt{\langle A^T A\mathbf{x}, \mathbf{x}\rangle} = \sqrt{\langle \mathbf{x}, \mathbf{x}\rangle} = \|\mathbf{x}\|.$$

Therefore for $\mathbf{x}_1, \mathbf{x}_2 \in \mathbb{R}^n$, $\|A\mathbf{x}_1 - A\mathbf{x}_2\| = \|x_1 - x_2\|$. □

4.21.3. Let (X, d_X) be a metric space and Y be a non-empty set. Let $\varphi: X \to Y$ be a bijection. Then we can transfer the metric from X to Y in the following obvious way:

$$d_Y(y_1, y_2) := d_X(x_1, x_2), \text{ where } y_1 = \varphi(x_1), y_2 = \varphi(x_2).$$

Then clearly the spaces (X, d_X) and (Y, d_Y) are isometric.

Let us now prove the mighty.

Theorem 4.22 *Let (X, d) be a metric space. Then there exists a metric space $(\widetilde{X}, \widetilde{d})$ and a map $\pi: X \to \widetilde{X}$ such that:*

i) $(\widetilde{X}, \widetilde{d})$ is a complete metric space.
ii) $\pi(X) \subset \widetilde{X}$ is dense.
iii) π is an isometry.
iv) $(\widetilde{X}, \widetilde{d})$ is unique.

Proof

Step 1: *Construction of* $\widetilde{X}$:

Let $\mathcal{F} := \{\{x_k\} \subset X : \{x_k\} \text{ is Cauchy}\}$ be the set of all Cauchy sequences in X. Call two Cauchy sequences $\{x_k\}, \{y_k\} \in \mathcal{F}$ *equivalent* iff $\lim_{k\to\infty} d(x_k, y_k) = 0$, i.e. define a relation $\sim$ on $\mathcal{F}$ by

$$\{x_k\} \sim \{y_k\} \iff \lim_{k\to\infty} d(x_k, y_k) = 0.$$

Clearly, if $\{x_k\}, \{y_k\}, \{z_k\} \in \mathcal{F}$, then $\{x_k\} \sim \{x_k\}$, $\{x_k\} \sim \{y_k\} \implies \{y_k\} \sim \{x_k\}$ and $\{x_k\} \sim \{y_k\}$, $\{y_k\} \sim \{z_k\}$ together imply $\{x_k\} \sim \{z_k\}$. Hence, $\sim$ is an equivalence relation on $\mathcal{F}$.

Consider the set of all equivalence classes $\mathcal{F}/\sim$ thus generated. Call it $\widetilde{X}$, i.e. $\widetilde{X}$ is the set of all equivalent Cauchy sequences in X.

Observe that the elements of $\widetilde{X}$ are of two types. An element of first type is an equivalence class of convergent Cauchy sequences in X. If $\{x_k\}$ is a Cauchy sequence converging to a point $x \in X$, then the equivalent class $[\{x_k\}]$ contains the constant sequence $\{x\} = \{x, x, x, \dots\}$. In this case, we will denote the class $[\{x_k\}]$ by $[\{x\}]$. And as for every point $x \in X$, we can consider the constant Cauchy sequence $\{x\}$, it makes sense to define the map $\pi : X \to \widetilde{X}$ by $\pi(x) = [\{x\}]$.

On the other hand, an element of the second type in $\widetilde{X}$ is an equivalence class which does not correspond to any element from X; it consists of equivalent Cauchy sequences which do not converge to any point of X.

Step 2: *Definition of* $\widetilde{d}$:

Choose arbitrary elements $\widetilde{\mathbf{x}} = [\{x_k\}]$, $\widetilde{\mathbf{y}} = [\{y_k\}]$ from $\widetilde{X}$. Then for indices k, p,

$$\begin{aligned} d(x_k, y_k) &\le d(x_k, x_p) + d(x_p, y_p) + d(y_p, y_k) \\ \text{or, } d(x_k, y_k) - d(x_p, y_p) &\le d(x_k, x_p) + d(y_p, y_k). \end{aligned}$$

Interchanging k and p, we obtain

$$|d(x_p, y_p) - d(x_k, y_k)| \le d(x_p, x_k) + d(y_p, y_k) \to 0 \text{ as } k, p \to \infty.$$

Hence, $\{d(x_k, y_k)\}$ is a Cauchy sequence of real numbers, and since $\mathbb{R}$ is complete, $\lim_{k\to\infty} d(x_k, y_k)$ exists.

Now we show that $\lim_{k\to\infty} d(x_k, y_k)$ does not depend on the choices of $\{x_k\}$ and $\{y_k\}$ in $\widetilde{\mathbf{x}}$ and $\widetilde{\mathbf{y}}$, respectively.

Suppose $\{x'_k\} \in \widetilde{\mathbf{x}}$ and $\{y'_k\} \in \widetilde{\mathbf{y}}$. Then $d(x'_k, x_k) \to 0$ and $d(y'_k, y_k) \to 0$ as $k \to \infty$. Hence,

$$d(x_k, y_k) \leq d(x_k, x'_k) + d(x'_k, y'_k) + d(y'_k, y_k).$$

$$\text{Hence, } |d(x_k, y_k) - d(x'_k, y'_k)| \leq d(x_k, x'_k) + d(y_k, y'_k) \to 0 \text{ as } k \to \infty,$$

i.e. $\lim_{k\to\infty} d(x_k, y_k) = \lim_{k\to\infty} d(x'_k, y'_k)$.
Hence, we can define $\widetilde{d}: \widetilde{X} \times \widetilde{X} \to \mathbb{R}_{\geq 0}$ by

$$\text{for } \widetilde{\mathbf{x}} = [\{x_k\}], \widetilde{\mathbf{y}} = [\{y_k\}] \in \widetilde{X}, \quad \widetilde{d}(\widetilde{\mathbf{x}}, \widetilde{\mathbf{y}}) = \lim_{k\to\infty} d(x_k, y_k).$$

Step 3: *$\widetilde{d}$ is a metric on $\widetilde{X}$:*
For arbitrary elements $\widetilde{\mathbf{x}} = [\{x_k\}]$, $\widetilde{\mathbf{y}} = [\{y_k\}]$, and $\widetilde{\mathbf{z}} = [\{z_k\}]$ from $\widetilde{X}$, we have:

(a) $\widetilde{d}(\widetilde{\mathbf{x}}, \widetilde{\mathbf{y}}) \geq 0$, and $\widetilde{d}(\widetilde{\mathbf{x}}, \widetilde{\mathbf{y}}) = 0 \iff d(x_k, y_k) \to 0 \iff \{x_k\} \sim \{y_k\} \iff \widetilde{\mathbf{x}} = \widetilde{\mathbf{y}}$.
(b) $\widetilde{d}(\widetilde{\mathbf{x}}, \widetilde{\mathbf{y}}) = \lim_{k\to\infty} d(x_k, y_k) = \lim_{k\to\infty} d(y_k, x_k) = \widetilde{d}(\widetilde{\mathbf{y}}, \widetilde{\mathbf{x}})$.
(c) $d(x_k, y_k) \leq d(x_k, z_k) + d(z_k, y_k) \implies \widetilde{d}(\widetilde{\mathbf{x}}, \widetilde{\mathbf{y}}) \leq \widetilde{d}(\widetilde{\mathbf{x}}, \widetilde{\mathbf{z}}) + \widetilde{d}(\widetilde{\mathbf{z}}, \widetilde{\mathbf{y}})$.

Hence $\widetilde{d}$ is a metric.

Step 4: *$\pi: X \to \widetilde{X}$ is an isometry:*
If for points $x, y \in X$, $\widetilde{\mathbf{x}} = \pi(x) = [\{x\}]$, $\widetilde{\mathbf{y}} = \pi(y) = [\{y\}]$, then

$$\widetilde{d}(\pi(x), \pi(y)) = \widetilde{d}(\widetilde{\mathbf{x}}, \widetilde{\mathbf{y}}) = \lim_{k\to\infty} d(x, y) = d(x, y).$$

Hence $\pi: X \to \widetilde{X}$ is an isometry. This proves (*iii*).
Step 5: *$\pi(X)$ is dense in $\widetilde{X}$:*
Clearly $\widetilde{A} := \pi(X)$ is the subset of elements of $\widetilde{X}$ which correspond to points in X. To show $\widetilde{A}$ is dense in $\widetilde{X}$, we show that every point in $\widetilde{X} \setminus \widetilde{A}$ is a limit point of $\widetilde{A}$.
At this point, let us agree to some notations to be used going further. We are going to consider sequences both in X and in $\widetilde{X}$*. As a rule of thumb, we are going to denote a sequence in X by $\{x_k\}$, with the index being k, whereas a sequence in $\widetilde{X}$ will be denoted by $\{\mathbf{x}^{(n)}\}$, with the index being n, which is placed as a bracketed superscript for distinguishing purpose. By the nature of the space $\widetilde{X}$, each term $\mathbf{x}^{(n)}$ denotes a class of equivalent Cauchy sequences in X which will generally be denoted as*

$$\mathbf{x}^{(n)} = [\{x_k^{(n)}\}_k] = [\{x_1^{(n)}, x_2^{(n)}, \ldots, x_k^{(n)}, \ldots\}].$$

Now, going back to the problem in hand, for arbitrarily chosen $\widetilde{\mathbf{x}} = [\{x_k\}] \in \widetilde{X} \setminus \widetilde{A}$, define a sequence $\{\widetilde{\mathbf{x}}^{(n)}\} \subset \widetilde{A}$ by

$$\widetilde{\mathbf{x}}^{(n)} = [\{x_n, x_n, x_n, \ldots\}],$$

Fig. 4.9 Construction of $\widetilde{\mathbf{x}}^{(n)}$

$$\begin{array}{rcl} \widetilde{\mathbf{x}}^{(1)} & = & [\{x_1,\ x_1,\ x_1,\ \ldots,\ x_1,\ \ldots\}] \\ \widetilde{\mathbf{x}}^{(2)} & = & [\{x_2,\ x_2,\ x_2,\ \ldots,\ x_2,\ \ldots\}] \\ \vdots & & \vdots \\ \widetilde{\mathbf{x}}^{(n)} & = & [\{x_n,\ x_n,\ x_n,\ \ldots,\ x_n,\ \ldots\}] \\ \vdots & & \vdots \\ \hline \downarrow & & \downarrow \\ \widetilde{\mathbf{x}} & = & [\{x_1,\ x_2,\ x_3,\ \ldots,\ x_k,\ \ldots\}] \end{array}$$

i.e. $\widetilde{\mathbf{x}}^{(n)}$ is the equivalence class containing the constant sequence $\{x_n, x_n, x_n, \ldots\}$ from X. In the notations agreed above, for each k, $x_k^{(n)} = x_n$.
Then

$$\widetilde{d}(\widetilde{\mathbf{x}}^{(n)}, \widetilde{\mathbf{x}}) = \lim_{k\to\infty} d(x_k^{(n)}, x_k) = \lim_{k\to\infty} d(x_n, x_k).$$

But as $\{x_k\} \subset X$ is a Cauchy sequence, for arbitrarily chosen $\epsilon > 0$, $\exists N_1 \in \mathbb{N}$ such that $\forall n, k \geq N_1$, $d(x_n, x_k) < \epsilon$. Hence

$$\forall n \geq N_1, \quad \widetilde{d}(\widetilde{\mathbf{x}}^{(n)}, \widetilde{\mathbf{x}}) \leq \epsilon.$$

Consequently $\widetilde{\mathbf{x}} = \lim_{n\to\infty} \widetilde{\mathbf{x}}^{(n)}$, i.e. $\widetilde{\mathbf{x}}$ is a limit point of $\widetilde{A}$. Hence $\pi(X) = \widetilde{A}$ is dense in $\widetilde{X}$. This proves *(ii)*.

Step 6: $(\widetilde{X}, \widetilde{d})$ *is complete:*

We first prove that every Cauchy sequence in $\widetilde{A}$ converges in $\widetilde{X}$ and complete the proof using *(ii)*.
Let $\{\widetilde{\mathbf{x}}^{(n)}\} \subset \widetilde{A}$ be a Cauchy sequence. Then each term $\widetilde{\mathbf{x}}^{(n)}$ looks as in Fig. 4.9. Therefore, by the definition of $\widetilde{d}$, we see that for every pair of indices m, n of the sequence $\{x_k\} \subset X$,

$$d(x_m, x_n) = \widetilde{d}(\widetilde{\mathbf{x}}^{(m)}, \widetilde{\mathbf{x}}^{(n)}).$$

Hence $\{x_k\} \subset X$ is Cauchy. Consequently, $\widetilde{\mathbf{x}} := [\{x_k\}] \in \widetilde{X}$. Therefore, for each index n, we have

$$\widetilde{d}(\widetilde{\mathbf{x}}^{(n)}, \widetilde{\mathbf{x}}) = \lim_{k\to\infty} d(x_k^{(n)}, x_k) = \lim_{k\to\infty} d(x_n, x_k).$$

Hence by the same argument as in Step 5, we conclude that

$$\widetilde{\mathbf{x}}^{(n)} \to \widetilde{\mathbf{x}}.$$

This proves that every Cauchy sequence in $\widetilde{A}$ converges in $\widetilde{X}$.
Now let $\{\widetilde{\mathbf{y}}^{(n)}\}$ be a Cauchy sequence in $\widetilde{X}$. So for arbitrarily chosen $\epsilon > 0$, $\exists N_2 \in \mathbb{N}$, such that $\forall m, n \geq N_2$, $\widetilde{d}(\widetilde{\mathbf{y}}^{(m)}, \widetilde{\mathbf{y}}^{(n)}) < \epsilon$. Again, as $\widetilde{A}$ is dense in $\widetilde{X}$,

for each index n, $\exists \widetilde{\mathbf{x}}^{(n)} \in \widetilde{A}$ such that $\widetilde{d}(\widetilde{\mathbf{y}}^{(n)}, \widetilde{\mathbf{x}}^{(n)}) < \epsilon$. Hence $\forall m, n \geq N_2$,

$$\widetilde{d}(\widetilde{\mathbf{x}}^{(m)}, \widetilde{\mathbf{x}}^{(n)}) \leq \widetilde{d}(\widetilde{\mathbf{x}}^{(m)}, \widetilde{\mathbf{y}}^{(m)}) + \widetilde{d}(\widetilde{\mathbf{y}}^{(m)}, \widetilde{\mathbf{y}}^{(n)}) + \widetilde{d}(\widetilde{\mathbf{y}}^{(n)}, \widetilde{\mathbf{x}}^{(n)}) < 3\epsilon.$$

This shows that $\{\widetilde{\mathbf{x}}_n\} \subset \widetilde{A}$ is Cauchy. But by the first part of this step, $\{\widetilde{\mathbf{x}}_n\}$ is convergent in $\widetilde{X}$. Call $\widetilde{\mathbf{x}} = \lim_{n\to\infty} \widetilde{\mathbf{x}}_n$. Then for every $m, n \geq N_2$, we have

$$\widetilde{d}(\widetilde{\mathbf{x}}^{(m)}, \widetilde{\mathbf{y}}^{(n)}) \leq \widetilde{d}(\widetilde{\mathbf{x}}^{(m)}, \widetilde{\mathbf{y}}^{(m)}) + \widetilde{d}(\widetilde{\mathbf{y}}^{(m)}, \widetilde{\mathbf{y}}^{(n)}) < 2\epsilon.$$

Letting $m \to \infty$, we obtain that $\forall n \geq N_2, \widetilde{d}(\widetilde{\mathbf{x}}, \widetilde{\mathbf{y}}^{(n)}) \leq 2\epsilon$. Hence $\lim_{n\to\infty} \widetilde{\mathbf{y}}^{(n)} = \widetilde{\mathbf{x}}$. This proves (i).

Step 7: $(\widetilde{X}, \widetilde{d})$ *is unique:*

Let (Y, d) be another metric space such that $X \subset Y$ is everywhere dense in Y and Y is complete. We will show that Y is isometric to $\widetilde{X}$.

For arbitrary $\widetilde{\mathbf{x}} \in \widetilde{X}$, let $\widetilde{\mathbf{x}} = [\{x_k\}]$. Then $\{x_k\}$ is a Cauchy sequence in X, and hence in Y. Since Y is complete, $\{x_k\}$ converges to some $\mathbf{y} \in Y$. Define $\varphi\colon \widetilde{X} \to Y$ by

$$\varphi(\widetilde{\mathbf{x}}) = \mathbf{y}.$$

To see φ is well-defined, observe that if $\widetilde{\mathbf{x}} = [\{x_k\}] = [\{x'_k\}]$, then

$$\lim d(x_k, x'_k) = 0 \implies \lim x_k = \lim x'_k.$$

We will show that φ is a bijective isometry.

To see φ is one to one, for $\widetilde{\mathbf{x}}^{(1)}$ and $\widetilde{\mathbf{x}}^{(2)}$ from $\widetilde{X}$, assume $\varphi(\widetilde{\mathbf{x}}^{(1)}) = \varphi(\widetilde{\mathbf{x}}^{(2)})$. Let $\widetilde{\mathbf{x}}^{(1)} = [\{x_k^{(1)}\}]$ and $\widetilde{\mathbf{x}}^{(2)} = [\{x_k^{(2)}\}]$. Then the hypothesis implies that in Y, $\lim x_k^{(1)} = \lim x_k^{(2)}$. But this, in turn, implies that in X, $d(x_k^{(1)}, x_k^{(2)}) \to 0$, i.e. $\widetilde{\mathbf{x}}^{(1)} = \widetilde{\mathbf{x}}^{(2)}$ in $\widetilde{X}$. This shows φ is one to one.

Now, to see φ is onto, consider $\mathbf{y} \in Y$ arbitrarily. As X is dense in Y, we can find a sequence $\{x_k\} \subset X$ converging to $\mathbf{y}$. Then for the element $\widetilde{\mathbf{x}} = [\{x_k\}] \in \widetilde{X}$, $\varphi(\widetilde{\mathbf{x}}) = \mathbf{y}$. Hence φ is onto. Consequently φ is a bijection.

To see φ an isometry is easy, as for $\mathbf{y}^{(1)} = \varphi(\widetilde{\mathbf{x}}^{(1)})$ and $\mathbf{y}^{(2)} = \varphi(\widetilde{\mathbf{x}}^{(2)})$,

$$\widetilde{d}(\widetilde{\mathbf{x}}^{(1)}, \widetilde{\mathbf{x}}^{(2)}) = \lim_{k\to\infty} d(x_k^{(1)}, x_k^{(2)}) = d(\mathbf{y}^{(1)}, \mathbf{y}^{(2)}). \qquad \text{[by Exercise 3.23]}$$

This proves (iv), and hence the theorem.

■

Example 4.22.1 We have already established that the set of rational numbers $\mathbb{Q}$ is incomplete with respect to the usual metric. The completion of $\mathbb{Q}$ yields the set of real numbers $\mathbb{R}$. This fact provides an alternative method for constructing the real numbers, distinct from the usual axiomatic development.

Now, let us replicate the completion process outlined in the previous proof for these two well-known spaces. Since we already know that $\mathbb{Q}$ is incomplete, we construct $\mathbb{R}$ as follows:

Step 1: *Identifying rational numbers within* $\mathbb{R}$: In the yet-to-be-constructed set of real numbers, each rational number r is represented not only by the constant sequence $\{r, r, r, \dots\}$ but also by any rational sequence that converges to r. For instance, all of the following sequences:

$$\{0, 0, 0, \dots\}, \quad \left\{1, -\tfrac{1}{2}, \tfrac{1}{3}, -\tfrac{1}{4}, \dots, \right\}, \quad \left\{1, \tfrac{1}{4}, \tfrac{1}{6}, \dots\right\}$$

are identified with 0 as a real number. In the notation used in the previous proof, 0 as a rational number is denoted by "0", whereas 0 as a real number is represented by the symbol $\widetilde{\mathbf{0}}$.
Similarly, the sequences

$$\{1, 1, 1, \dots\}, \quad \{0.9, 0.99, 0.999, \dots\}$$

are both identified with 1 as a real number and are denoted by $\widetilde{\mathbf{1}}$.

Step 2: *Handling rational Cauchy sequences without rational limits*: Now, consider rational Cauchy sequences that do not converge to any rational number. To incorporate these into our construction, we introduce a partitioning process: Sequences that exhibit the same asymptotic behaviour are grouped together.
For instance, the sequences

$$x_n = \left(1 + \frac{1}{n}\right)^n \text{ and } y_n = 1 + \frac{1}{1!} + \frac{1}{2!} + \dots + \frac{1}{n!}$$

belong to the same equivalence class, but this class differs from those containing the sequences

$$u_n = \frac{[10^n\sqrt{2}]}{10^n} \quad \text{and} \quad v_n = \frac{[10^n\sqrt{3}]}{10^n}.$$

By classifying all such non-convergent rational Cauchy sequences into distinct groups, we form an extended set that includes both the original rational numbers (represented as limits of convergent rational Cauchy sequences) and new elements arising from these partitions. Each element of this set is a particular real number.

Step 3: *Establishing the metric on* $\mathbb{R}$: The arguments in the above proof conclude that there exists a metric on this set of all real numbers such that:

- When restricted to the rationals, the new metric agrees with the existing usual metric on $\mathbb{Q}$.
- The set of all rationals is dense in the set of all reals.
- There is only one such space with the above two properties.

Identifying this constructed space with our familiar set $\mathbb{R}$ and verifying its standard properties is a separate mathematical endeavour. This approach provides an alternative perspective on the construction of $\mathbb{R}$, see [5] for more details.

Exercises: 4.3

4.20. Let X and Y be two isometric spaces. If X is complete, show that Y is also complete.

4.21. Let (X, d) be a metric space and A be a dense subset of X such that every Cauchy sequence in A converges in X. Show that X is complete.

4.22. Does there exist a metric d (other than the discrete metric) on the set $\mathbb{Q}$ of all rational numbers so that $(\mathbb{Q}, d)$ is complete?

Chapter 5
Continuity

Continuity lies at the core of analysis, serving as a fundamental concept around which the entire study of topology has evolved. British mathematician J J Sylvester, in his inaugural presidential address to the Mathematical and Physical Section of the British Association for the Advancement of Science in Exeter (August 1869), eloquently expressed the centrality of continuity:

> Geometry formerly was the chief borrower from arithmetic and algebra, but it has since repaid its obligations with abundant usury; and if I were asked to name, in one word, the pole star round which the mathematical firmament revolves, the central idea which pervades the whole corpus of mathematical doctrine, I should point to Continuity as contained in our notions of space, and say, it is this, it is this!

This chapter will explore the concept of continuity in depth. The first section develops the idea through various definitions, supported by numerous examples. The second section introduces the stronger notion of uniform continuity and highlights its advantages. In the third section, we examine how a special class of continuous functions, known as homeomorphisms, preserve the topologies of spaces. Finally, we conclude the chapter with Banach's contraction principle and its application in guaranteeing solutions to certain differential equations.

5.1 Definitions and Examples

To explore the fundamental concept of continuity, it is useful to build the idea from its very core. We begin our discussion within the framework of classical real analysis to preserve the essence of intuition.

The literal meaning of the word "continuous" naturally leads us to think of a function as *continuous* if its graph has no breaks—that is, if one can draw the graph without lifting the tip of the pen from the paper. Let us attempt to formulate this idea mathematically.

S. Paul, *Metric Spaces*, University Texts in the Mathematical Sciences,
https://doi.org/10.1007/978-981-96-9259-0_5

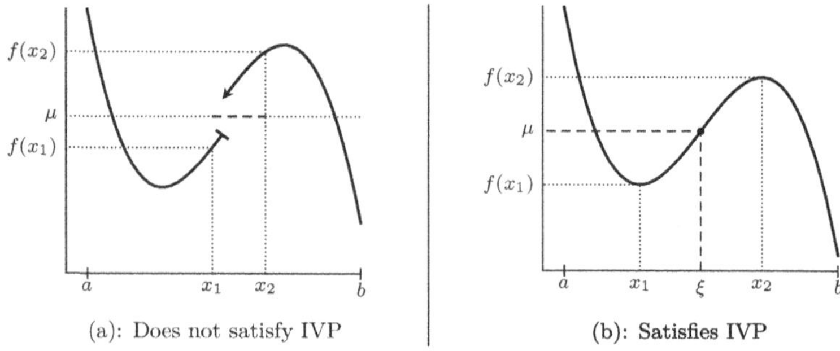

Fig. 5.1 Generic real-valued functions having breaks and no breaks in the graphs

To do so, we start by considering a function defined on a closed and bounded interval $[a, b]$. This choice allows us to begin drawing the graph at a specific point $(a, f(a))$ and finish at another point $(b, f(b))$. As illustrated in Fig. 5.1a, if f has a break in its graph, then there exist two values, $f(x_1)$ and $f(x_2)$, such that some intermediate value μ, lying between $f(x_1)$ and $f(x_2)$, is never attained by f at any point between x_1 and x_2.

Negating this statement leads to a precise mathematical formulation of our intuition: A function f has no breaks in its graph if it satisfies:

Definition 5.1 (Intermediate Value Property (IVP)) A function $f\colon [a, b] \to \mathbb{R}$ is said to satisfy the *intermediate value property* on $[a, b]$ if for every x_1, x_2 satisfying $a \leq x_1 < x_2 \leq b$ and for every μ between $f(x_1)$ and $f(x_2)$, there exists a point $\xi \in (x_1, x_2)$, such that $f(\xi) = \mu$.

Seen graphically, f satisfies IVP if every horizontal line $y = \mu$ sitting between the lines $y = f(x_1)$ and $y = f(x_2)$ intersects the graph at some point (ξ, μ), where ξ lies between x_1 and x_2.

Thus, IVP seems to capture the intuitive idea of continuity. However, can we take this as the definition of continuous functions? Unfortunately, no. The "no-break" interpretation of continuity, as captured by IVP, is too weak. There exist functions—ones that are not particularly pathological—that satisfy IVP but still fail to align with our deeper intuitions about continuity. The following examples illustrate this limitation:

- Consider the function $f\colon [-1, 1] \to \mathbb{R}$ given by

$$f(x) = \begin{cases} \sin\frac{1}{x}, & x \neq 0 \\ 0, & x = 0 \end{cases}.$$

  It is a good exercise to check that f satisfies IVP. However, it remains ambiguous whether there is a break in its graph around the point $(0, 0)$. See Fig. 5.2a.

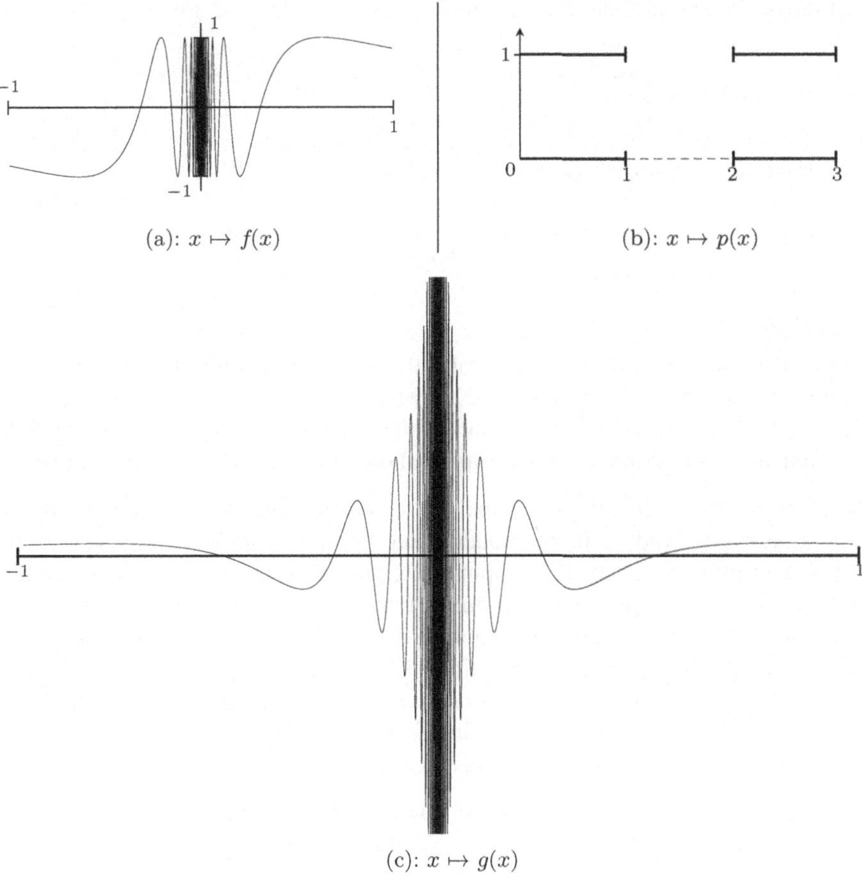

Fig. 5.2 Troubles with functions satisfying IVP

- The above example can be worsened by choosing the function $g\colon [-1, 1] \to \mathbb{R}$ given by

$$g(x) = \begin{cases} \frac{1}{x} \sin \frac{1}{x}, & x \neq 0 \\ 0, & x = 0 \end{cases}.$$

  See Fig. 5.2c. Beyond merely satisfying the IVP and raising ambiguity about the presence of a break at $x = 0$, the function g is also unbounded. This is particularly unsettling, as it contradicts the intuition that continuous functions over closed and bounded intervals should themselves be bounded.
- An expected consequence from our "no-break" notion of continuity is that the sum of two continuous functions should also be continuous. In other words, if two functions have no breaks in their graphs, their sum should likewise be free

of breaks. However, both the functions $h_1, h_2\colon [-1, 1] \to \mathbb{R}$ given by

$$h_1(x) = \begin{cases} \sin\frac{1}{x}, & x \neq 0 \\ 0, & x = 0 \end{cases}, \quad \text{and} \quad h_2(x) = \begin{cases} -\sin\frac{1}{x}, & x \neq 0 \\ 1, & x = 0 \end{cases},$$

satisfy IVP, but their sum $h\colon [-1, 1] \to \mathbb{R}$

$$h(x) := h_1(x) + h_2(x) = \begin{cases} 0, & x \neq 0 \\ 1, & x = 0 \end{cases}$$

does not!

- IVP does not support the "no-break" theory for functions defined over non-interval domains. For instance, the constant function $p(x) = 1$ defined over $A = [0, 1] \cup [2, 3]$ satisfies IVP, although it has a break in its graph. See Fig. 5.2b. To talk about continuity *only* over interval domains seems overtly restrictive.

Therefore, the above discussion compels us to abandon IVP as the defining criterion for continuity. Instead, a more modern and rigorous approach is based on the controlling-output-by-controlling-input principle. This perspective asserts that a small change in the input of a continuous function results in only a small change in the output. For an analogy, when cooking a dish that tastes less salty than desired, one adds salt gradually to adjust the saltiness to the right level. Here, the "saltiness" represents the output, which changes *continuously* in response to the input, "salt". Moreover, even the slightest change in saltiness can be controlled by adding an appropriately small amount of salt. This idea leads to:

Definition 5.2 Let $D \subset \mathbb{R}$ be non-empty and $c \in D$. Let $f\colon D \to \mathbb{R}$ be a map. f is said to be *continuous* at c if for every $\epsilon > 0$, there exists a $\delta > 0$ such that for every point $x \in (c - \delta, c + \delta) \cap D$, we will have $|f(x) - f(c)| < \epsilon$.

A continuous function, as defined above, satisfies IVP when restricted to an interval, remains bounded over a closed and bounded set, can be defined on non-interval domains, and allows for algebraic operations. This definition effectively resolves all the ambiguities discussed earlier.

However, it also leads to some surprising cases where pathologically defined functions turn out to be continuous at certain unusual points. For instance, consider the function defined by $f\colon \mathbb{R} \to \mathbb{R}$ by

$$f(x) = \begin{cases} x, & \text{when } x \text{ is rational} \\ -x, & \text{when } x \text{ is irrational} \end{cases}.$$

See Fig. 5.3. Then f is continuous at 0, which baffles our intuition.

Let us take a moment to discuss the role of ϵ in the definition of continuity. In analysis, ϵ conventionally represents a measure of how close one quantity is to another. It also carries a silent tag of "however small" in a very literal sense.

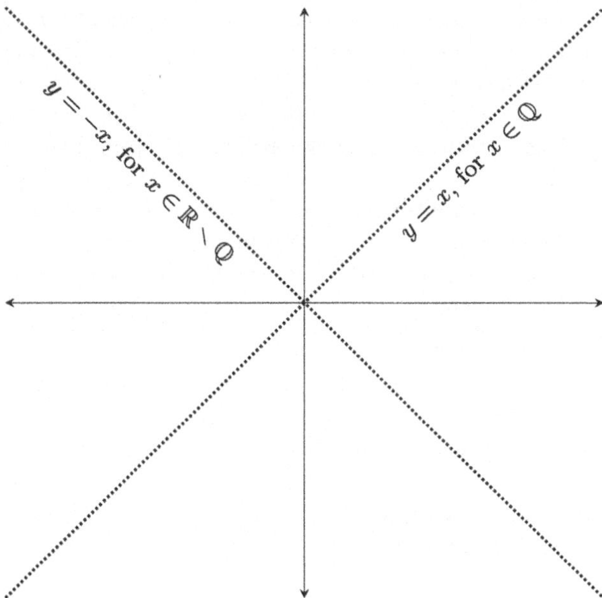

Fig. 5.3 f is continuous at 0

From a practical standpoint, ϵ often serves as an "error" parameter. Consider a mason constructing a single-storey building, where she needs to use the value of $\sqrt{2}$. She completes her job using the approximation 1.4, and the building stands tall through an earthquake. However, an engineer working on a high rise must account for additional forces such as wind, compelling her to use a more precise value, no cheaper than 1.414. A construction company designing a sophisticated cantilever bridge may need an approximation as precise as 1.414213, while manufacturing an aircraft could require 1.414213562. For a spacecraft, even greater precision, such as 1.414213562373, might be necessary.[1]

Notably, in none of these cases, do we use the exact value of $\sqrt{2}$, as no physical measuring scale can capture it perfectly. Instead, we choose an acceptable tolerable error for each scenario: 0.1 for the one-storey building, 0.001 for the high rise, and so on. The parameter ϵ represents this tolerable error. If we remain within the $\pm\,\epsilon$ range of the desired value, the approximation is considered sufficiently accurate.

Similarly, in the definition of continuity, ϵ signifies the permissible error in the output. Think of a function f as a manufacturing process, where an input x is transformed into an output $f(x)$. We certify this process as "continuous" if the output error can be controlled by restricting variations in the input. Since measuring the input x is also subject to inaccuracies—due to limitations in our measuring

[1] Disclaimer: The approximations used here are for illustrative purpose only and *do not* reflect actual engineering standards.

instruments—we must account for its deviation from the true value. The definition of continuity ensures that if we are *a priori* given a tolerable output error ϵ, we can wisely select an appropriate input scale with a controlled variation δ to keep the error in $f(x)$ within ϵ. Thus, a continuous manufacturing process is one that can produce outputs to any required degree of accuracy by appropriately controlling the inputs.

Beyond a few counter-intuitive examples, Definition 5.2 encapsulates our refined notion of continuity. Furthermore, it enriches the entire field by yielding profound and fundamental results. We now extend this definition to the broader setting of a general metric space to obtain:

Definition 5.3 Let (X, d_X) and (Y, d_Y) be two metric spaces and $a \in X$. A function $f: X \to Y$ is said to be *continuous* at a if for every $\epsilon > 0$, there exists $\delta > 0$, such that for each point x from the ball $B_X(a; \delta)$, the functional value $f(x)$ lies in the ball $B_Y(f(a); \epsilon)$, i.e. for every point $x \in X$ with $d_X(x, a) < \delta$, we will have $d_Y(f(x), f(a)) < \epsilon$.

Using notational jargon, we write:
$f: X \to Y$ is *continuous* at $a \in X$ if

$$\forall \epsilon > 0, \ \exists \delta > 0, \ \text{such that } \forall x \in B_X(a; \delta), \ \text{we have } d_Y(f(x), f(a)) < \epsilon.$$

Remarks

(a) Let A be a non-empty subset of X and f is defined on A. The continuity of f at a point $a \in A$ is determined by applying the above definition considering A as a subspace of X. Thus, $f: A \to Y$ is said to be continuous at a point $a \in A$ if for every $\epsilon > 0$, there exists $\delta > 0$ such that for every point $x \in B_A(x; \delta)$, we have $d_Y(f(x), f(a)) < \epsilon$, where $B_A(x; \delta) := B(x; \delta) \cap A$ is the δ neighbourhood of x in the subspace A.

(b) Let A be a non-empty subset of a metric space X. A point $c \in A$ is called an *isolated point* of A if for some $\delta > 0$, $B(c; \delta) \cap A = \{c\}$. If we switch to real analysis again, for an isolated point c in the domain of a function f, the point $C(c, f(c))$ is also an isolated point on the graph. There is no path passing through C. However, for convenience, we declare a function to be continuous at an isolated point. We carry the same convention in the metric spaces as well. This fact also follows from the definition. If c is an isolated point of the domain A, there is a $\delta > 0$ so that $B(c; \delta) \cap A = \{c\}$. Hence for any $\epsilon > 0$ and for each $x \in B(c; \delta) \cap A$:

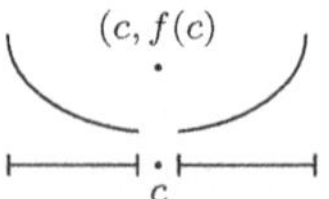

$$|f(x) - f(c)| = |f(c) - f(c)| = 0 < \epsilon.$$

(c) In the above definition, δ depends on both ϵ chosen and the concerned point a. That is why this is often called as *pointwise continuity*.
(d) If for a given $\epsilon > 0$, there is one $\delta > 0$ which works in the definition of continuity, then every δ' with $0 < \delta' < \delta$ will also work. Thus δ is far from being unique.
(e) If you look geometrically at the image of an interval under a continuous function, you will observe that it has morphed the shape of the straight line segment by stretching or squishing, but without tearing or piercing. This particular description of continuity will make our lives easier in higher levels of analysis. Let us look at the following illustration:

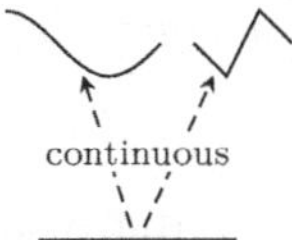

Let $D = [0, 2]$, and $f, g\colon D \to \mathbb{R}$ be defined as follows:

$$f(x) = 2x \qquad \text{and} \qquad g(x) = \begin{cases} x - 1, & \text{if } 0 \le x < 1 \\ x, & \text{if } 1 \le x \le 2 \end{cases}.$$

Let $c \in D$ be arbitrary. Fix $\epsilon > 0$ arbitrarily. Then

$$|f(x) - f(c)| < \epsilon \impliedby |2x - 2c| < \epsilon \impliedby |x - c| < \tfrac{\epsilon}{2}.$$

Choose $\delta = \frac{\epsilon}{2}$. Then for every point $x \in B(c; \delta) \cap D$, $|f(x) - f(c)| < \epsilon$. Hence, f is continuous at c, and as $c \in D$ is arbitrary, f is continuous over D. On the other hand, g is discontinuous at 1. Because, for the choice of $\epsilon = 0.5$, for any $0 < \delta < 1$, the point $x_\delta = 1 - \frac{\delta}{2}$ lies in $B(1; \delta) \cap D$, but $g(x_\delta) = -\frac{\delta}{2}$ which lies outside $(1 - \epsilon, 1 + \epsilon) = (f(1) - \epsilon, f(1) + \epsilon)$.
Now look at Figs. 5.4a and b. The function f stretches the domain interval $[0, 2]$, whereas the function g tears it at the point 1 and places the left part 1 distance apart. And as we proved, f is continuous, whereas g is discontinuous.

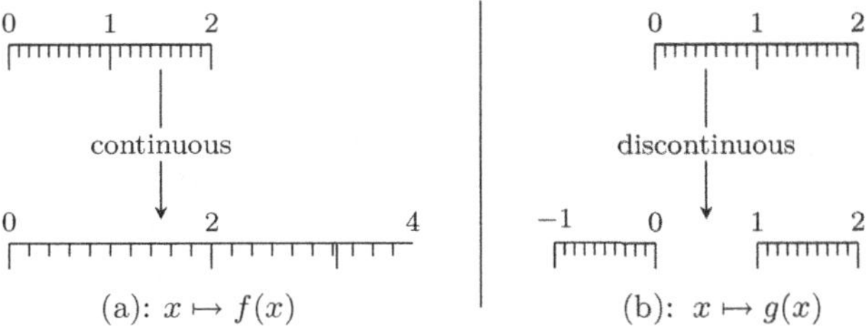

Fig. 5.4 Topological intuition of continuous functions

Before we see examples of continuous functions over different spaces, let us know some equivalent criteria for a function to be continuous.

Theorem 5.4 (Equivalent Definitions of Continuity) *Let (X, d_X) and (Y, d_Y) be two metric spaces and $f: X \to Y$. Then the followings are equivalent:*

i) *f is continuous on X.*
ii) Sequential criterion: *For each point $x \in X$ and for every sequence $\{x_n\} \subset X$ converging to x, the sequence $\{f(x_n)\}$ converges to $f(x)$ in Y.*
iii) *For every closed set $F \subset Y$, $f^{-1}(F)$ is closed in X.*
iv) *For every open set $G \subset Y$, $f^{-1}(G)$ is open in X.*
v) *For every point $x \in X$ and an open neighbourhood V of $f(x)$, there exists an open neighbourhood U of x, such that $f(U) \subset V$.*

Proof

- $(i) \implies (ii)$: Choose $x \in X$ arbitrarily and let $\{x_n\} \subset X$ converge to x. Fix $\epsilon > 0$ arbitrarily. As f is continuous at x, there exists $\delta > 0$ such that for every point $x' \in B(x; \delta)$, we will have $d_Y(f(x'), f(x)) < \epsilon$. However, as $x_n \to x$, we can find $k \in \mathbb{N}$ such that for each index $n \geq k$, $d_X(x_n, x) < \delta$, i.e. $\forall n \geq k$, $x_n \in B(x; \delta)$. Therefore, for the indices $n \geq k$, $d_Y(f(x_n), f(x)) < \epsilon$. Since $\epsilon > 0$ is arbitrary, $f(x_n) \to f(x)$.
- $(ii) \implies (iii)$: Let $F \subset Y$ be closed. If $f^{-1}(F)$ is empty, we are done. Otherwise, choose $x \in \overline{f^{-1}(F)}$ arbitrarily. We will conclude our argument by showing $x \in f^{-1}(F)$.
Let the sequence $\{x_n\} \subset f^{-1}(F)$ converge to x. Call $y_n = f(x_n)$. Then $\{y_n\} \subset F$. Therefore, by hypothesis (ii) of sequential criterion, $y_n = f(x_n) \to f(x)$. Hence, $f(x) \in \overline{F}$. But as F is closed, $\overline{F} = F$. Consequently, $f(x) \in F$, i.e. $x \in f^{-1}(F)$.
- $(iii) \implies (iv)$: Since $G \subset Y$ is open, $Y \setminus G$ is closed. Hence $f^{-1}(Y \setminus G)$ is closed. But $f^{-1}(Y \setminus G) = X \setminus f^{-1}(G)$. Hence $f^{-1}(G)$ is open in X.
- $(iv) \implies (v)$: Since $V \subset Y$ is open, $f^{-1}(V)$ is open in X. Call $U = f^{-1}(V)$. Thus, as $f(x) \in V$, the point x must belong to U. Also, $f(U) = f[f^{-1}(V)] \subset V$ (Fig. 5.5).
- $(v) \implies (i)$: Choose $x \in X$ arbitrarily. For $\epsilon > 0$ fixed arbitrarily, choose $V = B_Y(f(x); \epsilon)$. Then the corresponding open set $U \subset X$ contains x and $f(U) \subset$

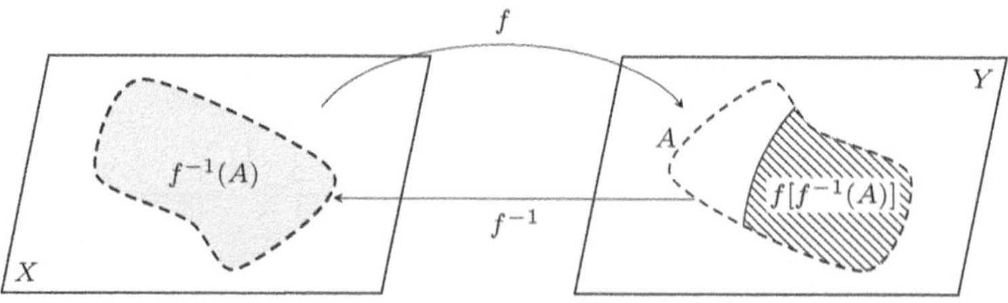

Fig. 5.5 Behaviour of a function and its inverse images

V. Choose $\delta > 0$ so that $B_X(x;\delta) \subset U$. Hence, $f(B_X(x;\delta)) \subset f(U) \subset V$, i.e. for every point $x' \in B_X(x;\delta)$, we have $d_Y(f(x'), f(x)) < \epsilon$. This completes the proof.

■

Examples

5.4.1. A constant map from one metric space to another is continuous.
Proof. Let the map $f\colon X \to Y$ be defined for each point $x \in X$ as

$$f(x) = k,$$

where k is a constant in Y. Then for any closed set $F \subset Y$, $f^{-1}(F)$ is either X or $\varnothing$ depending respectively on whether k belongs or does not belong to F. In either case, $f^{-1}(F)$ is closed. Consequently, f is continuous. □

5.4.2. The identity map $x \mapsto x$ on a metric space is continuous.
Proof. Let the map $f\colon X \to X$ be defined for each point $x \in X$ as

$$f(x) = x.$$

Choose $\epsilon > 0$ arbitrarily. Then for the choice $\delta = \epsilon$, every point $x' \in B(x;\delta)$ produces $f(x') = x' \in B(x;\delta) = B(x;\epsilon)$. Therefore, f is continuous at x. □

5.4.3. Let X be a discrete metric space. Then any function on X is continuous.
Proof. Let Y be another metric space and $f\colon X \to Y$. Let $G \subset Y$ be open. By Example 3.3.3, $f^{-1}(G)$ is always open in X. Consequently, f is continuous on X. □

5.4.4. The projection of a point $\mathbf{x} = (x_1, \ldots, x_n)$ in $\mathbb{R}^n$ onto its kth coordinate is given by the *projection map* $\pi_k\colon \mathbb{R}^n \to \mathbb{R}$ defined as $\pi_k(\mathbf{x}) = x_k$. Projection maps are all continuous.
Proof. Let $\{\mathbf{x}^{(m)}\}_m$ be a sequence in $\mathbb{R}^n$ converging to $\mathbf{x}$. Let the coordinates of each term be given as

$$\mathbf{x}^{(m)} = (x_1^{(m)}, \ldots, x_n^{(m)}).$$

Then by Exercise 3.20, for each coordinate k, the sequence $x_k^{(m)} \to x_k$. But $x_k^{(m)} = \pi_k(\mathbf{x}^{(m)})$. Hence by sequential criterion, π_k is continuous at $\mathbf{x}$. □

5.4.5. Fix a point a in a metric space X. Then the map $x \mapsto d(a, x)$ is continuous.
Proof. By miscellaneous result 2.1, we know that for points $x, x' \in X$,

$$|d(a,x) - d(a,x')| \le d(x,x').$$

Therefore, for any $\epsilon > 0$ chosen arbitrarily, $d(x,x') < \epsilon \implies |d(a,x) - d(a,x')| < \epsilon$. This concludes the map is continuous at x. Hence the proof. □

5.4.6. The vector addition map $\mathbb{R}^n \times \mathbb{R}^n \to \mathbb{R}^n$ given by $(\mathbf{x}, \mathbf{y}) \mapsto \mathbf{x} + \mathbf{y}$ is continuous in the product metric. Also, the scalar multiplication map $\mathbb{R} \times \mathbb{R}^n \to \mathbb{R}^n$ given by $(\alpha, \mathbf{x}) \mapsto \alpha\mathbf{x}$ is continuous.

Proof. Fix a point $(\mathbf{x}, \mathbf{y}) \in \mathbb{R}^n \times \mathbb{R}^n$ arbitrarily and let $\{(\mathbf{x}^{(m)}, \mathbf{y}^{(m)})\}_m$ be a sequence of points in $\mathbb{R}^n \times \mathbb{R}^n$ converging to $(\mathbf{x}, \mathbf{y})$ in the product metric d. Then

$$\begin{aligned} & (\mathbf{x}^{(m)}, \mathbf{y}^{(m)}) \to (\mathbf{x}, \mathbf{y}) \\ \Longrightarrow \quad & d\left((\mathbf{x}^{(m)}, \mathbf{y}^{(m)}), (\mathbf{x}, \mathbf{y})\right) \to 0 \\ \Longrightarrow \quad & \max\left\{\|\mathbf{x}^{(m)} - \mathbf{x}\|_n, \|\mathbf{y}^{(m)} - \mathbf{y}\|_n\right\} \to 0 \\ \Longrightarrow \quad & \|\mathbf{x}^{(m)} - \mathbf{x}\|_n \to 0 \quad \text{and} \quad \|\mathbf{y}^{(m)} - \mathbf{y}\|_n \to 0 \\ \Longrightarrow \quad & \mathbf{x}^{(m)} \to \mathbf{x} \quad \text{and} \quad \mathbf{y}^{(m)} \to \mathbf{y} \\ \Longrightarrow \quad & \mathbf{x}^{(m)} + \mathbf{y}^{(m)} \to \mathbf{x} + \mathbf{y}. \end{aligned}$$

Therefore, by sequential criterion, the vector addition map is continuous at $(\mathbf{x}, \mathbf{y})$.

In a similar fashion, we can show that the scalar multiplication map is also continuous. □

Following are couple of results which help us to create more continuous functions over specific spaces.

Proposition 5.5 *Let X be a metric space, and the maps $f, g: X \to \mathbb{R}$ are continuous at a point $x \in X$. Let $k \in \mathbb{R}$ be a constant. Then the maps kf, $f + g$, fg are continuous at x. Also, $\frac{f}{g}$ is continuous at x, provided $g(x) \neq 0$.*

Proof Consider a sequence $\{x_n\} \subset X$ converging to x. Then

$$\begin{aligned} (kf)(x_n) &= k \times f(x_n) \\ (f \pm g)(x_n) &= f(x_n) \pm g(x_n) \\ (fg)(x_n) &= f(x_n)g(x_n) \\ \left(\frac{f}{g}\right)(x_n) &= \frac{f(x_n)}{g(x_n)}. \end{aligned}$$

As both f and g are continuous, the sequences $f(x_n) \to f(x)$ and $g(x_n) \to g(x)$. For the quotient function $\frac{f}{g}$, we may assume $g(x_n) \neq 0$. This is a legitimate assumption; as $g(x_n) \to g(x)$ and $g(x) \neq 0$, all but only finitely many terms of the sequence $\{g(x_n)\}$ are non-zero. Therefore, by the algebra of limits of sequences of real numbers, the above sequences in the left-hand sides converge to the respective limits of the sequences in the right-hand sides. Hence, the proof. ■

Remarks

(a) The above proposition holds true if f, g are complex-valued continuous functions as well.
(b) Let $\mathcal{C}(X, \mathbb{R})$ and $\mathcal{C}(X, \mathbb{C})$ denote the set of all real-valued and complex-valued continuous functions on X, respectively. Then $\mathcal{C}(X, \mathbb{R})$ or $\mathcal{C}(X, \mathbb{C})$ form a vector space over $\mathbb{R}$ or $\mathbb{C}$, respectively, under obvious operations.

Proposition 5.6 *Let X, Y, Z be metric spaces. Let $f: X \to Y$ be continuous at $x \in X$ and $g: Y \to Z$ be continuous at $y = f(x)$. Then the composite map $g \circ f: X \to Z$ is continuous at x.* [This is often said as: "Continuous map of a continuous map is a continuous map".]

Proof Consider a sequence $\{x_n\} \subset X$ converging to x. As f is continuous at x, the sequence $\{f(x_n)\}$ converges to $f(x)$ in Y. Again, as g is continuous at $f(x)$, the sequence $\{g(f(x_n))\}$ converges to $g(f(x))$ in Z, i.e. $(g \circ f)(x_n) \to (g \circ f)(x)$. Therefore, by sequential criterion, $g \circ f$ is continuous at x. ■

Examples

5.6.1. The quantity $2x^3 + 4x - 5$ is a real polynomial of degree 3 in one variable. Similarly, $x^2y^3 - x^3 + 2xy - 2$ is a polynomial of degree 5 in two variables, whereas $xyz + yz + zx + xy$ is of degree 3 in three variables. In general, a real polynomial of degree r in n variables $x_1, \ldots, x_n$ takes the form

$$p(x_1, \ldots, x_n) = \sum_{\substack{0 \le i_1 + \cdots + i_n \le r \\ i_k \in \{0,1,\ldots,r\}}} a_{i_1 \ldots i_n} x_1^{i_1} \ldots x_n^{i_n},$$

where $a_{i_1 \ldots i_n}$'s are all real numbers, and for some occurrence of $i_1, \ldots, i_n$, $i_1 + \cdots + i_n = r$. The above form implies that a polynomial in n variables is a function $p: \mathbb{R}^n \to \mathbb{R}$. Now, for points $\mathbf{x} = (x_1, \ldots, x_n) \in \mathbb{R}^n$, Example 5.4.4 suggests that we can write

$$p(\mathbf{x}) = p(x_1, \ldots, x_n) = \sum_{\substack{0 \le i_1 + \cdots + i_n \le r \\ i_k \in \{0,1,\ldots,r\}}} a_{i_1 \ldots i_n} [\pi_1(\mathbf{x})]^{i_1} \ldots [\pi_n(\mathbf{x})]^{i_n}.$$

As all the π_k's are continuous, by Proposition 5.5, we can conclude that p is continuous, i.e. every polynomial function is continuous.

5.6.2. Let $\mathcal{M}_n(\mathbb{R})$ denote the set of all $n \times n$ real matrices. Identify $\mathcal{M}_n(\mathbb{R})$ with $\mathbb{R}^{n^2}$, equipped with the usual metric. Let σ be a permutation on the set $\{1, 2, \ldots, n\}$ of n elements, and let S_n be the collection of all such permutations. Then the *determinant* of a matrix $A = [\![a_{ij}]\!] \in \mathcal{M}_n(\mathbb{R})$ is given by

$$\det A = \sum_{\sigma \in S_n} (\operatorname{sgn} \sigma) a_{1\sigma(1)} \ldots a_{n\sigma(n)},$$

where

$$\operatorname{sgn}\sigma = \begin{cases} 1, & \text{if } \sigma \text{ is an even permutation} \\ -1, & \text{if } \sigma \text{ is an odd permutation} \end{cases}.$$

Therefore, the determinant function $\det\colon \mathcal{M}_n(\mathbb{R}) \to \mathbb{R}$ is a polynomial of degree n in n^2 many variables $\{a_{ij} : 1 \leq i, j \leq n\}$. Consequently, the determinant is a continuous function.

Miscellaneous Results

5.1 Show that the set of all invertible matrices in $\mathcal{M}_n(\mathbb{R})$ is open.

Proof. The set $GL_n(\mathbb{R})$ of all invertible matrices can be written as

$$GL_n(\mathbb{R}) = \det{}^{-1}(\mathbb{R} \setminus \{0\}).$$

Since the determinant map $\det\colon \mathcal{M}_n(\mathbb{R}) \to \mathbb{R}$ is continuous, and the set $\mathbb{R} \setminus \{0\}$ is open in $\mathbb{R}$, $GL_n(\mathbb{R})$ is open in $\mathcal{M}_n(\mathbb{R})$. □

5.2 Let $f, g\colon X \to \mathbb{R}$ be two continuous functions on a metric space X. Show that the map $\varphi\colon X \to \mathbb{R}^2$ given by

$$\varphi(x) := (f(x), g(x))$$

is continuous. In general, if $f_1, \ldots, f_n$ are real-valued continuous functions on X, then the map $x \mapsto (f_1(x), \ldots, f_n(x))$ is a continuous map from X to $\mathbb{R}^n$.

Proof. Fix $x \in X$ arbitrarily, and let the sequence $\{x_m\}_m \subset X$ converge to x. As f and g are continuous at x, the sequence $\{f(x_m)\}$ converges to $f(x)$ and $\{g(x_m)\}$ converges to $g(x)$. Therefore, by Example 3.39.3, the sequence $\{(f(x_m), g(x_m))\}$ converges to $(f(x), g(x))$, i.e. $\{\varphi(x_m)\}$ converges to $\varphi(x)$. Consequently, φ is continuous at x. □

5.3 A square matrix A is called *orthogonal* if $AA^T = I$. Show that the set $\mathcal{O}_n(\mathbb{R})$ of all orthogonal matrices is a closed subset of $\mathcal{M}_n(\mathbb{R})$.

Proof. For an $n \times n$ matrix A, consider the map $\varphi\colon \mathcal{M}_n(\mathbb{R}) \to \mathcal{M}_n(\mathbb{R})$ given by

$$\varphi(A) = AA^T.$$

If A is given by $[\![a_{ij}]\!]_{n\times n}$, then by definition, the $(i, j)^{\text{th}}$ entry of $\varphi(A)$ is

$$\varphi(A)_{ij} = \sum_{k=1}^{n} a_{ik}a_{jk}.$$

Observe that $\sum_{k=1}^{n} a_{ik}a_{jk}$ is a quadratic polynomial in the entries of A. By Example 5.6.1, each $\varphi(A)_{ij}$ is therefore a continuous function. But in that case, using the above example, we can conclude that $\varphi(A) = [\![\varphi(A)_{ij}]\!]_{n\times n}$

is continuous. Furthermore, by definition,

$$\mathcal{O}_n(\mathbb{R}) = \varphi^{-1}(\{I\}).$$

As a finite point set is always closed in a metric space (Example 3.20.5), $\{I\}$ is closed in $\mathcal{M}_n(\mathbb{R})$, and so is $\varphi^{-1}(\{I\})$. Hence the proof. □

5.4 A square matrix $A \in \mathcal{M}_n(\mathbb{R})$ is called *nilpotent* if $A^k = \mathbf{0}$ for some positive integer k, where $\mathbf{0}$ is the $n \times n$ null matrix. Show that the set $\mathcal{N}_n(\mathbb{R})$ of all nilpotent matrices is a closed subset of $\mathcal{M}_n(\mathbb{R})$.

Proof. Let us agree to use the index k to denote the smallest positive integer for which $A^k = \mathbf{0}$. We claim that $k \leq n$.

This is true, because $A^k = \mathbf{0}$ implies all the eigenvalues of A^k are 0. Hence, all the eigenvalues of A are 0. Consequently, the characteristic polynomial of A is $\Phi(x) := (-1)^n x^n$. Therefore, by Cayley-Hamilton theorem,

$$\Phi(A) = (-1)^n A^n = \mathbf{0},$$

i.e. $A^n = \mathbf{0}$. Hence $k \leq n$.

Now, by Exercise 5.8(b), for any positive integer k, the map $A \mapsto A^k$ is continuous. Also, the singleton $\{\mathbf{0}\}$ is closed in $\mathcal{M}_n(\mathbb{R})$. If φ denotes the function $\varphi(A) = A^k$, then the set

$$N_k := \{A \in \mathcal{M}_n(\mathbb{R}) : A^k = \mathbf{0}\} = \varphi^{-1}(\{\mathbf{0}\})$$

is closed. Thus, the set of all $n \times n$ nilpotent matrices given by

$$\mathcal{N}_n(\mathbb{R}) = \bigcup_{k=1}^{n} N_k$$

is a finite union of closed sets and hence is closed. □

5.5 Let (X, d) be a metric space. Consider the distance function $d \colon X \times X \to \mathbb{R}$. Equip $X \times X$ with the product metric. Show that $d(\cdot, \cdot)$ is a continuous map.

Proof. For arbitrary points (x, y) and (x', y') from $X \times X$, we have

$$\begin{aligned}
|d(x, y) - d(x', y')| &= |d(x, y) - d(x', y) + d(x', y) - d(x', y')| \\
&\leq |d(x, y) - d(x', y)| + |d(x', y) - d(x', y')| \\
&\leq d(x, x') + d(y, y') \quad \text{[by Miscellaneous result 2.1]} \\
&\leq 2d'((x, y), (x', y')),
\end{aligned}$$

where d' is the product metric on $X \times X$. For $\epsilon > 0$ chosen arbitrarily, choose $\delta = \frac{\epsilon}{2}$. Then

$$d'((x, y), (x', y')) < \delta \implies |d(x, y) - d(x', y')| < \epsilon.$$

Consequently, d is continuous at (x, y). Since $(x, y) \in X \times X$ is arbitrary, this completes the proof. □

Exercise: 5.1

5.1. Let X, Y be two metric spaces and $f: X \to Y$. Show that TFAE:

(a) f is continuous on X.
(b) For every $B \subset Y$, $f^{-1}(B^\circ) \subset [f^{-1}(B)]^\circ$. ($A^\circ$ denotes the interior of A.)
(c) For every $A \subset X$, $f(\overline{A}) \subset \overline{f(A)}$.
(d) For every $B \subset Y$, $\overline{f^{-1}(B)} \subset f^{-1}(\overline{B})$.

5.2. Let X, Y be two metric spaces and $f: X \to Y$ be continuous.

(a) Let $Z = f(X)$ be this subspace of Y. Show that the function $f: X \to Z$ remains continuous.
(b) Let $\varnothing \neq A \subset X$. Show that the restricted function $f\big|_A: A \to Y$ is also continuous.

5.3. Let X, Y be metric spaces. Fix $y_0 \in Y$. Show that the inclusion map $i_X: X \to X \times Y$ given by $x \mapsto (x, y_0)$ is continuous in the product metric.

5.4. Let $f: \mathbb{N} \to X$ be a sequence in the metric space X. Considering $\mathbb{N}$ as a subspace of $\mathbb{R}$ in the usual metric, show that f is continuous.

5.5. Let X, Y, and Z be metric spaces. Let $\varphi: X \to Y \times Z$, where $\varphi = (\varphi_1, \varphi_2)$. Here $\varphi_1: X \to Y$ and $\varphi_2: X \to Z$ are the coordinate functions of φ. Show that φ is continuous iff both φ_1 and φ_2 are continuous functions.

5.6. Let $f: \mathbb{R}^2 \to \mathbb{R}$ be continuous. Show that the map $g: \mathbb{R}^2 \to \mathbb{R}$ given by

$$g(x, y) := f(x + y, x - y)$$

is also continuous.

5.7. Show that the map $A \mapsto A^T$ from $\mathcal{M}_n(\mathbb{R})$ to $\mathcal{M}_n(\mathbb{R})$ is continuous, where A^T is the transpose of the matrix A.

5.8. (a) The *standard inner product* on the vector space $\mathbb{R}^n$ over $\mathbb{R}$ is defined for points $\mathbf{x} = (x_1, \ldots, x_n)$ and $\mathbf{y} = (y_1, \ldots, y_n)$ from $\mathbb{R}^n$ as

$$\langle \mathbf{x}, \mathbf{y} \rangle = x_1 y_1 + x_2 y_2 + \cdots + x_n y_n.$$

Show that $\langle \cdot, \cdot \rangle$ is a continuous map.

(b) Use the above result to show that the map $A \mapsto A^2$ is continuous. [*Hint:* The $(i, j)^{\text{th}}$ entry of the matrix A^2 is the standard inner product of the ith row and j^{th} column of A.] In general, the map $A \mapsto A^k$ is continuous for every positive integer k.

5.9. Let X and Y be two metric spaces and $X = \bigcup_\alpha U_\alpha$, where each U_α is open in X. Let $f: X \to Y$ be such that for each α, $f\big|_{U_\alpha}$ is continuous on U_α. Show that f is continuous on X.

5.10. *Pasting lemma*: Let X be a metric space and $X = A \cup B$, where A and B are closed subsets of X. Let Y be another metric space and the functions $f: A \to Y$ and $g: B \to Y$ are continuous on A and B, respectively. If for each point $x \in A \cap B$, $f(x) = g(x)$, show that the function $h: X \to Y$ given by

$$h(x) = \begin{cases} f(x), \text{ when } x \in A \\ g(x), \text{ when } x \in B \end{cases}$$

is continuous on X. What happens when both A and B are open in X?

5.11. For any two real numbers x and y, define two metrics d and d' on the set of all real numbers as follows: $d(x, y) = 3|x - y|$; and

$$d'(x, y) = \begin{cases} |x - y|, \; xy > 0 \\ 1 + |x - y|, \; xy < 0 \end{cases} \text{ and } d'(x, 0) = d'(0, x)$$

$$= \begin{cases} |x|, \; x \geq 0 \\ 1 + |x|, \; x < 0 \end{cases}.$$

For brevity of notations, denote the space of all real numbers with the usual metric by $\mathbb{R}$, the space of all real numbers with the metric d by X, and the space of all real numbers with the metric d' by X'. Let f denote the map $x \mapsto 2x$, $x \in \mathbb{R}$. Show that $f: \mathbb{R} \to X$ is continuous everywhere, whereas $f: \mathbb{R} \to X'$ is not. Explain the results from the light of the remark (e) on page 177.

5.12. Let X, Y be metric spaces and $f: X \to Y$ be continuous and onto. If $D \subset X$ is dense in X, show that $f(D)$ is also dense in Y.

5.13. Let X, Y be metric spaces and $D \subset X$ be dense. Let $f, g: X \to Y$ be continuous functions such that for points $x \in D$, $f(x) = g(x)$. Show that $f(x) = g(x)$ for every point $x \in X$.

5.14. Let A be a closed and bounded subset of $\mathbb{R}^n$ and $f: A \to \mathbb{R}$ be continuous. Show that f is bounded on A. [*Hint:* If not, $\exists \{\mathbf{x}^{(m)}\}_m \subset A$ such that $|f(\mathbf{x}^{(m)})| > m$. Use Exercise 3.21.]

(a) If A is not necessarily closed, does the result still hold true?
(b) If X is any arbitrary metric space and $A \subset X$ is closed and bounded, is every continuous map $f: A \to \mathbb{R}$ bounded on A?

5.15. If A is a closed set in a metric space X, show that there is a sequence $\{U_n\}$ of decreasing ($U_{n+1} \subset U_n$) open sets, such that $A = \bigcap_n U_n$. [*Hint:* The function $f(x) = d(x, A)$ is continuous on X. $f(x) = 0$ iff $x \in A$. Take $U_n = f^{-1}\left(-\frac{1}{n}, \frac{1}{n}\right)$.]

5.2 Uniform Continuity

See miscellaneous result 5.5. Here, the value of δ does not depend on the specific point (x, y). This means that if we switch to another point (x_1, y_1) and verify continuity using the ϵ-δ definition, the same $\delta = \frac{\epsilon}{2}$ will still be valid. However, this property does not hold in general.

For example, consider the function $f(x) = \frac{1}{x}$, which is continuous at every point in its domain $(0, \infty)$. Given an $\epsilon > 0$, a suitable choice of δ for a given point x is determined by

$$\delta(x) = x - \frac{1}{\frac{1}{x} + \epsilon}.$$

However, it is impossible to find a single positive δ that works uniformly for every x in $(0, \infty)$. Functions that do allow for such a uniform choice of δ exhibit a stronger form of continuity, known as uniform continuity. In this section, we explore this more robust notion of continuity.

Definition 5.7 Let (X, d_X) and (Y, d_Y) be metric spaces and $f: X \to Y$. f is said to be *uniformly continuous* on X if for arbitrary $\epsilon > 0$, we can find a $\delta > 0$, such that for each pair of points $x, y \in X$, $d_X(x, y) < \delta$ will imply $d_Y(f(x), f(y)) < \epsilon$.

In the realm of real analysis, uniform continuity provides a way to control the steepness of a function's graph. Given a positive number ϵ, imagine a ring—more precisely, a thimble open on both sides—with a fixed diameter ϵ and some positive width δ (see Fig. 5.6b). A function is uniformly continuous if one can slide this ring along the entire graph of the function, always keeping it perfectly vertical, such

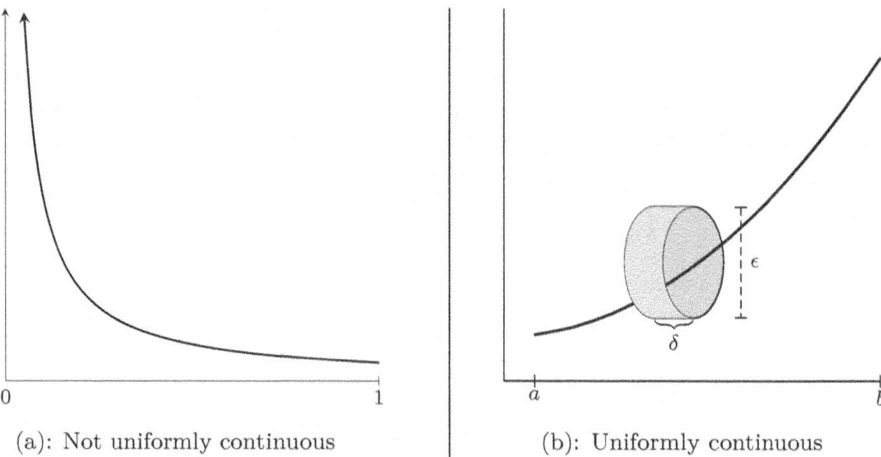

Fig. 5.6 Generic illustration of uniform continuity

that the graph passes *through* the ring without ever touching its boundaries. This geometric intuition captures the essence of uniform continuity: The function's rate of change remains controlled in a way that a single choice of δ works for the entire domain, rather than being dependent on specific points.

Pointwise continuity is a *local* property, ensuring that a function behaves predictably in the vicinity of a given point. In contrast, uniform continuity offers a *global* perspective, maintaining control over the function's behaviour across the entire domain. Issues arise when one attempts to extend local continuity to infer global properties, often leading to counter-intuitive results.

For example, pointwise continuity guarantees that outputs remain close when inputs are sufficiently close. Similarly, a Cauchy sequence concerns itself with the mutual proximity of its terms. It seems reasonable to expect that the image of a Cauchy sequence under a continuous function would also be a Cauchy sequence. Surprisingly, this is not always the case. Consider again the function $f(x) = \frac{1}{x}$ on $(0, \infty)$, and the Cauchy sequence $\left\{\frac{1}{n}\right\}$. The sequence of function values is then $\{n\}$, which is clearly not Cauchy.

This failure highlights a weakness of pointwise continuity in preserving key properties of sequences. Uniform continuity remedies this limitation by providing a stronger, more robust framework that aligns better with our natural intuitions over the entire domain.

Let us now see a proof of this result.

Theorem 5.8 *A uniformly continuous function maps Cauchy sequences to Cauchy sequences.*

Proof Let $f: X \to Y$ be uniformly continuous and $\{x_n\}$ be a Cauchy sequence in X. Choose $\epsilon > 0$ arbitrarily. By uniform continuity of f, we can find $\delta > 0$ such that for any pair of points $x, x' \in X$ with $d_X(x, x') < \delta$, we will have $d_Y(f(x), f(x')) < \epsilon$. Now, as $\{x_n\}$ is Cauchy, there exists a positive integer k such that for indices $m, n \geq k$, we have $d_X(x_m, x_n) < \delta$. Hence for every $m, n \geq k$, $d_Y(f(x_m), f(x_n)) < \epsilon$. As $\epsilon > 0$ is arbitrary, this implies $\{f(x_n)\}$ is Cauchy in Y. ■

Examples

5.8.1. Any function defined over a discrete space is uniformly continuous.
The proof is left for the reader as an exercise.

5.8.2. The function $x \mapsto x^2$ is uniformly continuous over $[0, a]$ for any positive number a but is not so over $[0, \infty)$.
Proof. The first part is left as exercise. Call $f(x) := x^2$. To show f is not uniformly continuous over $[0, \infty)$, consider the sequences $\{x_n\}$ and $\{y_n\}$, where $x_n = n$ and $y_n = n + \frac{1}{n}$. Then, $|x_n - y_n| = \frac{1}{n} \to 0$. But

$$|f(x_n) - f(y_n)| = \left|n^2 - \left(n + \frac{1}{n}\right)^2\right| = 2 + \frac{1}{n^2} \not\to 0.$$

Hence by Exercise 5.18, f cannot be uniformly continuous over $[0, \infty)$. □

Remark This example explicitly shows that uniform continuity depends upon the choice of domain of the function.

5.8.3. Let X, Y be two metric spaces. A map $f: X \to Y$ is called *Lipschitz* if there is a constant $L > 0$ (called a *Lipschitz constant*) of f such that for every pair of points $x, x' \in X$,

$$d_Y(f(x_1), f(x_2)) \leq L d_X(x_1, x_2).$$

A Lipschitz function is always uniformly continuous.

5.8.4. Let X, Y be metric spaces and $f: X \to Y$ be an isometry. Then f is Lipschitz and hence uniformly continuous.

The second significant contribution of uniform continuity is to be able to extend the definition of a function. It has far-reaching consequences, and we see a couple of them after proving the result.

Theorem 5.9 *Let X be a metric space and $A \subset X$ be dense in X. Let Y be a complete metric space. If $f: A \to Y$ is uniformly continuous on A, then f can be extended uniquely to a uniformly continuous function $g: X \to Y$.*

Proof If $A = X$, the proof is obvious. So assume $A \subsetneq X$. Let us define g suitably. For a point $x \in A$, we define $g(x) := f(x)$. For points $x \in X \setminus A$, we proceed in the following way: As A is dense in X, there is a sequence of points $\{a_n\}$ from A converging to x. Since $\{a_n\}$ is convergent, it is Cauchy. Since f is uniformly continuous over A, the sequence $\{f(a_n)\}$ is Cauchy in Y. As Y is complete, $\{f(a_n)\}$ is convergent. Therefore, for points $x \in X \setminus A$, let $g(x) := \lim f(a_n)$.

We first show that g thus obtained is well-defined. That is to show that for points $x \in X \setminus A$, the definition of g depends only on the point x and not on the sequence $\{a_n\}$. Let $\{b_n\} \subset A$ be another sequence converging to x. Then $d_X(a_n, b_n) \to 0$, and since f is uniformly continuous on A, by Exercise 5.18, $d_Y(f(a_n), f(b_n)) \to 0$ as well; i.e. $\lim f(a_n) = \lim f(b_n)$. Hence g is well-defined.

Next we show that g is uniformly continuous. Since f is uniformly continuous on A, for any chosen $\epsilon > 0$, there exists a $\delta > 0$ such that for points $a, a' \in A$, $d_X(a, a') < \delta$ implies $d_Y(f(a), f(a')) < \frac{\epsilon}{3}$. Now choose arbitrary points $x, x' \in X$ with $d_X(x, x') < \frac{\delta}{3}$. As A is dense in X, we get sequences $\{a_n\}$ and $\{a'_n\}$ in A converging to x and x', respectively. By definition of g, $g(x) = \lim f(a_n)$ and $g(x') = \lim f(a'_n)$. Hence for the concerned ϵ and δ, we can find a large enough $k \in \mathbb{N}$ such that

$$d_X(a_k, x) < \frac{\delta}{3}, \qquad d_X(a'_k, x') < \frac{\delta}{3}$$
$$d_Y(g(x), f(a_k)) < \frac{\epsilon}{3}, \qquad d_Y(g(x'), f(a'_k)) < \frac{\epsilon}{3}.$$

Therefore

$$d_X(a_k, a'_k) < d_X(a_k, x) + d_X(x, x') + d_X(x', a'_k) < \tfrac{\delta}{3} + \tfrac{\delta}{3} + \tfrac{\delta}{3} = \delta,$$

and hence $d_Y(f(a_k), f(a'_k)) < \frac{\epsilon}{3}$. Consequently,

$$\begin{aligned} d_Y(g(x), g(x')) &\le d_Y(g(x), f(a_k)) + d_Y(f(a_k), f(a'_k)) + d_Y(f(a'_k), g(x')) \\ &< \tfrac{\epsilon}{3} + \tfrac{\epsilon}{3} + \tfrac{\epsilon}{3} = \epsilon. \end{aligned}$$

Since $x, x' \in X$ are arbitrary, g is uniformly continuous on X.
The uniqueness of g is easy to prove and is left as an exercise. ■

Application of Theorem 5.9: Defining the Map $x \mapsto a^x$
Mathematics from elementary school has taught us simple definitions of indices like $3^2 := 3 \times 3$ or $5^3 := 5 \times 5 \times 5$. But, can we make sense of $2^{\sqrt[4]{3}}$? Precisely asked, for a fixed positive number a, can we define the map $x \mapsto a^x$ for every real value of x? It turns out that the answer is "yes" and the above theorem helps us in the process. Let us go through the successive steps of this manipulation.

Step 1: For a fixed positive number a, define

$$a^0 = 1.$$

Step 2: For any $a > 0$ and a positive integer n, define

$$a^n = \underbrace{a \times \cdots \times a}_{n \text{ times}}.$$

Step 3: For any $a > 0$ and a positive integer n, define

$$a^{-n} = \left(\frac{1}{a}\right)^n.$$

Thus we have defined a^m for every integer m.

Step 4: A natural way to define a^r for any rational $r = \frac{m}{n}$, where $m \in \mathbb{Z}$ and $n \in \mathbb{N}$ is

$$a^{m/n} = (a^{1/n})^m.$$

However, for this, we need to make sense of $a^{1/n}$. Let us now accomplish that.

Claim For any $a > 0$ and a positive integer n, there exists a unique real number $b > 0$ such that $b^n = a$.

Proof of Claim Consider the continuous function $f:[0,\infty)\to[0,\infty)$ given by $f(x)=x^n$. By Archimedean property of $\mathbb{R}$, we can find a positive integer A such that $A>a$. Hence $f(A)=A^n>a$. Also, $f(0)=0<a$. Hence by the intermediate value property, there exists a positive real number b between 0 and A such that $b^n=a$.
To see the uniqueness of b, assume that for $c>0$, $c^n=a$ as well. Then

$$\begin{aligned} & b^n = c^n \\ \implies & b^n - c^n = 0 \\ \implies & (b-c)(b^{n-1}+b^{n-2}c+\cdots+bc^{n-2}+c^{n-1}) = 0 \\ \implies & b = c.. \\ & [\text{as } b^{n-1}+b^{n-2}c+\cdots+bc^{n-2}+c^{n-1}>0] \end{aligned}$$

This proves our claim.

Denote b by $a^{1/n}$. Thus we have defined a^x for every rational x.

Step 5: We now wish to extend the definition of a^x for all real x. For that, we must first show that $x\mapsto a^x$ is uniformly continuous over $\mathbb{Q}$. However, that is not true. We tackle this by restricting our domain. Eventually, we prove that:

Claim Fix $a>0$ and a positive integer N arbitrarily. The map $x\mapsto a^x$ is uniformly continuous over $[-N,N]\cap\mathbb{Q}$.

Proof of Claim For arbitrary points $x, x+h\in[-N,N]\cap\mathbb{Q}$,

$$|a^{x+h}-a^x| = a^x|a^h-1| \le a^N|a^h-1|.$$

As $h\to 0$, $a^h\to 1$, and consequently, $|a^{x+h}-a^x|\to 0$. Writing y for $x+h$, we get that for a given $\epsilon>0$, there exists $\delta>0$ such that for arbitrary points $x, y\in[-N,N]\cap\mathbb{Q}$,

$$|x-y|<\delta \implies |a^x-a^y|<\epsilon.$$

Consequently, $x\mapsto a^x$ is uniformly continuous over $[-N,N]\cap\mathbb{Q}$. This proves our claim.

Thus by the last theorem, the function $x\mapsto a^x$ is well-defined for every point $x\in[-N,N]$ and is uniformly continuous there.
Since N is arbitrary, a^x is defined for all real x.

Remark Observe that the function $x\mapsto a^x$ is *not* uniformly continuous over $\mathbb{R}$. It is only pointwise continuous. Only when we restrict to a bounded domain $[-N,N]$, the map becomes uniformly continuous.

Application of Theorem 5.9: Alternate Proof of Theorem 4.22 (Completion of a Metric Space)

We first prove the following proposition as a consequence of Theorem 5.9:

Proposition 5.10 *A completion of a metric space is unique.*

Proof Let (Y_1, d_1) and (Y_2, d_2) be two completions of a metric space (X, d). We will show Y_1 and Y_2 must be isometric spaces.
According to Definition 4.21, there exist isometries

$$\varphi_1: X \to Y_1 \quad \text{and} \quad \varphi_2: X \to Y_2$$

so that $\varphi_1(X)$ is dense in Y_1 and $\varphi_2(X)$ is dense in Y_2. As an isometry is always injective, the maps $\varphi_1: X \to \varphi_1(X)$ and $\varphi_2: X \to \varphi_2(X)$ are bijections. Consequently, the map $\psi: \varphi_1(X) \to \varphi_2(X)$ given by $\psi := \varphi_2 \circ \varphi_1^{-1}$ is a well-defined bijection.

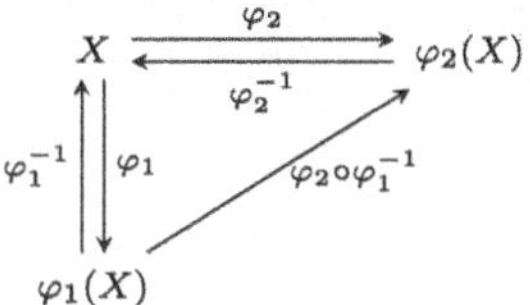

Also, for arbitrary points $y, y' \in \varphi_1(X)$, if $y = \varphi_1(x)$ and $y' = \varphi_1(x')$, $(x, x' \in X)$, then we have

$$\begin{aligned}
d_2(\psi(y), \psi(y')) &= d_2\big((\varphi_2 \circ \varphi_1^{-1})(\varphi_1(x)), (\varphi_2 \circ \varphi_1^{-1})(\varphi_1(x'))\big) \\
&= d_2(\varphi_2(x), \varphi_2(x')) \\
&= d(x, x') && [\text{as } \varphi_2 \text{ is an isometry}] \\
&= d_1(\varphi_1(x), \varphi_2(x')) && [\text{as } \varphi_1 \text{ is an isometry}] \\
&= d_1(y, y').
\end{aligned}$$

This shows that ψ is an isometry as well. Therefore, ψ is uniformly continuous (cf. Example 5.8.4).
As $\varphi_1(X)$ is dense in Y_1, by Theorem 5.9, ψ admits an extension $\widetilde{\psi}$ to all of Y_1. We intend to show this $\widetilde{\psi}$ is the required bijective isometry between Y_1 and Y_2.
We start by showing $\widetilde{\psi}$ is an isometry. Choose points $y, y' \in Y_1$ arbitrarily. Let $\{y_n\}$ and $\{y'_n\}$ be two sequences in $\varphi_1(X)$ converging to y and y', respectively. As in the proof of Theorem 5.9, $\widetilde{\psi}$ will be defined for y and y' as

$$\widetilde{\psi}(y) = \lim \psi(y_n) \quad \text{and} \quad \widetilde{\psi}(y') = \lim \psi(y'_n).$$

Recall that by miscellaneous result 5.5, any metric is a continuous function in both of its variables. Therefore,

$$
\begin{aligned}
d_2(\widetilde{\psi}(y), \widetilde{\psi}(y')) &= d_2\left(\lim_n \psi(y_n), \lim_m \psi(y'_m)\right) \\
&= \lim_{m,n} d_2(\psi(y_n), \psi(y'_m)) && \text{[as } d_2 \text{ is continuous]} \\
&= \lim_{m,n} d_1(y_n, y'_m) && \text{[as } \psi \text{ is an isometry]} \\
&= d_1\left(\lim_n y_n, \lim_m y'_m\right) && \text{[as } d_1 \text{ is continuous]} \\
&= d_1(y, y').
\end{aligned}
$$

This shows that $\widetilde{\psi}$ is an isometry, hence injective. Therefore, it only remains to show that $\widetilde{\psi}$ is surjective as well, i.e. we want to show $\widetilde{\psi}(Y_1) = Y_2$.
Now, as $\widetilde{\psi}$ is an isometry, and Y_1 is complete, by Exercise 4.20, $\widetilde{\psi}(Y_1)$ is a complete subspace of Y_2. Therefore, $\overline{\widetilde{\psi}(Y_1)} = \widetilde{\psi}(Y_1)$. But, as $\psi : \varphi_1(X) \to \varphi_2(X)$ is a bijection, we have

$$\varphi_2(X) = \psi(\varphi_1(X)) = \widetilde{\psi}(\varphi_1(X)) \subset \widetilde{\psi}(Y_1).$$

However, as $\varphi_2(X)$ is dense in Y_2, we have

$$Y_2 = \overline{\varphi_2(X)} \subset \overline{\widetilde{\psi}(Y_1)} = \widetilde{\psi}(Y_1) \subset Y_2.$$

Thus, $\widetilde{\psi}(Y_1) = Y_2$. This completes the proof. ■

With this proposition, we now give:

Alternate Proof of Theorem 4.22 Let (X, d) be a metric space. Fix a point $x_0 \in X$. Consider the space $\mathcal{B}(X)$ of all real-valued bounded continuous functions on X w.r.t. the metric

$$d_\infty(f, g) = \sup_{x \in X} |f(x) - g(x)|.$$

By Exercise 4.10, $\mathcal{B}(X)$ is complete. Define a map $F : X \to \mathcal{B}(X)$ given by $F(x) = f_x$, where

$$f_x(y) = d(y, x) - d(y, x_0); \qquad y \in X.$$

We first confirm that each f_x thus defined is a member of $\mathcal{B}(X)$. For any $x \in X$ fixed arbitrarily, the map $y \mapsto d(x, y)$ is a real-valued continuous map on X (cf. Example 5.4.5). Hence, so is $y \mapsto f_x(y)$. Also, for any point $y \in X$,

$$|f_x(y)| = |d(y, x) - d(y, x_0)| \le d(x, x_0).$$

As x and x_0 are fixed in X, the above relation shows that f_x is bounded. Consequently, for any $x \in X$, $f_x \in \mathcal{B}(X)$.
We will now show that F is an isometry. Observe that for any $x, x' \in X$,

$$\begin{aligned} d_\infty(F(x), F(x')) &= \sup_{y \in X} |f_x(y) - f_{x'}(y)| \\ &= \sup_{y \in X} |d(y, x) - d(y, x_0) - d(y, x') + d(y, x_0)| \\ &= \sup_{y \in X} |d(y, x) - d(y, x')| \\ &= d(x, x'). \end{aligned}$$

[as both $d(y, x), d(y, x') > 0$ and $d(y, x') = 0 \iff y = x'$.]

Therefore, F is an isometry. Consequently, F is uniformly continuous as well.
Consider the closure of the image $\overline{F(X)}$ in $\mathcal{B}(X)$. Now, $\overline{F(X)}$ is a closed set in the complete metric space $\mathcal{B}(X)$. Hence by Theorem 4.6, $\overline{F(X)}$ is complete. Again, $\overline{F(X)}$ contains an isometric image $F(X)$ of X which is also dense in $\overline{F(X)}$. Therefore, according to Definition 4.21, $\overline{F(X)}$ is a completion of X.
By the previous proposition, any two completions of the same space are isometric. This completes the proof. ■

Miscellaneous Results

5.6. Let X be a metric space and $A \subset X$. For $x \in X$, define $f: X \to \mathbb{R}$ by

$$f(x) = d(x, A) := \inf\{d(x, a) : a \in A\}.$$

Show that f is uniformly continuous over X.
Proof. Let $x_1, x_2 \in X$ be arbitrary. Then

$$\begin{aligned} f(x_1) &= d(x_1, A) = \inf\{d(x_1, a) : a \in A\}, \\ f(x_2) &= d(x_2, A) = \inf\{d(x_2, a) : a \in A\}. \end{aligned}$$

But for any point $a \in A$, $d(x_1, a) \leq d(x_1, x_2) + d(x_2, a)$. Hence

$$\begin{aligned} & \inf\{d(x_1, a) : a \in A\} \leq d(x_1, x_2) + \inf\{d(x_2, a) : a \in A\} \\ \implies & f(x_1) \leq d(x_1, x_2) + f(x_2) \\ \implies & f(x_1) - f(x_2) \leq d(x_1, x_2). \end{aligned}$$

Interchanging x_1 and x_2, we get $f(x_2) - f(x_1) \leq d(x_1, x_2)$. Therefore,

$$|f(x_1) - f(x_2)| \leq d(x_1, x_2).$$

Consequently, f is Lipschitz and hence uniformly continuous. □

Remark Since f is continuous on X, the set $f^{-1}(\{0\})$ is closed. But by miscellaneous result 3.3, we know that $f^{-1}(\{0\}) = \overline{A}$. Therefore, if A is a non-empty closed subset of X, then for the aforementioned map f, $A = f^{-1}(\{0\})$. Thus for a given closed set $A \subset X$, there always exists a continuous function f on X such that $f(x) = 0 \iff x \in A$. This observation leads to the following result:

Lemma 5.11 (Urysohn's Lemma) *Let X be a metric space and A, B be two non-empty disjoint closed subsets of X. Then there exists a continuous function $f: X \to \mathbb{R}$ such that $f = 0$ on A and $f = 1$ on B.*

Proof Consider the map $f: X \to \mathbb{R}$ given by

$$f(x) = \frac{d(x, A)}{d(x, A) + d(x, B)}.$$

To show f is well-defined, observe that the denominator $d(x, A)+d(x, B)$ becomes 0 iff $d(x, A) = 0 = d(x, B)$. But by miscellaneous result 3.3, this would imply x belongs to both $\overline{A}$ and $\overline{B}$ simultaneously. As A and B are closed sets, $\overline{A} = A$ and $\overline{B} = B$. As a result, $x \in A \cap B$, a contradiction to the hypothesis that A and B are disjoint. Hence, f is well-defined.
As the maps $x \mapsto d(x, A)$ and $x \mapsto d(x, B)$ are continuous, by the algebra of continuous functions listed in Proposition 5.5, f is continuous over X. Also, for points $x \in A$, $f(x) = \frac{0}{0+d(x,B)} = 0$, and for $x \in B$, $f(x) = \frac{d(x,A)}{d(x,A)+0} = 1$. This completes the proof. ■

Corollary 5.11.1 *Any two disjoint closed sets in a metric space can be separated by disjoint open sets (Fig. 5.7).*

Proof Let A, B be two disjoint closed sets in a metric space X. Then by Urysohn's lemma, there exists a continuous function $f: X \to \mathbb{R}$ such that $f = 0$ on A and $f = 1$ on B. Choose $0 < \epsilon < \frac{1}{2}$. Then $B(0; \epsilon)$ and $B(1; \epsilon)$ are disjoint open sets in $\mathbb{R}$. As f is continuous, both $f^{-1}(B(0; \epsilon))$ and $f^{-1}(B(1; \epsilon))$ are disjoint open sets in X and $A \subset f^{-1}(B(0; \epsilon))$, while $B \subset f^{-1}(B(1; \epsilon))$. ■

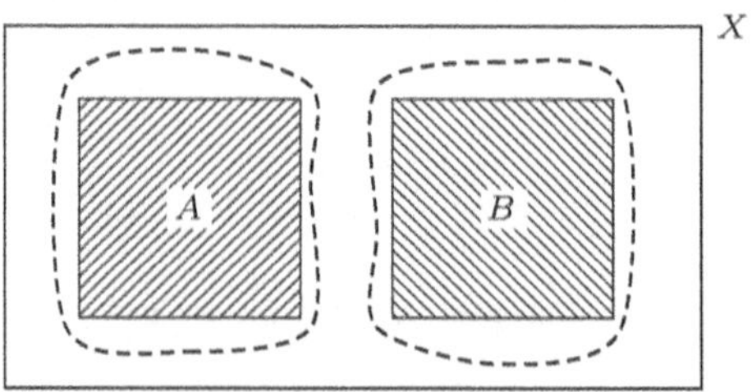

Fig. 5.7 Any two disjoint closed sets in a metric space can be separated by disjoint open sets

Exercise: 5.2

5.16. Show that any function defined over a discrete space is uniformly continuous.
5.17. For any positive number a, show that the function $x \mapsto x^2$ is uniformly continuous over $[0, a]$.
5.18. Let X, Y be metric spaces and $f: X \to Y$. Show that f is uniformly continuous on X if and only if for any two sequences $\{x_n\}$ and $\{y_n\}$ from X, $d_X(x_n, y_n) \to 0$ implies $d_Y(f(x_n), f(y_n)) \to 0$.
5.19. Let X, Y be metric spaces and $f: X \to Y$ be such that f maps Cauchy sequences in X to Cauchy sequences in Y. Show that f is continuous on X. Is f also uniformly continuous? Justify your answer.
5.20. Show that a linear map from $\mathbb{R}^m$ to $\mathbb{R}^n$ is uniformly continuous.
5.21. Show that a uniformly continuous function of a uniformly continuous function is again a uniformly continuous function.

5.3 Homeomorphism

While dealing with abstract entities, we often need a way to recognise fundamentally equivalent objects, even if they are represented differently. Different notations may be used to describe the same concept. For example, the numbers one, two, and three can be written as "1", "2", and "3" in Hindu-Arabic numerals or as "I", "II", and "III" in Roman numerals. Despite the difference in representation, the underlying numerical values remain the same. If we learn to multiply two by three using Hindu-Arabic numerals, we do not need to relearn it in Roman numerals—the result remains valid, only expressed differently.

This ability to recognise equivalent structures reduces effort and allows results to be translated across different notational systems. In algebra, such a correspondence is called an isomorphism; in metric spaces, an isometry (more precisely, an *onto* isometry); and in topology, a homeomorphism.[2] Let us now build a definition for the last one.

Let X and Y be two non-empty sets and let $\mathcal{T}_X$ and $\mathcal{T}_Y$ be the topologies (the collection of all open subsets) on X and Y. Suppose that the topological space $(X, \mathcal{T}_X)$ is structurally identical to $(Y, \mathcal{T}_Y)$. Now, in outset for this to happen, the sets X and Y must be equivalent to one another. In other words, there must exist a bijection $f: X \to Y$. In addition, the open sets of X must also be in a one-to-one correspondence with the open sets of Y, i.e. f must map every open set in X to an open set in Y, and conversely, must pull back every open set in Y to an open set in X (Fig. 5.8). The first condition translates to that the inverse map f^{-1} (as f is a bijection, f^{-1} exists) from Y to X is continuous, whereas the second

[2] This is a rare case in mathematics where spelling matters: Hom*eo*morphism $\neq$ Hom*o*morphism.

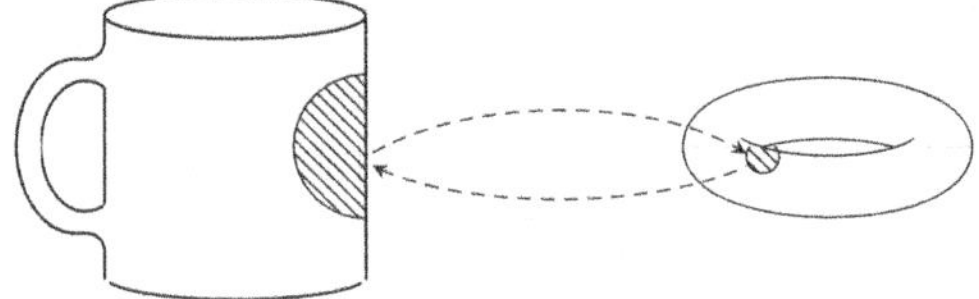

Fig. 5.8 Homeomorphic spaces

condition compels the forward map f from X to Y to be continuous as well. Thus, summarising this entire paragraph, we arrive at:

Definition 5.12 Let X, Y be metric spaces and $f: X \to Y$ be a bijection. f is called a *homeomorphism* between X and Y if both the maps f and f^{-1} are continuous. In that case, the spaces X and Y are called *homeomorphic* spaces.

In the theory of metric or topological spaces, a property of the space is called a *topological property* if that is preserved under a homeomorphism. In other words, a topological property is expressible solely in terms of the open sets of the space. While we will explore the fundamental topological properties of metric spaces in later chapters, it is worth noting that certain properties do not qualify as topological properties. The following list highlights a few such examples:

- **Distance:** Distance between two points may not remain the same under a homeomorphism. Consider the space of all real numbers $\mathbb{R}$ with the usual metric. Let $f: \mathbb{R} \to \mathbb{R}$ be defined by $f(x) = 2x$. It is a matter of a trivial exercise to check that f is a homeomorphism from $\mathbb{R}$ into itself. But for any two points $x, x' \in \mathbb{R}$, the distance between $f(x)$ and $f(x')$

$$|f(x) - f(x')| = 2|x - x'|$$

is twice the distance between x and x'. Thus, *distance is not a topological property.*
- **Boundedness:** Consider the metric spaces $X = (0, 1]$ and $Y = [1, \infty)$ with usual metrics and the map $f: X \to Y$ given by $x \mapsto \frac{1}{x}$. Then f is a continuous bijection from X to Y with the continuous self-inverse. Consequently, f is a homeomorphism. But X is a bounded space, while Y is not. Hence, *boundedness is not a topological property.*
- **Completeness:** Consider the same examples of X and Y as above. The discussions in the previous chapter have shown us X is incomplete, while Y is complete. Thus, *completeness is not a topological property.*

Examples (Worked Out)

5.12.1. Any two non-trivial closed and bounded intervals in $\mathbb{R}$ are homeomorphic.
Proof. Let $[a, b]$ and $[c, d]$ be any two closed and bounded intervals in $\mathbb{R}$. Define $f: [a, b] \to [c, d]$ by

$$f(x) = \frac{d-c}{b-a}x - \frac{ad-bc}{b-a}.$$

Then f is increasing, $f(a) = c$, $f(b) = d$, and f is a polynomial of degree 1. Therefore, f is bijective, continuous, and its inverse (which is also a polynomial of degree 1) is also continuous. Consequently, f is homeomorphism between $[a, b]$ and $[c, d]$. □

5.12.2. The function $x \mapsto \log x$ is a homeomorphism between $(0, \infty)$ and $\mathbb{R}$ w.r.t. the usual metric.

The proof is left for the reader as an exercise.

5.12.3. **Stereographic projection**: Let S^n denote the unit sphere in $\mathbb{R}^{n+1}$ given by

$$S^n := \left\{ \mathbf{x} = (x_1, \ldots, x_n, x_{n+1}) \in \mathbb{R}^{n+1} : \|\mathbf{x}\| = 1 \right\}.$$

Let $\mathbf{p} = (\underbrace{0, 0, \ldots, 0}_{n \text{ times}}, 1) \in \mathbb{R}^{n+1}$ be the north pole. Then $U := S^n \setminus \{\mathbf{p}\}$ is homeomorphic to $\mathbb{R}^n$.

Proof. Included in Theorem B.1. □

5.12.4. Any two isometric spaces are homeomorphic.

Proof. Let (X, d_X) and (Y, d_Y) be two isometric spaces and $\varphi: X \to Y$ be an onto isometry. Then φ is a bijection. Also, by definition of isometry, for any two points $x, x' \in X$,

$$d_X(x, x') = d_Y(\varphi(x), \varphi(x')).$$

Hence for any sequence $\{x_n\} \subset X$ converging to a point $x \in X$,

$$d_X(x_n, x) \to 0 \iff d_Y(\varphi(x_n), \varphi(x)) \to 0.$$

This proves continuity of both φ and φ^{-1}. Thus φ is a homeomorphism as well. □

Exercise: 5.3

5.22. Show that the following spaces are homeomorphic w.r.t. their usual metrics:

(a) $(0, \infty)$ and $\mathbb{R}$.
(b) Any two non-empty open intervals (bounded or unbounded) in $\mathbb{R}$.
(c) $\mathbb{R}$ and the parabola $y = x^2$ in $\mathbb{R}^2$.
(d) A circle and an ellipse in $\mathbb{R}^2$.
(e) $\mathbb{R}^n$, and the equator hyperplane

$$\{(x_1, \ldots, x_n, x_{n+1}) \in \mathbb{R}^{n+1} : x_{n+1} = 0\}$$

embedded in $\mathbb{R}^{n+1}$, where n is any positive integer.

5.23. Let X, Y, Z be three metric spaces so that X is homeomorphic to Y and Y is homeomorphic to Z. Show that X is homeomorphic to Z.
5.24. Show that two metrics d and d' on a set X are equivalent iff the identity map from (X, d) to (X, d') is a homeomorphism. [*Hint:* Use Theorem 3.40.]
5.25. Let X and Y be two homeomorphic spaces. Consider the product spaces X^n and Y^n (defined as the products of n copies of X and Y, respectively). Show that X^n and Y^n are homeomorphic as well. [*Hint*: If $f: X \to Y$ is a homeomorphism, then $\varphi(x_1, \ldots, x_n) := (f(x_1), \ldots, f(x_n))$ is a homeomorphism from X^n to Y^n.]
5.26. Show that separability is a topological property, i.e. homeomorphic image of a separable space is separable. How is separability expressed only in terms of open sets?
5.27. If a metric space X is homeomorphic to a metric space Y via a homeomorphism f and $A \subset X$, then show that $X \setminus A$ is homeomorphic to $Y \setminus f(A)$ via the same homeomorphism f restricted to $X \setminus A$.

5.4 Contraction Mappings and Its Applications

In this section, we explore a special class of functions that reduce the distance between any two points. A useful analogy is the process of compressing a city into its map, where distances shrink proportionally. Remarkably, under certain conditions, these functions leave a specific point unchanged within the space. This property has significant applications, particularly in finding solutions to certain types of differential equations.

We start with the formal definition of this distance-diminishing function.

Definition 5.13 Let (X, d) be a metric space. A map $\varphi: X \to X$ is called a *contraction* if there exists a positive real number $\alpha < 1$ such that for each pair of points x, x' from X,

$$d(\varphi(x), \varphi(x')) \leq \alpha\, d(x, x').$$

Remark A contraction is always Lipschitz. Hence uniformly continuous.

Examples

5.13.1. Any Lipschitz function with Lipschitz constant $L < 1$ is a contraction map.
5.13.2. Let $f: [a, b] \to \mathbb{R}$ be a differentiable function with its derivative always bounded by a constant k with $0 < k < 1$. Then f is a contraction.
Proof. By Lagrange's MVT, for any choice of points $x, x' \in [a, b]$, we can find a point ξ between x and x' such that

$$\frac{f(x) - f(x')}{x - x'} = f'(\xi).$$

$$Therefore,$$

$$\begin{aligned}|f(x) - f(x')| &= |f'(\xi)||x - x'| \\ &\leq k|x - x'|,\end{aligned}$$

showing f is a contraction. □

5.13.3. Fix $1 \leq p \leq \infty$. For a point $\mathbf{x} = \{x_n\} \in \ell_p$, define

$$\varphi(\mathbf{x}) = \frac{1}{2}\mathbf{x} = \left\{\frac{1}{2}x_n\right\}.$$

Then $\varphi: \ell_p \to \ell_p$ is a contraction.

Proof. Choose another point $\mathbf{y} = \{y_n\} \in \ell_p$ arbitrarily. Then

$$\begin{aligned}d_p(\varphi(\mathbf{x}), \varphi(\mathbf{y})) &= \left[\sum_{n=1}^{\infty}\left|\frac{x_n}{2} - \frac{y_n}{2}\right|^p\right]^{\frac{1}{p}} \\ &= \frac{1}{2}\left[\sum_{n=1}^{\infty}|x_n - y_n|^p\right]^{\frac{1}{p}} \\ &= \frac{1}{2}d_p(\mathbf{x}, \mathbf{y}),\end{aligned}$$

and for $p = \infty$,

$$\begin{aligned}d_\infty(\varphi(\mathbf{x}), \varphi(\mathbf{y})) &= \sup_n\left|\frac{x_n}{2} - \frac{y_n}{2}\right| \\ &= \frac{1}{2}\sup_n|x_n - y_n| \\ &= \frac{1}{2}d_\infty(\mathbf{x}, \mathbf{y}).\end{aligned}$$

Consequently, φ is a contraction. □

Definition 5.14 Let X be a metric space and $\varphi: X \to X$ be any map from X to itself. A point $\xi \in X$ is called a *fixed point* of f if $\varphi(\xi) = \xi$.

We then have the following theorem in a *complete* metric space.

Theorem 5.15 (Banach's Contraction Principle) *Let (X, d) be a complete metric space and $\varphi: X \to X$ be a contraction. Then φ has a unique fixed point.*

Proof Since φ is a contraction, there exists $0 < \alpha < 1$ such that for any pair of points $x, x' \in X$,

$$d(\varphi(x), \varphi(x')) \leq \alpha\, d(x, x').$$

Fix an arbitrary point $x_0 \in X$. Define $x_1 := \varphi(x_0)$ and $x_{n+1} = \varphi(x_n)$, for $n \geq 1$. First we see that $\{x_n\}$ is a Cauchy sequence. For any positive integer p,

$$\begin{aligned} d(x_{p+1}, x_p) &= d(\varphi(x_p), \varphi(x_{p-1})) \\ &\leq \alpha\, d(x_p, x_{p-1}) \\ &= \alpha\, d(\varphi(x_{p-1}), \varphi(x_{p-2})) \\ &\leq \alpha^2\, d(x_{p-1}, x_{p-2}) \\ &\vdots \\ &\leq \alpha^p\, d(x_1, x_0). \end{aligned}$$

Hence for any index n of the sequence $\{x_n\}$, using triangle inequality, we obtain

$$\begin{aligned} d(x_{n+p}, x_n) &\leq d(x_{n+p}, x_{n+p-1}) + d(x_{n+p-1}, x_{n+p-2}) + \cdots + d(x_{n+1}, x_n) \\ &\leq (\alpha^{n+p-1} + \alpha^{n+p-2} + \cdots + \alpha^n) d(x_1, x_0) \\ &= \alpha^n (\alpha^{p-1} + \alpha^{p-2} + \cdots + 1) d(x_1, x_0) \\ &\leq \frac{\alpha^n}{1-\alpha} d(x_1, x_0). \end{aligned}$$

As $0 < \alpha < 1$, $\alpha^n \to 0$ as $n \to \infty$. As a result, $d(x_{n+p}, x_n) \to 0$, showing that $\{x_n\}$ is Cauchy. Since X is complete, $\{x_n\}$ is convergent. Let $\lim\limits_{n\to\infty} x_n = \xi$. Using continuity of φ, we see that

$$\varphi(\xi) = \varphi\left(\lim_{n\to\infty} x_n\right) = \lim_{n\to\infty} \varphi(x_n) = \lim_{n\to\infty} x_{n+1} = \xi.$$

Hence ξ is a fixed point of φ.
To see ξ is unique, let for $\eta \in X$, $\varphi(\eta) = \eta$. Then

$$d(\xi, \eta) = d(\varphi(\xi), \varphi(\eta) \leq \alpha\, d(\xi, \eta).$$

This is only possible when $d(\xi, \eta) = 0$. Consequently, $\xi = \eta$. ■

Let us revisit the analogy of the city and its map. Consider both as subspaces of $\mathbb{R}^2$. Since they are closed, they are also complete. Let φ be the function that assigns each location in the city to its corresponding point on the map. Suppose the inverse scaling factor of the map is $\alpha : 1$, i.e. that the distance between two places p_1 and p_2 in the city and the corresponding points $\varphi(p_1)$ and $\varphi(p_2)$ on the map satisfy the relation

$$\|\varphi(p_1) - \varphi(p_2)\| = \alpha \|p_1 - p_2\|.$$

Since the map is much smaller than the city, we may assume $0 < \alpha \ll 1$, and thus $0 < 2\alpha < 1$. Therefore, $\|\varphi(p_1) - \varphi(p_2)\| \leq 2\alpha \|p_1 - p_2\|$, establishing φ as a contraction. Consequently, by the above theorem, if the map is spread on the ground somewhere in that city, there will be exactly one point on the map staying precisely on top of the corresponding point of the city it represents.

Remarks

(a) The theorem of Banach's contraction principle is also known as Banach's fixed point theorem.

(b) A map $\varphi: X \to X$ is called a *weak contraction map* if for all pair of points $x, x' \in X$,

$$d(\varphi(x), \varphi(x')) \leq d(x, x'),$$

where the equality occurs only if $x = x'$. A weak contraction map does not satisfy the contraction principle in general. For example, let $X = [1, \infty)$ and

$$\varphi(x) = x + \frac{1}{x}, \quad x \in X.$$

Then for points $x, x' \in [1, \infty)$,

$$|\varphi(x) - \varphi(x')| = |x - x'| \left(1 - \frac{1}{xx'}\right) < |x - x'|.$$

However, φ does not admit a fixed point in X. See Fig. 5.9.
However, *if* a weak contraction map admits a fixed point, it is unique. The proof is easy and is left for the reader as an exercise.

The following corollary extends the last theorem for a larger class of functions.

Corollary 5.15.1 *Let X be a complete metric space and $\varphi: X \to X$ be continuous. For $n \in \mathbb{N}$, define $\varphi^{(n)}$ inductively by $\varphi^{(1)} = \varphi$, and for $n \geq 1$, $\varphi^{(n+1)} = \varphi \circ \varphi^{(n)}$. Let for some positive integer k, $\varphi^{(k)}$ be a contraction. Then φ has a unique fixed point.*

Proof Fix $x_0 \in X$ arbitrarily. Denote $\widetilde{\varphi} = \varphi^{(k)}$. Construct the sequence $\{x_n\}$ by

$$x_1 := \widetilde{\varphi}(x_0) \quad \text{and} \quad x_{n+1} := \widetilde{\varphi}(x_n)$$

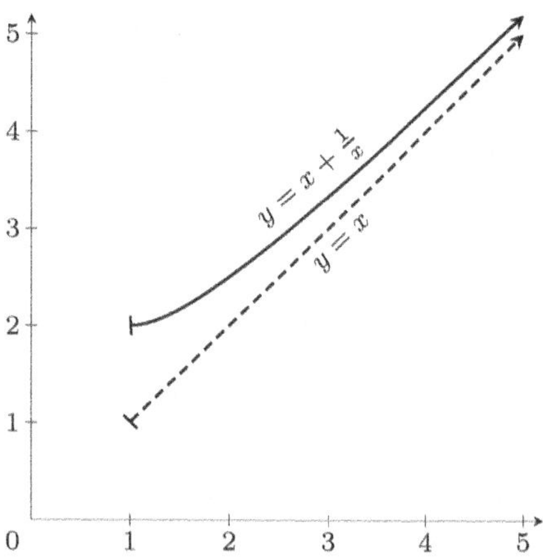

Fig. 5.9 $x \mapsto x + \frac{1}{x}$ does not admit a fixed point in $[1, \infty)$

as in the proof of contraction principle. By the arguments presented in that proof, for any choice of the starting point x_0, the sequence $\{x_n\}$ will converge to a unique point $\xi \in X$, and also, $\widetilde{\varphi}(\xi) = \xi$. We will show $\varphi(\xi) = \xi$ as well.
Observe here that for any $n \geq 1$,

$$\begin{aligned} x_{n+1} &= \widetilde{\varphi}(x_n) = \varphi^{(k)}(x_n) = \varphi^{(k)}[\widetilde{\varphi}(x_{n-1})] \\ &= \varphi^{(k)}[\varphi^{(k)}(x_{n-1})] = \varphi^{(2k)}(x_{n-1}). \end{aligned}$$

Thus, in general, $x_n = \varphi^{(nk)}(x_0)$. Therefore by the fact mentioned above, for any point $x \in X$, the sequence $\{\varphi^{(nk)}(x)\}$ converges to ξ. Clubbing this information with the fact that φ is continuous, we note that

$$\begin{aligned} \varphi(\xi) &= \varphi\left(\lim_{n\to\infty} x_n\right) = \lim_{n\to\infty} \varphi(x_n) \\ &= \lim_{n\to\infty} \varphi[\varphi^{(nk)}(x_0)] = \lim_{n\to\infty} \varphi^{(nk)}[\varphi(x_0)] = \xi.. \qquad [\text{choosing } x = \varphi(x_0)] \end{aligned}$$

To see the uniqueness of ξ, observe that for any index $n \geq 1$, $\varphi^{(n)}(\xi) = \xi$. Hence, for another fixed point η of φ, choosing $n = k$, we have

$$d(\xi, \eta) = d(\varphi^{(k)}(\xi), \varphi^{(k)}(\eta)) \leq \alpha\, d(\xi, \eta); , \qquad (0 < \alpha < 1)$$

which can only be true if $\xi = \eta$. This completes the proof. ■

We can conclude more in the above corollary. We proved above that for any choice $x \in X$, the sequence $\{\varphi^{(nk)}(x)\}$ converges to ξ. In fact, the sequence $\{\varphi^{(n)}(x)\}$ converges to ξ as well. By division algorithm, for any n fixed arbitrarily, let $n = mk + r$, where $m \geq 0$ is an integer, and $r \in \{0, 1, \ldots, k-1\}$. Also, $m \to \infty$ as

$n \to \infty$. Therefore,

$$\lim_{n\to\infty} \varphi^{(n)}(x) = \lim_{m\to\infty} \varphi^{(mk+r)}(x) = \lim_{m\to\infty} \varphi^{(mk)}[\varphi^{(r)}(x)] = \xi.$$

Example 5.15.1 Consider the space $\mathcal{C}\left[0, \frac{\pi}{2}\right]$ with the sup metric denoted by d. For any function $f \in \mathcal{C}\left[0, \frac{\pi}{2}\right]$, define $\varphi\colon \mathcal{C}\left[0, \frac{\pi}{2}\right] \to \mathcal{C}\left[0, \frac{\pi}{2}\right]$ by

$$[\varphi(f)](x) = \int_0^x f(t)\sin t\, dt.$$

If we choose $f(x) = x + 1$ and $g(x) = x$, then

$$d(f, g) = \sup_{x\in[0,\frac{\pi}{2}]} |f(x) - g(x)| = 1,$$

and also,

$$\begin{aligned} d(\varphi(f), \varphi(g)) &= \sup_{x\in[0,\frac{\pi}{2}]} \left|[\varphi(f)](x) - [\varphi(g)](x)\right| \\ &= \sup_{x\in[0,\frac{\pi}{2}]} \left|\int_0^x (f(t) - g(t)) \sin t\, \mathrm{d}t\right| \\ &= \sup_{x\in[0,\frac{\pi}{2}]} |1 - \cos x| = 1. \end{aligned}$$

Therefore, φ is not a contraction.
However, for arbitrary $f, g \in \mathcal{C}\left[0, \frac{\pi}{2}\right]$, at each point $x \in \left[0, \frac{\pi}{2}\right]$, we see that

$$\begin{aligned} |[\varphi(f)](x) - [\varphi(g)](x)| &= \left|\int_0^x (f(t) - g(t)) \sin t\, \mathrm{d}t\right| \\ &\le \sup_{t\in[0,x]} |f(t) - g(t)| \left|\int_0^x \sin t\, \mathrm{d}t\right| \\ &\le d(f, g)(1 - \cos x). \end{aligned}$$

Therefore, using linearity of φ, we obtain

$$\begin{aligned} \left|[\varphi^{(2)}(f)](x) - [\varphi^{(2)}(g)](x)\right| &= [\varphi[\varphi(f) - \varphi(g)]](x) \\ &= \left|\int_0^x [\varphi(f)(t) - \varphi(g)(t)] \sin t\, \mathrm{d}t\right| \end{aligned}$$

$$\leq \left| \int_0^x d(f,g)(1-\cos t)\sin t \, dt \right|$$
$$= d(f,g) \left| \int_0^x (1-\cos t)\sin t \, dt \right|$$
$$= d(f,g) \left| \frac{1}{2}(1-\cos x)^2 \right|.$$
[using the substitution $z = 1 - \cos t$]

$$d(\varphi^{(2)}(f), \varphi^{(2)}(g)) = \sup_{x\in[0,\frac{\pi}{2}]} \left| [\varphi^{(2)}(f)](x) - [\varphi^{(2)}(g)](x) \right|$$
$$= \frac{1}{2} d(f,g).$$

Consequently, $\varphi^{(2)}$ is a contraction. This is an example of a map φ which is not a contraction but $\varphi^{(2)}$ is.

Miscellaneous Results

5.7 Fix real numbers a, b with $0 < b < 1$. Consider the metric space $\mathcal{C}_{\sup}[0,b]$. Let

$$X := \{f \in \mathcal{C}[0,b] : f(0) = a\}.$$

For $f \in X$, define $\varphi \colon X \to X$ by

$$[\varphi(f)](x) = a + \int_0^x |f(t)| \, dt, \quad 0 \leq x \leq b.$$

Show that φ is a contraction. What is its fixed point?

Proof. We first check that φ is well-defined. For arbitrary $f \in X$, $\int_0^x |f|$ is a continuous function, and therefore, so is $\varphi(f)$. Also, $[\varphi(f)](0) = a$. Consequently, $\varphi(f) \in X$.

Next we prove that φ is a contraction. Observe that for $f, g \in X$,

$$\|\varphi(f) - \varphi(g)\|_\infty = \sup_{x\in[0,b]} \left| [\varphi(f)](x) - [\varphi(g)](x) \right|$$
$$= \sup_{x\in[0,b]} \left| \int_0^x |f(t)| \, dt - \int_0^x |g(t)| \, dt \right|$$
$$= \sup_{x\in[0,b]} \left| \int_0^x (|f(t)| - |g(t)|) \, dt \right|$$

$$
\begin{aligned}
&\leq \sup_{x\in[0,b]} \int_0^x \big||f(t)| - |g(t)|\big|\, \mathrm{d}t \\
&\leq \sup_{x\in[0,b]} \int_0^x |f(t) - g(t)|\, \mathrm{d}t \\
&\leq \sup_{x\in[0,b]} \left[\sup_{t\in[0,x]} |f(t) - g(t)| \int_0^x \mathrm{d}t \right] \\
&= \sup_{x\in[0,b]} \left[x \cdot \sup_{t\in[0,x]} |f(t) - g(t)| \right] \\
&\leq b\, \|f - g\|_\infty .
\end{aligned}
$$

As $0 < b < 1$, φ is a contraction on X.
We conclude about the fixed point of φ by showing X is complete. Consider the map $F\colon \mathcal{C}[0, b] \to \mathbb{R}$ defined by

$$F(f) = f(0).$$

Fix $f \in \mathcal{C}[0, b]$ arbitrarily and let $\{f_n\} \subset \mathcal{C}[0, b]$ be a sequence converging to f. Then $F(f_n) = f_n(0)$, and thus the sequence $\{F(f_n)\}$ converges to $f(0) = F(f)$, showing that F is continuous. Observe that we can write

$$X = F^{-1}(\{a\}).$$

Since the set $\{a\}$ is closed in $\mathbb{R}$, and F is continuous, X is a closed subset of $\mathcal{C}[0, b]$, by Theorem 5.4(*iii*). But by Example 4.5.7, $\mathcal{C}[0, b]$ is complete. Hence by Theorem 4.6, X is complete.
Hence by contraction principle, there exists a unique function $f \in X$ such that $\varphi(f) = f$. Now

$$\varphi(f) = f \implies f(x) = a + \int_0^x |f(t)|\, \mathrm{d}t.$$

Therefore, by the fundamental theorem of Integral Calculus, $f' = |f|$ on $[0, b]$.

□

5.4.1 Picard's Existence Theorem

The contraction mapping principle can be used to obtain a general result about the existence of a unique solution to the differential equation of the form

$$\frac{dy}{dx} = f(x, y)$$

satisfying certain conditions. For that purpose, let us first establish the following framework.

Definition 5.16 Let (x_0, y_0) be a point fixed in $\mathbb{R}^2$ and $R := \{(x, y) \in \mathbb{R}^2 : |x - x_0| \leq a,\ |y - y_0| \leq b\}$, where $a, b > 0$, be a closed and bounded rectangle in $\mathbb{R}^2$. Let $f: R \to \mathbb{R}$ be continuous. A real-valued function φ defined on an interval I is said to be a *solution of the initial value problem*

$$\frac{dy}{dx} = f(x, y), \quad y(x_0) = y_0 \tag{5.1}$$

if for each point $x \in I$, we have $(x, \varphi(x)) \in R$, $\varphi'(x) = f(x, \varphi(x))$, and $\varphi(x_0) = y_0$.

Definition 5.17 Let (x_0, y_0) be a point fixed in $\mathbb{R}^2$ and $R := \{(x, y) \in \mathbb{R}^2 : |x - x_0| \leq a,\ |y - y_0| \leq b\}$, where $a, b > 0$, be a closed and bounded rectangle in $\mathbb{R}^2$. Let $f: R \to \mathbb{R}$ be continuous. A real-valued function φ defined on an interval I is said to be a *solution of the integral equation*

$$y(x) = y_0 + \int_{x_0}^{x} f(t, y(t))\, dt \tag{5.2}$$

if for each point $x \in I$, we have $(x, \varphi(x)) \in R$ and $\varphi(x) = y_0 + \int_{x_0}^{x} f(t, \varphi(t))\, dt$.

Remark An immediate application of fundamental theorem of integral calculus shows that φ is a solution of (5.1) iff φ is a solution of (5.2). Details of the proof are left as an exercise.

Lemma 5.18 *Let $[a, b]$ and $[l, m]$ be two closed and bounded intervals in $\mathbb{R}$. Consider $\mathcal{C}[a, b]$ with sup metric. Let X denote the subspace of $\mathcal{C}[a, b]$ containing the functions which map $[a, b]$ into $[l, m]$, i.e.*

$$X := \{f \in \mathcal{C}[a, b] : f([a, b]) \subset [l, m]\}.$$

Then X is a complete subspace of $\mathcal{C}[a, b]$.

Proof We only require to show X is closed in $\mathcal{C}[a, b]$. By Example 4.5.7, $\mathcal{C}[a, b]$ is complete, and thus by Theorem 4.6, X will be complete as well.

Let f be a limit point of X. Then we can find a sequence $\{f_n\} \subset X$ such that $f_n \to f$. Therefore, for any $\epsilon > 0$ chosen arbitrarily, there will exist a positive integer k such that for each $n \geq k$,

$$\sup_{x \in [a,b]} |f_n(x) - f(x)| < \epsilon,$$

i.e. f_n's are converging to f uniformly. This implies, as all f_n's are continuous, f is also so. Since each $f_n \in X$, continuity of f in turn will imply that for every $n \in \mathbb{N}$ and for each point $x \in [a, b]$,

$$l \leq f_n(x) \leq m \implies l \leq f(x) \leq m.$$

Consequently, $f \in X$, showing X is closed. ■

Definition 5.19 Let R be a non-empty subset of $\mathbb{R}^2$ and $f: R \to \mathbb{R}$. f is said to satisfy the *Lipschitz condition in the y-variable uniformly in the x-variable* if there exists a constant $L > 0$ such that for any value of x, and for each pair of y_1, y_2,

$$|f(x, y_1) - f(x, y_2)| \leq L|y_1 - y_2|, \quad \text{whenever } (x, y_1), (x, y_2) \in R.$$

Similarly, we can have the definition of *Lipschitz condition in the x-variable uniformly in the y-variable.*[3]

With this framework, we now proceed to state and prove main theorem of this subsection, viz.:

Theorem 5.20 (Picard's Existence Theorem) *Let (x_0, y_0) be a point in $\mathbb{R}^2$. Let*

$$R = \{(x, y) \in \mathbb{R}^2 : |x - x_0| \leq a, |y - y_0| \leq b\},$$

where $a, b > 0$, be a closed and bounded rectangle in $\mathbb{R}^2$ and $f: R \to \mathbb{R}$ be continuous. If f satisfies the Lipschitz condition in the y-variable uniformly in the x-variable, then there exists a $\delta > 0$ and a unique function $\varphi: [x_0 - \delta, x_0 + \delta] \to \mathbb{R}$ such that φ is the solution to the initial value problem

$$\frac{dy}{dx} = f(x, y), \quad y(x_0) = y_0.$$

[3] If we fix the x-variable, say at x_0, $f(x_0, y)$ becomes a single-variable function in y. Now, when $f(x_0, y)$ becomes Lipschitz with some constant, say L_0, it certainly depends upon x_0. Furthermore, if we manage to find a uniform constant L to work for every choice of x_0, then f becomes Lipschitz in y-variable uniformly in x-variable.

Proof Call $I_0 = [x_0 - \delta, x_0 + \delta]$. By Exercise 5.30, it is enough to find a $\varphi \in \mathcal{C}(I_0)$ such that

$$\varphi(x) = y_0 + \int_{x_0}^{x} f(t, \varphi(t))\,\mathrm{d}t, \qquad x \in I_0. \tag{$*$}$$

From the Lipschitz condition on f, we obtain a constant $L > 0$ such that for any value of x, and for each pair of y_1, y_2,

$$|f(x, y_1) - f(x, y_2)| \leq L|y_1 - y_2|, \quad \text{whenever } (x, y_1), (x, y_2) \in R.$$

Now, as R is a closed and bounded subset of $\mathbb{R}^2$ and f is continuous on R, by Exercise 5.14, f is bounded on R. Let for $M > 0$,

$$|f(x, y)| \leq M$$

holds for every point $(x, y) \in R$. Choose $\delta > 0$ such that $L\delta < 1$ and

$$I_0 \times [y_0 - M\delta, y_0 + M\delta] = [x_0 - \delta, x_0 + \delta] \times [y_0 - M\delta, y_0 + M\delta] \subset R,$$

i.e. choose $0 < \delta < \min\left\{\frac{1}{L}, a, \frac{b}{M}\right\}$. Call Y to be the subspace of $\mathcal{C}(I_0)$ given by

$$Y = \{g \in \mathcal{C}(I_0) : |g(x) - y_0| \leq M\delta \text{ holds for every } x \in I_0\},$$

see Fig. 5.10. By Lemma 5.18, Y is complete. For $g \in Y$, define $T\colon Y \to Y$ by

$$[T(g)](x) = y_0 + \int_{x_0}^{x} f(t, g(t))\,\mathrm{d}t, \quad x \in I_0.$$

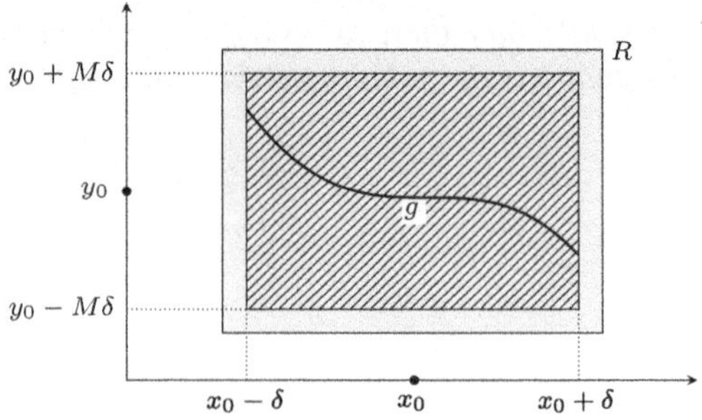

Fig. 5.10 Illustrative diagram of R, $I_0 \times [y_0 - M\delta, y_0 + M\delta]$, and a $g \in \mathcal{C}(I_0)$

To see T is well-defined, we first observe that T is continuous. Also, for each $x \in I_0$,

$$\begin{aligned}\left|[T(g)](x) - y_0\right| &= \left|\int_{x_0}^{x} f(t, g(t))\,\mathrm{d}t\right| \\ &\leq \int_{x_0}^{x} |f(t, g(t))|\,\mathrm{d}t \\ &\leq M\delta.\end{aligned}$$

Hence $T: Y \to Y$ is well-defined. We also observe that a fixed point φ of T will solve $(*)$.

Since Y is already known to be complete, to guarantee the existence of a fixed point of T, we only need to show that T is a contraction.

Consider $g, h \in Y$ arbitrarily. Hence for each point $x \in I_0$,

$$|g(x) - y_0| \leq M\delta \quad \text{and} \quad |h(x) - y_0| \leq M\delta.$$

Also, remember $d(g, h) = \sup\{|g(x) - h(x)| : x \in I_0\}$. Now

$$\begin{aligned}d(T(g), T(h)) &= \sup_{x \in I_0} \left|\int_{x_0}^{x} [f(t, g(t)) - f(t, h(t))]\,\mathrm{d}t\right| \\ &\leq \sup_{x \in I_0} \int_{x_0}^{x} \left|f(t, g(t)) - f(t, h(t))\right|\mathrm{d}t \\ &\leq \sup_{x \in I_0} \int_{x_0}^{x} L|g(t) - h(t)|\,\mathrm{d}t \\ &\leq d(g, h) \sup_{x \in I_0} \int_{x_0}^{x} L\,\mathrm{d}t \\ &\leq d(g, h)L\delta.\end{aligned}$$

But by the choice of δ, $L\delta < 1$. Hence T is a contraction and admits a fixed point φ. The uniqueness of φ is guaranteed by uniqueness of the fixed point of T. This completes the proof. ∎

Example 5.20.1 (Worked Out) Consider the initial value problem

$$\frac{dy}{dx} = x + y, \quad y(0) = 0.$$

By Exercise 5.30, φ is a solution of this initial value problem if and only if

$$\varphi(x) = \int_{0}^{x} [t + \varphi(t)]\,\mathrm{d}t.$$

As shown in the proof of Theorem 5.20, φ will be the fixed point of the contraction

$$[T(g)](x) = \int_0^x [t + g(t)]\, dt.$$

Again, as described in the proof of Theorem 5.15, $\varphi = \lim_{n\to\infty} T^{(n)}(g)$, where $T^{(n)}$ denotes n times iterative composition of T, i.e.

$$T^{(n)} = \underbrace{T \circ T \circ \cdots \circ T}_{n \text{ times}}.$$

Set $g(x) = 0$. Then

$$[T(g)](x) = \int_0^x t\, dt = \frac{x^2}{2!}.$$

Similarly,

$$[T^{(2)}(g)](x) = \int_0^x \left[t + \frac{t^2}{2!}\right] dt = \frac{x^2}{2!} + \frac{x^3}{3!}.$$

Continuing inductively, we obtain

$$[T^{(n)}(g)](x) = \frac{x^2}{2!} + \frac{x^3}{3!} + \cdots + \frac{x^n}{n!}.$$

Consequently,

$$\varphi(x) = \lim_{n\to\infty} [T^{(n)}(g)](x) = \lim_{n\to\infty} \left[\frac{x^2}{2!} + \frac{x^3}{3!} + \cdots + \frac{x^n}{n!}\right] = e^x - x - 1,$$

which is the required solution.

It is worthy to note that $e^x - x - 1$ is a solution of the given initial value problem on the whole of $\mathbb{R}$ and not merely on a bounded interval as permitted by Theorem 5.20.

5.28 Show that if a weak contraction map admits a fixed point, then it must be unique.

5.29 Show that the map φ on $\mathcal{C}_{\text{sup}}[0, 1]$ defined by

$$[\varphi(f)](x) = \int_0^x (x - t) f(t)\, dt, \quad x \in [0, 1],\ f \in \mathcal{C}[0, 1]$$

is a contraction. What is its fixed point?

5.30 Show that a real-valued function φ is a solution of Eq. (5.1) iff φ is a solution of Eq. (5.2).

Chapter 6
Compactness

In mathematics, where infinite sets are abundant, it is comforting to know that some of them exhibit properties similar to finite sets when placed in the right context. Finite sets possess several fundamental characteristics. Here are a few mentions:

- If we assign real numbers to the elements of a finite set, there will always be a highest and a lowest number. In formal mathematical terms, any real-valued function defined on a finite set attains its bounds.
- If we form a sequence using elements of a finite set, at least one element must appear infinitely often—i.e. that every sequence over a finite set contains a constant subsequence.
- If a finite set is covered by a collection of subsets, only a finite number of these subsets are needed to cover the entire set—i.e. every cover of a finite set has a finite subcover.

While these properties generally do not hold for infinite sets, certain infinite subsets do exhibit them when the underlying space is equipped with a topology. Such sets are called *compact* sets. However, there is a caveat: We must reinterpret these fundamental properties using the attributes available in topology. This leads to the following reformulations:

- Any real-valued *continuous* function defined on a compact set attains its bounds.
- Any sequence in a compact set admits a *convergent* subsequence.
- Any *open* cover of a compact set admits a finite subcover.

Thus, compact sets serve as a natural generalisation of finite sets. For a deeper exploration, readers may refer to [13].

In this chapter, we will study the concept of compactness and its implications in metric spaces. The first section will establish the various equivalent definitions of compactness. The second section will examine its consequences. The third section will explore the behaviour of continuous functions on compact sets. Finally, we will conclude the chapter with examples of compact sets in a couple of specific spaces.

S. Paul, *Metric Spaces*, University Texts in the Mathematical Sciences,
https://doi.org/10.1007/978-981-96-9259-0_6

6.1 Definitions

There are multiple equivalent definitions of compactness in a metric space. This section is divided into subsections, each presenting a necessary and sufficient condition (NASC).

The primary definition of a compact set is given by the third condition mentioned earlier: Every open cover of a compact set has a finite subcover. The first subsection will introduce an alternative definition derived logically from this description.

In real analysis, the well-known Heine-Borel theorem provides a convenient NASC for compactness in $\mathbb{R}$: A subset of $\mathbb{R}$ is compact if and only if it is closed and bounded. However, this result does not extend to arbitrary metric spaces. In the second subsection, we will examine why this is the case. As a result, we will establish an NASC for compactness in general metric spaces that aligns with the Heine-Borel theorem. Additional NASCs characterising compact sets will be discussed in later subsections.

Before proceeding, we formally define what it means for a set to be covered by a collection of subsets, laying the foundation for the precise definition of compactness.

Definition 6.1 Let X be a metric space. A collection $\mathcal{G}$ of open sets in X is called an *open cover* of X if all the sets in $\mathcal{G}$ together cover X, i.e.

$$X = \bigcup_{G \in \mathcal{G}} G.$$

Definition 6.2 Let X be a metric space and $\mathcal{G}$ be an open cover of X. If for some subcollection $\mathcal{G}'$ of $\mathcal{G}$, we have

$$X = \bigcup_{G \in \mathcal{G}'} G,$$

then $\mathcal{G}'$ is called a *subcover* of $\mathcal{G}$. If $\mathcal{G}'$ happens to contain finite or countable members, then $\mathcal{G}'$ is called a *finite subcover* or a *countable subcover* of $\mathcal{G}$, respectively.

Example 6.2.1 Let $X = (0, 1)$ and $\mathcal{G} = \left\{\left(\frac{1}{n}, 1 - \frac{1}{n}\right) : n \in \mathbb{N}, n \geq 3\right\}$. Then $\mathcal{G}$ is an open cover of X.

Definition 6.3 A metric space X is called *compact* if every open cover admits a finite subcover.

In the above example, the space $(0, 1)$ is not compact as the given cover $\mathcal{G}$ does not admit a finite subcover.

Remark Compactness of a metric space is a topological property as it is defined in terms of its open sets only.

Definition 6.4 Let X be a metric space and Y be a subset of X. Y is said to be a *compact subset* of X if Y is compact as a subspace of X, i.e. every cover of Y with sets *open in* Y admits a finite subcover.

Before this definition causes any inconvenience, we pacify it with the following:

Proposition 6.5 *Let X be a metric space and Y be a subset of X. Then Y is compact if and only if every cover of Y with sets open in X admits a finite subcover.*

Proof Let Y be compact and $\mathcal{G}$ be a cover of Y with sets open in X. Then, by Theorem 3.34, for each $G \in \mathcal{G}$, the set $G \cap Y$ is open in Y. Therefore the collection $\mathcal{G}' := \{G \cap Y : G \in \mathcal{G}\}$ is a cover of Y by sets open in Y. As Y is compact, $\mathcal{G}'$ admits a finite subcover, say $\{G_1 \cap Y, \ldots, G_n \cap Y\}$. Then

$$[G_1 \cap Y] \cup \cdots \cup [G_n \cap Y] = Y.$$

Thus $Y \subset G_1 \cup \cdots \cup G_n$. Therefore, $\{G_1, \ldots, G_n\}$ is a finite subcover from $\mathcal{G}$. Conversely, let every cover of Y with sets open in X admit a finite subcover of Y. Let $\mathcal{G}'$ be a cover of Y by sets open in Y. Then, by Theorem 3.34 again, for each $G' \in \mathcal{G}'$, there is a set G open in X, such that $G' = G \cap Y$. Gather all such open sets G in X and call the collection $\mathcal{G}$. Thus, as $\mathcal{G}'$ covers Y, so does $\mathcal{G}$. Since by hypothesis, $\mathcal{G}$ admits a finite subcover of Y, so does $\mathcal{G}'$. Hence Y is compact. ■

Remark In light of this proposition, the definition of a compact subset Y of a metric space X can unambiguously be stated as

> Y is said to be compact iff every open cover of Y admits a finite subcover.

6.1.1 Finite Intersection Property

Definition 6.6 Let X be a metric space. A collection $\mathcal{A}$ of subsets of X is said to have *finite intersection property* if every finite subcollection of $\mathcal{A}$ has non-empty intersection.

Example 6.6.1 Let $X = \mathbb{R}$ and $\mathcal{A} = \left\{\left(0, \frac{1}{n}\right) : n \in \mathbb{N}\right\}$. Then $\mathcal{A}$ satisfies finite intersection property.

With this definition onboard, the first equivalent definition of compact sets is recorded in the following:

Theorem 6.7 *A metric space X is compact if and only if every collection of closed sets in X satisfying the finite intersection property has a non-empty intersection.*

Proof Given a collection $\mathcal{G}$ of subsets of X, let

$$\mathcal{F} = \{X \setminus G : G \in \mathcal{G}\}$$

be the collection of their complements. Observe that $\mathcal{G}$ is a collection of open sets if and only if $\mathcal{F}$ is a collection of closed sets. We will use the symbols G and F to denote the mutual complements in this regard.

We have the following logical equivalences about the compactness of X.

X is compact:

$\iff$ Given any collection $\mathcal{G}$ of open subsets of X, if $\mathcal{G}$ covers X, then some finite subcollection $\mathcal{G}'$ of $\mathcal{G}$ covers X.

$\iff$ Given any collection $\mathcal{G}$ of open subsets of X, if no finite subcollection $\mathcal{G}'$ of $\mathcal{G}$ covers X, then $\mathcal{G}$ does not cover X.

$\iff$ Given any collection $\mathcal{G}$ of open subsets of X, and any finite subcollection $\mathcal{G}'$ of $\mathcal{G}$,

$$X \setminus \left[\bigcup_{G \in \mathcal{G}'} G \right] \neq \varnothing \implies X \setminus \left[\bigcup_{G \in \mathcal{G}} G \right] \neq \varnothing.$$

$\iff$ Given any collection $\mathcal{F}$ of closed subsets of X, and any finite subcollection $\mathcal{F}'$ of $\mathcal{F}$,

$$X \setminus \left[\bigcup_{F \in \mathcal{F}'} (X \setminus F) \right] \neq \varnothing \implies X \setminus \left[\bigcup_{F \in \mathcal{F}} (X \setminus F) \right] \neq \varnothing.$$

$\iff$ Given any collection $\mathcal{F}$ of closed subsets of X, and any finite subcollection $\mathcal{F}'$ of $\mathcal{F}$,

$$\bigcap_{F \in \mathcal{F}'} F \neq \varnothing \implies \bigcap_{F \in \mathcal{F}} F \neq \varnothing. \qquad \text{(using De Morgan's law)}$$

$\iff$ Any collection of closed subsets of X satisfying finite intersection property has a non-empty intersection.[1]

This completes the proof. ■

6.1.2 *Modification of Heine-Borel Theorem*

In real analysis, the mighty Heine-Borel theorem and its converse set up a remarkably easy and intuitive description of compact sets. We want to modify that

[1] $X \setminus F$ is the complement of F, and thus $\bigcup(X \setminus F)$ is the union of complements. By De Morgan's law, it is same as complement of the intersections. And finally, $X \setminus (\bigcup(X \setminus F))$ is the complement of the intersections.

here to fit into the context of metric spaces. We start with a recollection of their statements and the proofs.

Theorem 6.8 (Heine-Borel Theorem in $\mathbb{R}$) *Let $K \subset \mathbb{R}$. Then every open cover of K admits a finite subcover if and only if K is closed and bounded.*

Proof of the Sufficient Part Assume the contrary. Let the open cover $\mathcal{G}$ of K does not admit a finite subcover.

Since K is bounded, $K \subset [a, b]$ for some $a, b \in \mathbb{R}$. Call $c = \frac{a+b}{2}$. Then at least one of $[a, c] \cap K$ or $[c, b] \cap K$ must not be covered by finitely many open sets from $\mathcal{G}$. Choose that subinterval which is not covered by finitely many members from $\mathcal{G}$, and call it $I_1 = [a_1, b_1]$. Next, call $c_1 = \frac{a_1+b_1}{2}$. Then at least one of $[a_1, c_1] \cap K$ or $[c_1, b_1] \cap K$ must not be covered by finitely many open sets from $\mathcal{G}$. Choose that subinterval which is not covered by finitely many members from $\mathcal{G}$ and call it $I_2 = [a_2, b_2]$. Proceeding thus, we obtain a sequence of nested closed and bounded intervals $\{I_n = [a_n, b_n]\}$ such that for each index $n \in \mathbb{N}$:

i) $I_n \cap K$ is not covered by finitely many members from $\mathcal{G}$.
ii) $I_{n+1} \subset I_n$.
iii) $\lim\limits_{n\to\infty} \operatorname{diam}(I_n) = \lim\limits_{n\to\infty} \frac{b-a}{2^n} = 0$.

Since $\mathbb{R}$ is complete, by Cantor's intersection theorem, $\bigcap_n I_n$ is precisely a singleton set. Let $\bigcap_n I_n = \{\xi\}$. We first show that ξ is a limit point of K.

Choose $\epsilon > 0$ arbitrarily. Since $\operatorname{diam}(I_n) \to 0$, there exists a positive integer m such that $\operatorname{diam}(I_m) < \epsilon$. As $\xi \in I_m$, $I_m \subset B(\xi; \epsilon)$. But by construction, $I_m \cap K$ is not covered by finitely many members from $\mathcal{G}$. Hence I_m must contain infinitely many elements from K. Consequently, $B(\xi; \epsilon)$ also contains infinitely many elements from K. Since $\epsilon > 0$ is arbitrary, ξ is a limit point of K.

Since K is closed, $\xi \in K$.

As $\mathcal{G}$ is a cover of K, ξ is contained in some member G_0 from $\mathcal{G}$. But, as G_0 is open, for some $\delta > 0$, $B(\xi; \delta) \subset G_0$. Again, as $\operatorname{diam}(I_n) \to 0$, for some positive integer k, $\operatorname{diam}(I_k) < \delta$. Since $\xi \in I_k$, therefore $I_k \subset B(\xi; \delta)$. But in that case, $I_k \cap K \subset B(\xi; \delta) \subset G_0$, a contradiction; as by construction, $I_k \cap K$ is not covered by finitely many members from $\mathcal{G}$. Hence our assumption was wrong. Every open cover of K must admit a finite subcover. ■

Proof of the Necessary Part For any subset A of $\mathbb{R}$, the set $\{(-n, n) : n \in \mathbb{N}\}$ is an open cover of A. Therefore by hypothesis, this admits a finite subcover for K, say $\{(-n_k, n_k) : k = 1, \ldots, m\}$. Call $n_0 = \max\{n_k : k = 1, \ldots, m\}$. Then $K \subset (-n_0, n_0)$, showing that K is bounded.

To show K is closed, choose $y \in \mathbb{R} \setminus K$. For each $n \in \mathbb{N}$, construct the open sets

$$G_n := \left\{x \in \mathbb{R} : |x - y| > \frac{1}{n}\right\},$$

see Fig. 6.1. Then $\bigcup_{n\in\mathbb{N}} G_n = \mathbb{R} \setminus \{y\}$, and hence, $\mathcal{G} := \{G_n : n \in \mathbb{N}\}$ covers K. By hypothesis, $\mathcal{G}$ admits a finite subcover, $\{G_{n_k} : k = 1, \ldots, m\}$, say. Call

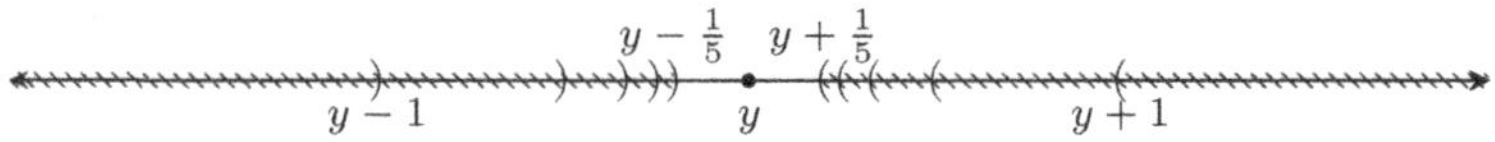

Fig. 6.1 Construction of G_n's

$n_0 = \max\{n_k : k = 1, \ldots, m\}$. Therefore,

$$K \subset \bigcup_{k=1}^{m} G_{n_k} = \left\{x \in \mathbb{R} : |x - y| > \frac{1}{n_0}\right\}.$$

Consequently, $B\left(y; \frac{1}{2n_0}\right) \cap K = \varnothing$, showing that y is not a limit point of K. Hence, $y \notin K$ implies y is not a limit point of K. Contrapositively, if y is a limit point of K, then $y \in K$, showing that K is closed. ■

Suppose we seek an NASC for compactness in metric spaces that parallels the Heine-Borel theorem in $\mathbb{R}$. While the necessary part holds, mere "closed and bounded"-ness do not suffice to guarantee compactness in an arbitrary metric space.

To address this, we take a twofold approach:

- Introduce a stronger notion of total boundedness
- Replace the criterion of closedness with the more suitable concept of *completeness*.

The justification for these refinements is provided in the following sub-subsections.

Compact Implies Closed and Bounded

Theorem 6.9 *A compact subset of a metric space is closed and bounded.*

A standard proof of this result is obtained by imitating the arguments from the proof of the necessary part of the last theorem.

Remark There is another sophisticated line of argument to prove the above theorem using the Hausdorff property of a metric space.

Let Y be a compact subset of the metric space X. We assume the part of showing Y is bounded. Thus we only show here that Y is closed.

If $Y = X$, we are done. Otherwise, fix a point $x \in X \setminus Y$ arbitrarily. Then for each point $y \in Y$, by Hausdorff property, there exists $\delta_y > 0$ such that $B(y; \delta_y) \cap B(x; \delta_y) = \varnothing$. Call

$$\mathcal{G} := \{B(y, \delta_y) \mid y \in Y,\ B(y; \delta_y) \cap B(x; \delta_y) = \varnothing\}.$$

Then $\mathcal{G}$ is a cover of Y, hence admitting a finite subcover $\{B\left(y_k, \delta_{y_k}\right) : k = 1, \ldots, m\}$, say (Fig. 6.2). Choose $\delta = \min\{\delta_{y_k} : k = 1, \ldots, m\} > 0$. We show

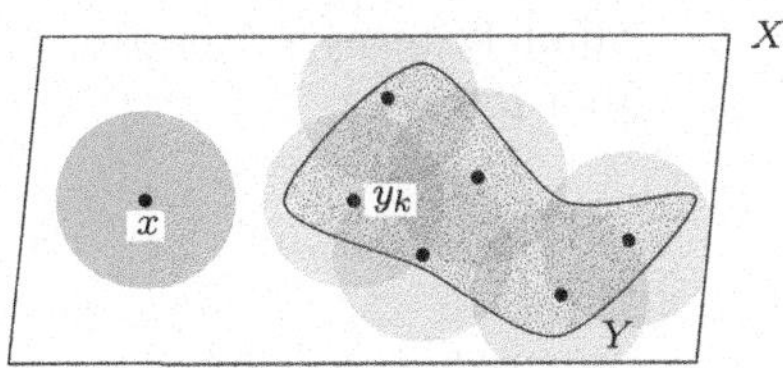

Fig. 6.2 A compact set is closed

that $Y \cap B(x; \delta) = \varnothing$. If $y' \in Y \cap B(x; \delta)$, then $y' \in B(y_k; \delta_{y_k}) \cap B(x; \delta)$ for some $k \in \{1, \ldots, m\}$, implying that $y' \in B(y_k; \delta_{y_k}) \cap B(x; \delta_{y_k})$, a contradiction. Hence $Y \cap B(x; \delta) = \varnothing$, and hence x is not a limit point of Y. Contrapositively, Y contains all its limit points.

Closed and Bounded Does Not Necessarily Imply Compact

Following examples show that a mere closed and bounded set in a metric space is not always compact.

Examples

6.9.1. An infinite subset of a discrete space is not compact.
Proof. Let X be an infinite set with the discrete metric, and $Y \subset X$ be an infinite subset. For every point $x \in X$, $B\left(x; \frac{1}{2}\right) = \{x\}$. Thus, every subset of X is open, and hence, simultaneously, closed. Also, $\operatorname{diam} Y = 1$, i.e. Y is bounded. However, the open cover $\{\{y\} : y \in Y\}$ of Y does not admit any finite subcover. Consequently, Y is not compact. □

6.9.2. The closed unit ball $B[\mathbf{0}; 1]$ in ℓ_p is not compact, where $\mathbf{0} = (0, 0, \ldots)$ is the constant 0 sequence in ℓ_p.
Proof. Call $X = B[\mathbf{0}; 1]$. Of course, $\operatorname{diam} X = 2$, and thus, X is bounded. We claim that the open cover $\mathcal{G} := \left\{B(\mathbf{x}; \frac{1}{2}) : \mathbf{x} \in X\right\}$ of X does not admit a finite subcover. To prove this claim, let us take notes of the following elements in X.
For $n \in \mathbb{N}$, denote the element $\{x_k^{(n)}\}_k$ in ℓ_p by $\mathbf{e}^{(n)}$, where

$$x_k^{(n)} := \begin{cases} 0, \text{ if } n \neq k \\ 1, \text{ if } n = k \end{cases}.$$

Then for each n, $d_p(\mathbf{e}^{(n)}, \mathbf{0}) = 1$ showing that $\mathbf{e}^{(n)} \in X$. Also, for $m \neq n$,

$$d_p(\mathbf{e}^{(m)}, \mathbf{e}^{(n)}) = 2^{1/p} > 1.$$

Assume, if possible, $\mathcal{G}$ admits a finite subcover $\mathcal{G}' = \left\{B(x_i; \frac{1}{2}) : i = 1, \ldots, l\right\}$ of X. Then some open set $B(\mathbf{x}_j; \frac{1}{2})$ from $\mathcal{G}'$ must contain infinitely many $\mathbf{e}^{(n)}$'s. But in that case, for any two distinct $\mathbf{e}^{(m)}$ and $\mathbf{e}^{(n)}$ from $B(\mathbf{x}_j; \frac{1}{2})$,

$$d_p(\mathbf{e}^{(m)}, \mathbf{e}^{(n)}) \leq d_p(\mathbf{e}^{(m)}, \mathbf{x}_j) + d_p(\mathbf{x}_j, \mathbf{e}^{(n)}) < 1,$$

a contradiction. Thus X is not compact. □

Introduce Total Boundedness

As "closed and bounded"-ness failed to guarantee compactness, we look towards total boundedness as a stronger version of boundedness. We start with the following:

Definition 6.10 Let X be a metric space and ϵ be a fixed positive number. A subset $A \subset X$ is said to be an *ϵ-net* for X if every point in X is within ϵ distance from a point in A, i.e. for each point $x \in X$, there exists a point $a \in A$, such that $d(x, a) < \epsilon$, which is again to say $X = \bigcup_{a \in A} B(a; \epsilon)$.

When A is finite, it is called a *finite ϵ-net* for X.

Definition 6.11 A metric space X is said to be *totally bounded* if for every $\epsilon > 0$, X admits a finite ϵ-net. A non-empty subset Y of X is *totally bounded* if Y is totally bounded as a subspace of X.

Remark To paraphrase Hermann Weyl (1885–1955), if a city is totally bounded, it can be guarded by a finite number of arbitrarily near-sighted policemen [4] (Fig. 6.3).

We prove the following result at the earliest, which will turn out to be very useful at times.

Lemma 6.12 *Let X be a totally bounded metric space and Y be a subspace of X. Then Y is also totally bounded.*

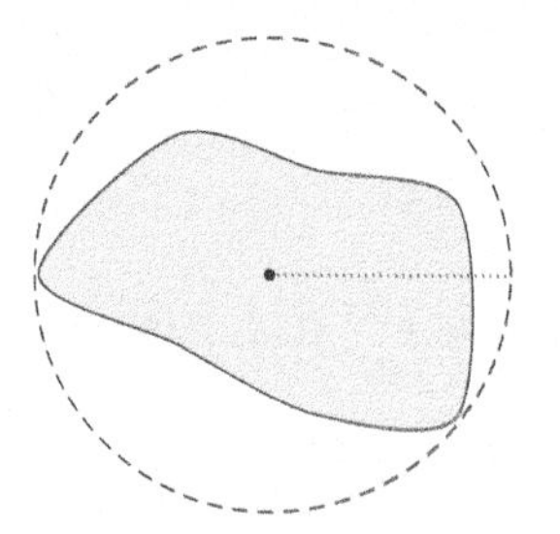

(a): *Bounded*: Entire space is contained in some ball of finite radius

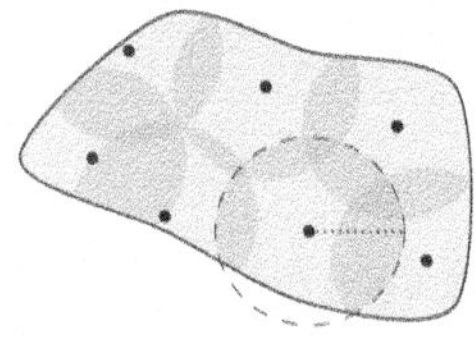

(b): *Totally bounded*: Entire space is contained in finitely many balls of arbitrarily small radius

Fig. 6.3 Bounded and totally bounded spaces

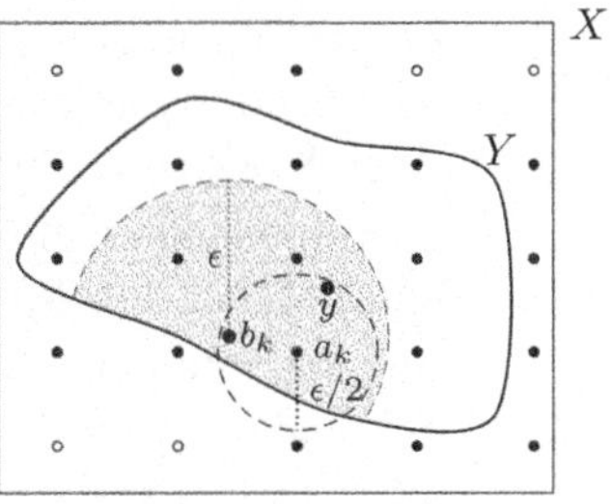

Fig. 6.4 Subspace of a totally bounded space is totally bounded

Proof Fix $\epsilon > 0$ arbitrarily. Let A be a finite $\frac{\epsilon}{2}$-net of X. Assume that for points $\{a_1, \ldots, a_n\}$ from A, Y intersects each of the balls $B(a_k; \frac{\epsilon}{2})$. See Fig. 6.4. Fix points $b_k \in B(a_k; \frac{\epsilon}{2}) \cap Y$, $k = 1, \ldots, n$. Then for any arbitrary point $y \in Y$, if $y \in B(a_k; \frac{\epsilon}{2})$ (as a point of X), then

$$d(y, b_k) \leq d(y, a_k) + d(a_k, b_k) < \frac{\epsilon}{2} + \frac{\epsilon}{2} = \epsilon.$$

Thus the set $B = \{b_1, \ldots, b_n\}$ is a finite ϵ-net for Y. As $\epsilon > 0$ is arbitrary, Y is totally bounded. ■

Now we will see that total boundedness is actually stronger than boundedness, i.e. a totally bounded space is always bounded, but the converse is not true. We prove the first part in the following:

Proposition 6.13 *A totally bounded space is always bounded.*

Proof Let X be totally bounded. Fix $\epsilon > 0$. There is a finite $A \subset X$ such that for each point $x \in X$, there exists a point $a \in A$ such that $d(x, a) < \epsilon$. As A is a finite set, $\text{diam}(A) < \infty$. Therefore for arbitrary points $x, x' \in X$, if a and a' are the points from A so that $d(x, a) < \epsilon$ and $d(x', a') < \epsilon$, then

$$d(x, x') \leq d(x, a) + d(a, a') + d(a', x') < \epsilon + \text{diam}(A) + \epsilon.$$

$$\text{diam}(X) = \sup\{d(x, x') : x, x' \in X\} \leq \text{diam}(A) + 2\epsilon.$$

Consequently, X is bounded. ■

To establish the second part, we utter Examples 6.9.1 and 6.9.2 again.

Examples

6.13.1. In Example 6.9.1, $\left\{B\left(x; \frac{1}{2}\right) : x \in X\right\}$ is an $\frac{1}{2}$-net for X which does not have a finite $\frac{1}{2}$-net.

6.13.2. Now consider Example 6.9.2 in ℓ_2. $B[\mathbf{0}; 1]$ is bounded but not totally bounded.

Proof. Construct the set $Y = \{\mathbf{e}^{(n)} : n \in \mathbb{N}\}$, where $\mathbf{e}^{(n)}$ are elements in ℓ_2 as defined in that example. Then Y is an infinite subset of $B[\mathbf{0}; 1]$. We will show Y does not admit a finite $\frac{1}{\sqrt{2}}$-net.

The arguments will be similar as used earlier. Assume, if possible, A is a finite $\frac{1}{\sqrt{2}}$-net for $B[\mathbf{0}; 1]$. Then for some point $\mathbf{a} \in A$, the ball $B(\mathbf{a}; \frac{1}{\sqrt{2}})$ contains infinitely many terms from Y. But for any two distinct elements $\mathbf{e}^{(m)}$ and $\mathbf{e}^{(n)}$ from Y, $d_2(\mathbf{e}^{(m)}, \mathbf{e}^{(n)}) = \sqrt{2}$. However, if they both belong to the same ball $B(\mathbf{a}; \frac{1}{\sqrt{2}})$, then

$$\begin{aligned} d_2(\mathbf{e}^{(m)}, \mathbf{e}^{(n)}) &\leq d_2(\mathbf{e}^{(m)}, \mathbf{a}) + d_2(\mathbf{a}, \mathbf{e}^{(n)}) \\ &< \frac{1}{\sqrt{2}} + \frac{1}{\sqrt{2}} = \sqrt{2}, \end{aligned}$$

a contradiction. Therefore, Y, and hence, $B[\mathbf{0}; 1]$ is not totally bounded. □

Remark From this example, we can conclude that the space ℓ_2 itself is not totally bounded as guaranteed by Lemma 6.12.

Before incorporating total boundedness as part of the sufficient conditions for compactness, we first ensure that this addition does not impose any extra constraints in the case of $\mathbb{R}$. The following lemma addresses this concern.

Lemma 6.14 *A bounded subset of* $\mathbb{R}$ *is totally bounded.*

Proof Let $S \subset \mathbb{R}$ be bounded. Enclose S in some interval $[a, b]$. Fix $\epsilon > 0$ arbitrarily. Choose $n \in \mathbb{N}$ such that $h = \frac{b-a}{n} < \epsilon$. For every $k \in \{1, \dots, n\}$, let a_k be any arbitrary fixed point in the interval $(a + (k-1)h, a + kh)$. Then

$$[a, b] \subset \bigcup_{k=1}^{n} B(a_k; \epsilon).$$

(See Fig. 6.5.) Thus $\{a_1, \dots, a_n\}$ is a finite ϵ-net for $[a, b]$. Therefore $[a, b]$ is totally bounded, so is S, by Lemma 6.12. ■

Now, let us address the necessary part. Does compactness imply the stronger notion of total boundedness in general? The following proposition affirms that.

Proposition 6.15 *A compact metric space is totally bounded.*

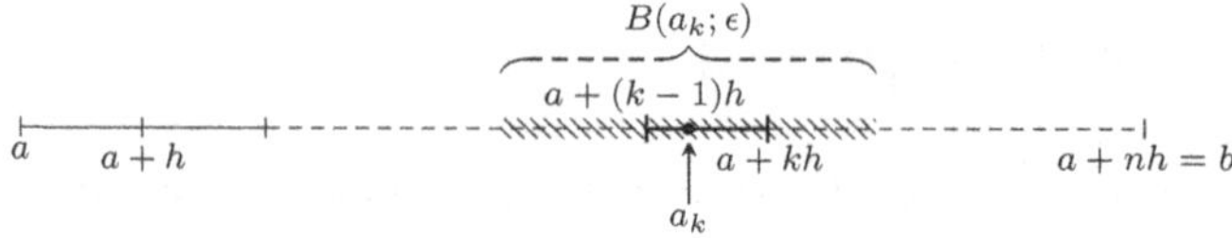

Fig. 6.5 Bounded subset of $\mathbb{R}$ is totally bounded

Proof Let X be a compact metric space. For arbitrarily chosen $\epsilon > 0$, consider the cover $\mathcal{G} = \{B(x; \epsilon) : x \in X\}$ of X. The finite subcover by $\mathcal{G}$ is a finite ϵ-net for X. Consequently, X is totally bounded. ∎

Not Only Closed, But *Complete*

Consider the metric space $X = (-1, 1)$ and the subset $Y = [0, 1)$ within X. The set Y is closed and bounded (and also totally bounded) in X, yet it is not compact.

This example highlights an important insight: Compactness in the Heine-Borel theorem relies not only on the closedness of the subset but also on the completeness of the underlying space $\mathbb{R}$. This observation follows from two key facts: (a) in the proof of the sufficient part of the theorem, we used Cantor's intersection theorem, which holds only in complete spaces; (b) a closed subspace of a complete space is itself complete.

Recognising this, we now establish the necessary condition.

Proposition 6.16 *A compact metric space is complete.*

Proof Let (X, d) be a compact metric space and $\{x_n\} \subset X$ be a Cauchy sequence. Suppose the range $A = \{x_n : n \in \mathbb{N}\}$ of the sequence has only finitely many points. Then $\{x_n\}$ has to admit a constant, and therefore, a convergent subsequence. Thus by Exercise 4.1, the Cauchy sequence $\{x_n\}$ will be convergent.
Therefore, WLOG, we may assume that $\{x_n\}$ does not have any constant subsequence, i.e. its range $A = \{x_n : n \in \mathbb{N}\}$ has infinitely many points. Assume, if possible, that $\{x_n\}$ does not converge to any point in X. Fix $x \in X$ arbitrarily. Then $\{x_n\}$ does not converge to x, i.e.

there exists $\epsilon_x > 0$ such that for every index $k \in \mathbb{N}$, there exists some violating index $n_k \geq k$ such that the term x_{n_k} from the sequence $\{x_n\}$ will satisfy $d(x_{n_k}, x) \geq 2\epsilon_x$. $\quad (*)$

However, as $\{x_n\}$ is Cauchy, for this $\epsilon_x > 0$, there exists some index $p_x \in \mathbb{N}$ such that whenever $m, n \geq p_x$, we would have $d(x_m, x_n) < \epsilon_x$. Choosing $k = p_x$ in $(*)$, we thus obtain that for every index $m \geq k(= p_x)$, $d(x_m, x_{n_k}) < \epsilon_x$. Therefore, using $(*)$, we obtain that for every $m \geq p_x$,

$$d(x_m, x) \geq d(x_{n_k}, x) - d(x_m, x_{n_k}) > 2\epsilon_x - \epsilon_x = \epsilon_x. \quad (**)$$

Hence, the ball $B(x; \epsilon_x)$ contains at most first $p_x - 1$ many terms of the sequence $\{x_n\}$. Thus for each point $x \in X$, the ball $B(x; \epsilon_x)$ contains only finitely many points from A.
Now we will exploit the fact that X is compact. The collection $\{B(x; \epsilon_x) : x \in X\}$ is a cover for X and thus admits a finite subcover. However, as each ball $B(x; \epsilon_x)$

contains only finitely many elements from A, existence of a finite subcover will imply that the A has only finitely many terms, a contradiction to the hypothesis. Therefore, a compact metric space is always complete. ■

Remark Since a compact metric space is complete, a contraction map defined over a compact space admits a unique fixed point by Banach's contraction principle.

Complete and Total Boundedness Guarantee Compactness

Finally, we arrive at the much anticipated:

Theorem 6.17 *A totally bounded, complete metric space is compact.*

Proof We imitate the technique used in the proof of Heine-Borel theorem with modifications for total boundedness. Let X be a totally bounded, complete metric space. Assume, if possible, that X is not compact. Then there is an open cover $\mathcal{G}$ of X which does not admit a finite subcover.
Since X is totally bounded, it is bounded as well. Hence for some $x_0 \in X$ and $r > 0$, $X = B(x_0; r)$. For each $n \in \mathbb{N}$, set $\delta_n = \frac{r}{2^n}$.
Since X is totally bounded, for δ_1, we get a finite set of points from X which makes a δ_1-net. By our hypothesis, there must exist a point $x_1 \in X$ such that the ball $B(x_1; \delta_1)$ is *not* contained in finitely many open sets from $\mathcal{G}$; otherwise, entire X will be contained in finitely many open sets from $\mathcal{G}$. As X is totally bounded, by Lemma 6.12, $B(x_1; \delta_1)$ is totally bounded as well. Therefore, we can find a finite δ_2-net for $B(x_1; \delta_1)$. By similar arguments, we obtain $x_2 \in B(x_1; \delta_1)$ such that the ball $B(x_2; \delta_2)$ is not contained in finitely many open sets from $\mathcal{G}$. Continuing thus, we get a sequence $\{x_n\}$ of points from X such that each $x_{n+1} \in B(x_n; \delta_n)$, and $B(x_n; \delta_n)$ is not contained in finitely many open sets from $\mathcal{G}$.
As X is complete, we want to use Cantor's intersection theorem. We choose the closed sets as $F_n := B[x_n; \delta_n]$. Thus, for the sequence $\{F_n\}$ of closed sets, we have the following facts:

i) For each $n \in \mathbb{N}$, $F_{n+1} \subset F_n$.
ii) No F_n is contained in finitely many open sets from $\mathcal{G}$.
iii) $\operatorname{diam} F_n = 2\delta_n = \frac{r}{2^{n-1}}$. Thus $\lim_{n\to\infty} \operatorname{diam} F_n = 0$.

Therefore, by Cantor's intersection theorem, there exists a unique point $\xi \in X$ such that $\bigcap_{n\in\mathbb{N}} F_n = \{\xi\}$. Let ξ be contained in the open set G from $\mathcal{G}$. Therefore, some δ-ball $B(\xi; \delta)$ is also contained in G (Fig. 6.6). But $\operatorname{diam} F_n \to 0$ implies that for some index $k \in \mathbb{N}$, $\operatorname{diam} F_k < \delta$. In that case, $F_k \subset B(\xi; \delta)$. (Recall that ξ is contained in each F_n.) Thus, $F_k \subset B(\xi; \delta) \subset G$, and this is a contradiction, as no F_n is contained in finitely many open sets from $\mathcal{G}$. This disproves our assumption. Consequently, X is compact. ■

Remark If X is a complete metric space and Y is a closed subset of X, by Theorem 4.6, Y is also complete. Thus, the nearest analogy to the Heine-Borel

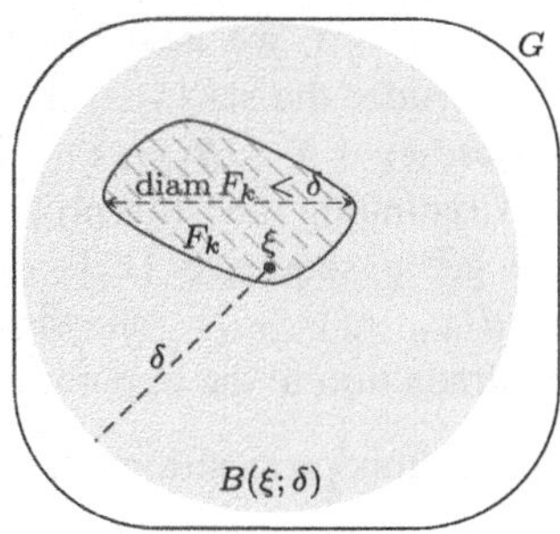

Fig. 6.6 Behaviour of F_k

theorem for an NASC for a subset in a complete metric space to be compact is given by the following:

Corollary 6.17.1 *Let X be a complete metric space. A subset Y of X is compact if and only if Y is closed and totally bounded.*

6.1.3 Sequential Compactness

In this subsection and the next, we discuss a couple of important and practical equivalent definitions of compactness in a metric space.

Definition 6.18 A metric space (X, d) is called *sequentially compact* if every sequence in X admits a convergent subsequence.

We shall prove that sequential compactness is equivalent to compactness. To see that, we start with the following:

Proposition 6.19 *A metric space X is totally bounded if and only if every sequence has a Cauchy subsequence.*

Proof Let X be totally bounded and $\{x_n\} \subset X$ be a sequence. WLOG, we may assume that $\{x_n\}$ does not admit a constant subsequence (in that case, the proof is trivial). Then the range of the sequence $A = \{x_n : n \in \mathbb{N}\}$ is an infinite set. Fix $\epsilon > 0$ arbitrarily. Since X is totally bounded, there is a finite $\frac{\epsilon}{2}$-net of X, say $\left\{B\left(a_k; \frac{\epsilon}{2}\right) : k = 1, \ldots, m\right\}$. Then at least one of $B\left(a_k; \frac{\epsilon}{2}\right)$ must contain infinitely many terms from the sequence $\{x_n\}$. Denote that infinite subsequence which is contained in one of the $B\left(a_k; \frac{\epsilon}{2}\right)$'s by $\{x_{n_p}\}_p$. Hence for any two values p_1, p_2 of p,

$$d\left(x_{n_{p_1}}, x_{n_{p_2}}\right) < \frac{\epsilon}{2} + \frac{\epsilon}{2} = \epsilon.$$

Hence $\{x_{n_p}\}_p$ is a Cauchy subsequence of $\{x_n\}$.

Conversely, let every sequence in X admit a Cauchy subsequence. Fix $\epsilon > 0$ arbitrarily. We need to produce a finite ϵ-net for X. Let $x_1 \in X$ be fixed arbitrarily. If

$B(x_1; \epsilon) = X$, we are done. Otherwise, choose $x_2 \in X \setminus B(x_1; \epsilon)$. Then $d(x_1, x_2) \geq \epsilon$. Consider the set $\{x_1, x_2\}$. If $B(x_1; \epsilon) \cup B(x_2; \epsilon) = X$, we are done. Otherwise, choose $x_3 \in X \setminus B(x_1; \epsilon) \cup B(x_2; \epsilon)$. Then $d(x_1, x_3) \geq \epsilon$ and $d(x_2, x_3) \geq \epsilon$.

Continuing thus, this process must terminate after finite steps. Because otherwise we get a sequence $\{x_n\} \subset X$ such that for any two distinct indices $m \neq n$, $d(x_m, x_n) \geq \epsilon$, showing that $\{x_n\}$ cannot have a Cauchy subsequence, a contradiction to the hypothesis. Hence X is totally bounded. ■

We use this proposition to prove the following:

Theorem 6.20 *A metric space is compact if and only if it is sequentially compact.*

Proof Let X be a compact metric space. Then by Propositions 6.15 and 6.16, X is totally bounded and complete. Let $\{x_n\}$ be any arbitrary sequence in X. The total boundedness of X guarantees that $\{x_n\}$ admits a Cauchy subsequence, by the previous lemma. Again, as X is complete, the Cauchy subsequence of $\{x_n\}$ will be convergent as well. Hence any arbitrary sequence $\{x_n\}$ in X admits a convergent subsequence. Therefore, X is sequentially compact.
Conversely, assume X is sequentially compact, i.e. every sequence in X has a convergent subsequence. As a convergent sequence is always Cauchy, every sequence in X will thus have a Cauchy subsequence. Hence by the previous lemma, X is totally bounded. To show X is complete, let $\{x_n\}$ be a Cauchy sequence in X. By sequential compactness, $\{x_n\}$ admits a convergent subsequence. But by Exercise 4.1 a Cauchy sequence with a convergent subsequence is convergent. Hence $\{x_n\}$ is convergent and X is complete. Hence by Theorem 6.17, X is compact. ■

6.1.4 Bolzano-Weierstrass Compactness

Definition 6.21 A metric space (X, d) is said to satisfy *Bolzano-Weierstrass property* if every infinite subset has a limit point in X.

Theorem 6.22 *A metric space is sequentially compact if and only if it satisfies Bolzano-Weierstrass property.*

Proof Let X satisfy Bolzano-Weierstrass' property. Let $\{x_n\} \subset X$ be a sequence. WLOG, we may assume that $\{x_n\}$ does not admit a constant subsequence (in that case, the proof is trivial). Then the range of the sequence $A = \{x_n : n \in \mathbb{N}\}$ is an infinite set and hence has a limit point, say ξ in X. However, by the definition of a limit point, for every positive integer k, we will find a term x_{n_k} from the sequence $\{x_n\}$ such that $d(x_{n_k}, \xi) < \frac{1}{k}$. Therefore $\lim_{k\to\infty} x_{n_k} = \xi$. Hence the proof.

Conversely, assume X is sequentially compact. Then from any infinite set A, we can extract a sequence, which, in turn, will have a convergent subsequence. Hence by the same argument as in the previous case, the subsequential limit will be a limit point of A. ■

Hereby, we summarise this entire section in the following:

Theorem 6.23 *Let X be a metric space. Then the followings are equivalent:*

i) X is compact: Every open cover of X has a finite subcover.
ii) Every collection of closed sets in X satisfying the finite intersection property has a non-empty intersection.
iii) X is complete and totally bounded.
iv) X is sequentially compact: Every sequence in X has a convergent subsequence.
v) X satisfies Bolzano-Weierstrass property: Every infinite subset of X has a limit point in X.

Examples

6.23.1. A closed and bounded interval $[a, b]$ in $\mathbb{R}$ is compact by Heine-Borel theorem.

6.23.2. For fixed $a \neq 0$, the parabola $P = \{(x, y) \in \mathbb{R}^2 : y^2 = 4ax\}$ is *not* compact.
Proof. Consider the sequence of points $\{\mathbf{x}^{(n)}\}$ from P, where

$$\mathbf{x}^{(n)} = (an^2, 2an), \quad n \in \mathbb{N}.$$

Then for each n, $\|\mathbf{x}^{(n)}\| \geq |a|n^2$. This shows $\{\mathbf{x}^{(n)}\}$ is an unbounded sequence. Thus P is unbounded as well. Consequently, P is not totally bounded (Proposition 6.13); and hence, not compact. □

6.23.3. The n-dimensional hyper-disc

$$D^n = \{\mathbf{x} = (x_1, \ldots, x_n) \in \mathbb{R}^n : \|\mathbf{x}\| \leq 1\}$$

is compact.
Proof. Consider a sequence $\{\mathbf{x}^{(m)}\}_m \subset D^n$, where $\mathbf{x}^{(m)} = (x_1^{(m)}, \ldots, x_n^{(m)})$. As D^n is bounded, $\{\mathbf{x}^{(m)}\}$ is also so. Moreover, as D^n is closed, by Bolzano-Weierstrass theorem in $\mathbb{R}^n$ (Exercise 3.21), $\{\mathbf{x}^{(m)}\}$ admits a subsequence converging to a point in D^n. Consequently, D^n is sequentially compact, and hence, compact. □

6.23.4. For the same reasons used above, the n-dimensional hyper-sphere

$$S^{n-1} = \{\mathbf{x} = (x_1, \ldots, x_n) \in \mathbb{R}^n : \|\mathbf{x}\| = 1\}$$

is compact.

6.23.5. Using the same arguments as in Example 6.23.4, we conclude that the ellipse

$$\left\{(x, y) \in \mathbb{R}^2 : \frac{x^2}{a^2} + \frac{y^2}{b^2} = 1\right\}$$

is compact.

6.23.6. ℓ_2 is *not* compact.

Proof. The subset Y of ℓ_2 as mentioned in Example 6.13.2 is not totally bounded. Therefore Lemma 6.12 concludes that ℓ_2 is not totally bounded as well, hence not compact. □

6.23.7. The metric space $\mathcal{C}_{\text{int}}[0, 1]$ with integral metric is incomplete (see Example 4.5.8), therefore not compact.

Exercises: 6.1

6.1. Show that a metric space X is compact iff every collection of closed sets with empty intersection admits a finite subcollection with empty intersection.
6.2. Use finite intersection property to show that $\mathbb{R}$ is not compact w.r.t. the usual Euclidean metric.
6.3. Show that a compact subset of a metric space is closed and bounded. [*Hint*: Imitate the proof of the necessary part of Heine-Borel theorem in $\mathbb{R}$.]
6.4. Show that the closed unit ball in $(\mathcal{C}[0, 1], \|\cdot\|_\infty)$ is not compact. [*Hint*: See [6], p. 92.]
6.5. Let A be a compact subset of a metric space X. Show that the derived set A' of A is also compact.
6.6. If S is a totally bounded subset of a metric space X, show that $\overline{S}$ is also so.
6.7. Show that the completion $\widetilde{X}$ of a metric space X is compact iff X is totally bounded.
6.8. Show that a subset D of a metric space X is dense iff D is an ϵ-net for every $\epsilon > 0$.
6.9. Show that a closed subset of a compact metric space is compact.
6.10. Find all compact subsets of a discrete metric space.
6.11. Let the metric space X satisfy the Bolzano-Weierstrass' property. Use Proposition 6.19 to provide an independent proof of the fact that X is totally bounded.
6.12. If A, B are compact subsets of $\mathbb{R}^n$, show that their sum

$$A + B := \{\mathbf{a} + \mathbf{b} : \mathbf{a} \in A, \mathbf{b} \in B\}$$

is compact as well.
6.13. A metric space X is called *countably compact* if every countable open cover of X has a finite subcover. Show that a metric space is compact iff it is countably compact.

6.2 Results and Consequences

The examples of compact and non-compact sets in Euclidean space $\mathbb{R}^n$ presented at the end of the last section may have led you to wonder: Is the Heine-Borel theorem also true in $\mathbb{R}^n$? The answer is Yes, and we will prove it in this section.

In fact, there are multiple proofs of this fundamental result. Our approach will be to first explore the consequences of compactness in general metric spaces. These developments will then allow us to derive progressively simpler proofs of the theorem.

For future reference, we formally state the result:

Theorem 6.24 (Heine-Borel Theorem in $\mathbb{R}^n$) *Let $K \subset \mathbb{R}^n$. Then every open cover of K admits a finite subcover if and only if K is closed and bounded.*

A brief disclaimer: The proof of the necessary direction is straightforward (as shown in Theorem 6.9). Therefore, this section will focus exclusively on proving the sufficient direction. Our first proof follows the same structure as the proof for $\mathbb{R}$ without relying on the machinery developed in the previous section. However, this approach requires a significant amount of brute force to establish the result!

(First) Proof of Heine-Borel Theorem in $\mathbb{R}^n$ As K is bounded, there are closed and bounded intervals $[a_k, b_k]$ for each coordinate $k = 1, \ldots, n$, so that $K \subset \prod_{k=1}^{n}[a_k, b_k]$. Denote this hyper-parallelepiped by

$$\mathbb{P} = \prod_{k=1}^{n}[a_k, b_k].$$

By miscellaneous result 3.1, $\mathbb{P}$ is closed.

Now assume K is not compact, i.e. there is an open cover $\mathcal{G}$ of K not having any finite subcover.

We start by observing that

$$\operatorname{diam} \mathbb{P} = \text{length of the diagonal of } \mathbb{P} = \sqrt{\sum_{k=1}^{n}(b_k - a_k)^2} := D, \text{ say.}$$

Fragment $\mathbb{P}$ into 2^n subcomponents by the perpendicular hyperplanes through the midpoints of each side. These hyperplanes are given by

$$x_1 = \frac{a_1 + b_1}{2};\ x_2 = \frac{a_2 + b_2}{2};\ \ldots\ ;\ x_n = \frac{a_n + b_n}{2}.$$

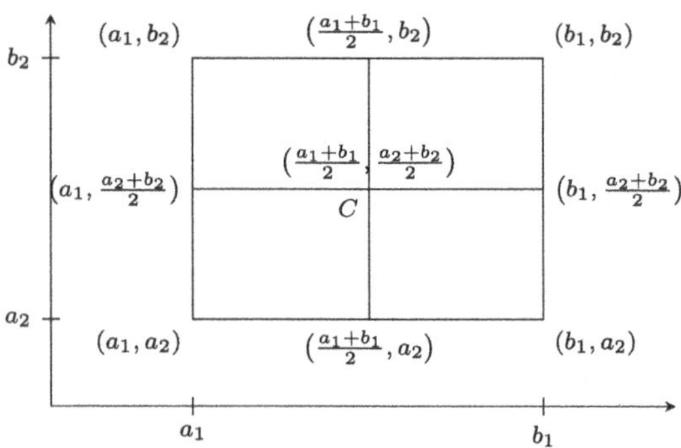

Fig. 6.7 Subdivisions of intervals in $\mathbb{R}^2$

They will all meet at the centroid of $\mathbb{P}$ given by

$$C = \left(\frac{a_1 + b_1}{2}, \frac{a_2 + b_2}{2}, \dots, \frac{a_n + b_n}{2}\right).$$

An illustrative figure for $\mathbb{R}^2$ is given in Fig. 6.7.

As $\mathcal{G}$ does not admit a finite subcover for K, at least one of these 2^n components of $K \cap \mathbb{P}$ cannot be covered by finitely many open sets from $\mathcal{G}$. Otherwise entire $K \cap \mathbb{P}$ will be covered by finitely many open sets. Choose one such subcomponent of $\mathbb{P}$ for which the intersection with K is not covered by finitely many open sets from $\mathcal{G}$. Call it $\mathbb{P}_1 := \prod_{k=1}^{n} \left[a_k^{(1)}, b_k^{(1)}\right]$. Then $\mathbb{P}_1$ is a closed subset of $\mathbb{P}$ and

$$\operatorname{diam} \mathbb{P}_1 = \sqrt{\sum_{k=1}^{n} \left(\frac{b_k - a_k}{2}\right)^2} = \frac{D}{2}.$$

Divide $\mathbb{P}_1$ into 2^n subcomponents by the perpendicular hyperplanes through the midpoints of each side. Again, at least one of these 2^n components of $K \cap \mathbb{P}_1$ cannot be covered by finitely many open sets from $\mathcal{G}$. Otherwise, entire $K \cap \mathbb{P}_1$ will be covered by finitely many open sets. Choose one such subcomponent of $\mathbb{P}_1$ for which the intersection with K is not covered by finitely many open sets, and call it $\mathbb{P}_2 := \prod_{k=1}^{n} \left[a_k^{(2)}, b_k^{(2)}\right]$. Then $\mathbb{P}_2$ is a closed subset of $\mathbb{P}_1$ and

$$\operatorname{diam} \mathbb{P}_1 = \sqrt{\sum_{k=1}^{n} \left(\frac{b_k^{(1)} - a_k^{(1)}}{2}\right)^2} = \sqrt{\sum_{k=1}^{n} \left(\frac{b_k - a_k}{2^2}\right)^2} = \frac{D}{2^2}.$$

Continuing thus, we get a sequence of closed sets $\{\mathbb{P}_n\}$ such that:

i) For each index $n \in \mathbb{N}$, $K \cap \mathbb{P}_n$ is *not* covered by finitely many open sets from $\mathcal{G}$.

ii) For each index $n \in \mathbb{N}$, $\mathbb{P}_{n+1} \subset \mathbb{P}_n$.

iii) $\operatorname{diam} \mathbb{P}_n = \frac{D}{2^n} \to 0$ as $n \to \infty$.

Since $\mathbb{R}^n$ is complete (Example 4.5.3), by Cantor's intersection theorem, $\bigcap_{n\in\mathbb{N}} \mathbb{P}_n$ is a singleton set, say $\{\xi\}$.

Now, for any arbitrary $\epsilon > 0$, $\operatorname{diam} \mathbb{P}_n \to 0$ implies that for some $i \in \mathbb{N}$, $\operatorname{diam} \mathbb{P}_i < \epsilon$. But as $\xi \in \mathbb{P}_i$, we must conclude

$$K \cap \mathbb{P}_i \subset \mathbb{P}_i \subset B(\xi; \epsilon).$$

Again, as $K \cap \mathbb{P}_i$ is not contained in finitely many open sets from $\mathcal{G}$, $K \cap \mathbb{P}_i$ must be an infinite set. Hence, $B(\xi; \epsilon)$ contains infinitely many elements from K. Since $\epsilon > 0$ is arbitrary, ξ is a limit point of K.

As K is closed, $\xi \in K$.

Therefore ξ is in some open set G_0 from $\mathcal{G}$. As G_0 is open, there exists $\delta > 0$ such that $B(\xi; \delta) \subset G_0$. However, $\operatorname{diam} \mathbb{P}_n \to 0$ implies again that for some index $j \in \mathbb{N}$, $\operatorname{diam} \mathbb{P}_j < \delta$. Therefore $\xi \in \mathbb{P}_j$ and $\operatorname{diam} \mathbb{P}_j < \delta$ together imply that $\mathbb{P}_j \subset B(\xi; \delta)$, which again eventually implies

$$K \cap \mathbb{P}_j \subset \mathbb{P}_j \subset B(\xi; \delta) \subset G_0.$$

But this is a contradiction, as by hypothesis, $K \cap \mathbb{P}_j$ is not covered by finitely many open sets from $\mathcal{G}$.

Hence, every open cover of K admits a finite subcover. ■

Now when we have experienced the crude way, let us look for more sophisticated routes, as agreed. We start with the following:

Proposition 6.25 *A subset of $\mathbb{R}^n$ is totally bounded if and only if it is bounded.*

Proof In Proposition 6.13, we have seen that a totally bounded metric space is always bounded. To see the converse in $\mathbb{R}^n$, let $S \subset \mathbb{R}^n$ be bounded. Then we can enclose S inside a hypercube $\mathbb{P} := \prod_{k=1}^{n} [a_k, b_k]$, where for each coordinate $k = 1, \ldots, n$, $b_k - a_k$ is the same constant, say r. Then

$$\operatorname{diam} \mathbb{P} = \text{length of the diagonal of } \mathbb{P} = \sqrt{\sum_{k=1}^{n} (b_k - a_k)^2} = r\sqrt{n}.$$

Choose $\epsilon > 0$ arbitrarily. By Archimedean property of $\mathbb{R}$, we find $m \in \mathbb{N}$ such that $m\epsilon > r\sqrt{n}$. Set $h = \frac{r}{m}$.

See Fig. 6.8 for an illustration in two dimensions. Divide each $[a_k, b_k]$ into m subintervals by inserting $m+1$ intermediate points at $a_k^{(i)} = a_k + ih$, $i = 0, 1, \ldots, m$. So, $a_k^{(0)} = a_k$ and $a_k^{(m)} = b_k$. Divide $\mathbb{P}$ into sub-hypercubes by the perpendicular hyperplanes through these intermediate points of each side (much like in the first

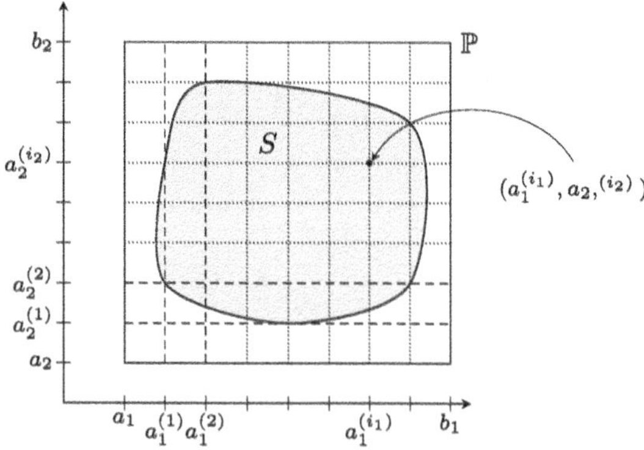

Fig. 6.8 Construction of finite ϵ-net for $\mathbb{P} \subset \mathbb{R}^2$

proof of this theorem). Then the divided sub-hypercubes will have their vertices at

$$\left(a_1^{(i_1)}, a_2^{(i_2)}, \ldots, a_n^{(i_n)}\right),$$

where each of the indices $i_1, \ldots, i_n$ will independently vary over the set $\{0, 1, \ldots, m\}$. We claim that the set

$$A = \left\{\left(a_1^{(i_1)}, a_2^{(i_2)}, \ldots, a_n^{(i_n)}\right): i_1, \ldots, i_n \in \{0, 1, 2, \ldots, m\}\right\}$$

of all these vertices thus generated is a finite ϵ-net for $\mathbb{P}$.

Let $(p_1, \ldots, p_n) \in \mathbb{P}$. Then for each coordinate $k = 1, \ldots, n$, we will find an index $j_k \in \{0, 1, \ldots, m-1\}$, such that

$$a_k + j_k h \leq p_k \leq a_k + (j_k + 1)h. \qquad (*)$$

For brevity, let us agree to use the following notations:

$$\begin{aligned} \mathbf{p} &= (p_1, \ldots, p_n), \\ \mathbf{a}^{(\mathbf{J})} &= \left(a_1^{(j_1)}, \ldots, a_n^{(j_n)}\right), \quad \text{and} \\ \mathbf{a}^{(\mathbf{J}+1)} &= \left(a_1^{(j_1+1)}, \ldots, a_n^{(j_n+1)}\right). \end{aligned}$$

Both $\mathbf{a}^{(\mathbf{J})}, \mathbf{a}^{(\mathbf{J}+1)}$ are vertex points coming from A. Thus,

$$\begin{aligned} d(\mathbf{p}, A) &\leq d\left(\mathbf{p}, \mathbf{a}^{(\mathbf{J}+1)}\right) \\ &= \|\mathbf{a}^{(\mathbf{J}+1)} - \mathbf{p}\| \\ &\leq \|\mathbf{a}^{(\mathbf{J}+1)} - \mathbf{a}^{(\mathbf{J})}\| && [\text{by}(*)] \end{aligned}$$

$$= \sqrt{\sum_{k=1}^{n}([a_k + (j_k+1)h] - [a_k + j_k h])^2}$$

$$= h\sqrt{n} = \frac{r\sqrt{n}}{m} < \epsilon.$$

Since $\mathbf{p} \in \mathbb{P}$ is arbitrary, A is a finite ϵ-net of $\mathbb{P}$. Since $\epsilon > 0$ is arbitrary, $\mathbb{P}$, and hence S, is totally bounded. ■

The above result enables us to have:

(Second) Proof of *Heine-Borel Theorem in* $\mathbb{R}^n$ Let $K \subset \mathbb{R}^n$ be closed and bounded. Then by the previous proposition, K is totally bounded. As $\mathbb{R}^n$ is complete (Example 4.5.3), and K is a closed subset of $\mathbb{R}^n$, it is complete as well (by Theorem 4.6). Therefore, by Theorem 6.17, K is compact, i.e. every open cover of K admits a finite subcover. ■

The third proof depends upon the fact that product of compact spaces is compact again. This fact is "*perhaps the most important single theorem in general topology*"[2] named as:

Theorem 6.26 (Tychonoff's Theorem) *The product of any non-empty class of compact sets is compact.*

We provide below a proof of a very soft version of this theorem.

Theorem 6.27 *If* (X, d_X) *and* (Y, d_Y) *are compact metric spaces, then* $X \times Y$ *is also compact with respect to the metric* $d := \sqrt{d_X^2 + d_Y^2}$.

Proof We will show that $X \times Y$ is sequentially compact w.r.t. the metric d. Let $\{(x_n, y_n)\} \subset X \times Y$ be a sequence. Then $\{x_n\}$ and $\{y_n\}$ are sequences in X and Y, respectively. Since X is compact, $\{x_n\}$ admits a convergent subsequence, say $\{x_{k_n}\}$. Then the subsequence $\{y_{k_n}\}$, being a sequence in the compact space Y, admits a convergent subsequence as well, say $\{y_{p_n}\}$. Then $\{x_{p_n}\}$, being a subsequence of the convergent sequence $\{x_{k_n}\}$, also converges. If $x_{p_n} \to x$ in X and $y_{p_n} \to y$ in Y, then for an arbitrarily chosen $\epsilon > 0$, there exists a positive integer n_0 such that for every $n \geq n_0$,

$$d_X(x_{p_n}, x) < \frac{\epsilon}{\sqrt{2}} \quad \text{and} \quad d_Y(y_{p_n}, y) < \frac{\epsilon}{\sqrt{2}}.$$

Hence, for every $n \geq n_0$,

$$d((x_{p_n}, y_{p_n}), (x, y)) = \sqrt{\left[d_X(x_{p_n}, x)\right]^2 + \left[d_Y(y_{p_n}, y)\right]^2} < \epsilon.$$

[2] Simmons, George F, pens this comment in his book [12].

Thus $\{(x_{p_n}, y_{p_n})\}$ is a convergent subsequence of $\{(x_n, y_n)\}$, showing that $X \times Y$ is sequentially compact, hence compact by Theorem 6.20. ■

Corollary 6.27.1 *If $(X_1, d_1), (X_2, d_2), \ldots, (X_n, d_n)$ are compact metric spaces, then the product metric space $X_1 \times \cdots \times X_n$ is also compact w.r.t. the metric $\sqrt{d_1^2 + \cdots + d_n^2}$.*

Remark Since compactness is a topological property and the above metric is equivalent to the product metric d_∞ (Exercise 3.25), the product space $\prod_{k=1}^n X_k$ is also compact in the d_∞ metric.

In the light of the above corollary, we have:

(Third) Proof of *Heine-Borel Theorem in $\mathbb{R}^n (n > 1)$* Since K is bounded, we get a set of closed and bounded intervals $\{[a_k, b_k] : k = 1, \ldots, n\}$, such that $K \subset \prod_{k=1}^n [a_k, b_k]$. Since each $[a_k, b_k]$ is closed and bounded in $\mathbb{R}$, it is compact. Therefore by the above corollary, $\prod_{k=1}^n [a_k, b_k]$ is compact in $\mathbb{R}^n$ w.r.t. the Euclidean metric. As K is a closed subset of the compact set $\prod_{k=1}^n [a_k, b_k]$, by Exercise 6.9, it is compact. ■

The same line of arguments of enclosing a given set in a compact (hyper)-cuboid, along with the equivalent conditions recorded in Theorem 6.23, also proves:

Theorem 6.28 (Bolzano-Weierstrass Theorem in $\mathbb{R}^n$) *Every infinite bounded subset of $\mathbb{R}^n$ has a limit point in $\mathbb{R}^n$.*

A mention-worthy result in this regard is given in the following:

Theorem 6.29 *A compact metric space is separable.*

Proof Let X be a compact metric space. Then X is totally bounded by the Proposition 6.15. Thus for every positive integer n, we obtain a finite set $A_n := \{x_{n_1}, x_{n_2}, \ldots, x_{n_{p(n)}}\} \subset X$ such that A_n is an $\frac{1}{n}$-net of X, i.e.

$$X = \bigcup_{j=1}^{p(n)} B\left(x_{n_j}; \frac{1}{n}\right).$$

Call $A = \bigcup_{n \in \mathbb{N}} A_n$. Since all A_n's are finite, a countable union of them will be countable again. Thus A is countable. We will show that A is dense in X.

Choose $x \in X$ arbitrarily. Fix $\delta > 0$ arbitrarily. Let for $k \in \mathbb{N}$, $\frac{1}{k} < \delta$. Thus by construction, $X = \bigcup_{j=1}^{p(k)} B\left(x_{k_j}; \frac{1}{k}\right)$, where the points x_{k_j} are coming from A_k, and therefore, from A. Let $x \in B\left(x_{k_{j_0}}; \frac{1}{k}\right)$ for some j_0. As $B\left(x_{k_{j_0}}; \frac{1}{k}\right) \subset B\left(x_{k_{j_0}}; \delta\right)$, and $\delta > 0$ is arbitrary, x must be in $\overline{A}$. Since $x \in X$ is arbitrary, A is dense in X. ■

Examples

6.29.1. Look at Examples 6.23.1 to 6.23.5. Their compactness or non-compactness can be directly (and much easily) be guaranteed by the Heine-Borel theorem.

6.29.2. The metric space ℓ_∞ is not compact because it is inseparable (see Example 3.47.6).

6.29.3. In the set of all $n \times n$ real matrices $\mathcal{M}_n(\mathbb{R})$, the set $\mathcal{O}_n(\mathbb{R})$ of all orthogonal[3] matrices, and the set $\mathcal{SO}_n(\mathbb{R})$ of all orthonormal matrices are compact.
Proof. By miscellaneous result 5.3, $\mathcal{O}_n(\mathbb{R})$ is a closed subset of $\mathcal{M}_n(\mathbb{R})$. Now, for any $n \times n$ orthogonal matrix A, if $\{R_1, \ldots, R_n\}$ denote the rows of A, then

$$(i, j)^{\text{th}} \text{ entry of } AA^T = \langle R_i, R_j \rangle = \begin{cases} 1, \text{ if } i = j \\ 0, \text{ if } i \neq j \end{cases}.$$

Thus, for each row, $\|R_i\|^2 = 1$, and hence $\|A\|^2 = \sum_{i=1}^n R_i^2 = n$. This shows that $\mathcal{O}_n(\mathbb{R})$ is bounded as well. Consequently, by Heine-Borel theorem, $\mathcal{O}_n(\mathbb{R})$ is compact.
Now, $\mathcal{SO}_n(\mathbb{R}) \subset \mathcal{O}_n(\mathbb{R})$ is also bounded. To see $\mathcal{SO}_n(\mathbb{R})$ closed, remember that determinant is a continuous function (Example 5.6.2) and

$$\mathcal{SO}_n(\mathbb{R}) = [\det{}^{-1}(\{1\})] \cap \mathcal{O}_n(\mathbb{R}).$$

Thus, it is a closed subset of $\mathcal{O}_n(\mathbb{R})$, and therefore, by Exercise 6.9, is compact. □

6.29.4. In $\mathcal{M}_n(\mathbb{R})$, let $GL_n(\mathbb{R})$ denote the set of all invertible matrices and $SL_n(\mathbb{R})$ denote the set of all matrices with determinant 1. Both $GL_n(\mathbb{R})$ and $SL_n(\mathbb{R})$ are unbounded subsets of $\mathcal{M}_n(\mathbb{R})$, hence not compact. For the same reason, the set of all nilpotent matrices is not compact.

We conclude this section with the useful Lebesgue covering lemma. Let us set up the relevant definition first.

Definition 6.30 Let X be a metric space and $\mathcal{G}$ be an open cover of X. A positive number δ is called a *Lebesgue number* of $\mathcal{G}$ if for every subset A of X with diam $A < \delta$, A can be contained in a single open set G_A from $\mathcal{G}$.

Let us resort to some crude intuitions to understand the situation (Fig. 6.9). Think of the space X as a closed rectangular region in the plane and $\mathcal{G}$ as a collection of open discs covering X. Notice that the discs in $\mathcal{G}$ cannot all be pairwise disjoint; otherwise, they would fail to cover the entire area of X.

---

[3] A square matrix A is *orthogonal* if $AA^t = I$. A is *orthonormal* if A is orthogonal and $\det A = 1$. *Note* that a matrix can have determinant 1 without being orthonormal, for instance, $\begin{pmatrix} 1 & 1 \\ 2 & 3 \end{pmatrix}$.

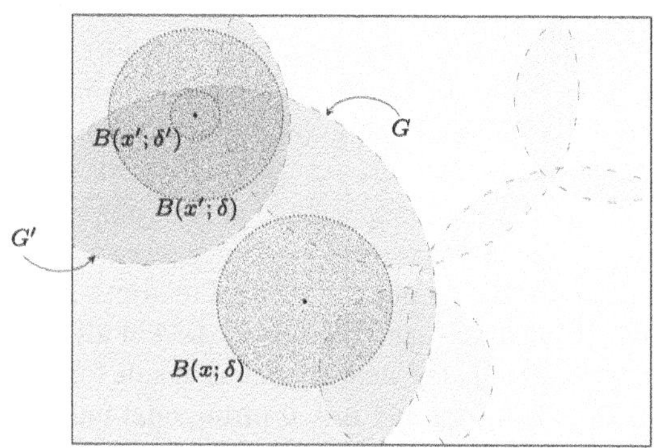

Fig. 6.9 Illustration of Lebesgue number

Now, each point $x \in X$ belongs to some open disc G from $\mathcal{G}$, i.e. there exists a ball of positive radius (say δ) centred at x that is entirely contained in G. If we consider another point x' within the same disc G, we can similarly find a ball of positive radius (say δ') centred at x' that fits within G. However, δ' may be much smaller than δ, so as we move from point to point, we cannot always guarantee a uniform positive radius.

However, if we consider the neighbouring discs of G and identify x' as a member of another disc G', it may turn out that the entire δ-radius ball $B(x'; \delta)$ fits inside G'. This process suggests that we do not always need to reduce the radius drastically to keep the ball contained within a single open disc. Thus, we can hope to find a uniform positive radius that works for every point in X. This uniform radius defines a Lebesgue number for $\mathcal{G}$.

Examples

6.30.1. Let $X = [0, 1]$. Call $G_0 := [0, \frac{1}{10})$ and $G_a := (\frac{a}{2}, 1]$, $a \in (0, 1]$. Then $G_0 \cup \{G_a : a \in (0, 1]\}$ is an open cover of X. Here $\frac{1}{10}$ is a Lebesgue number. (Prove it!)

6.30.2. Let $X = (0, 1)$ and $G_n := \left(\frac{1}{n}, 1\right)$. Consider the cover $\mathcal{G} := \{G_n : n \geq 2\}$ of X. If $\delta > 0$ is a Lebesgue number, then the set $\left(0, \frac{\delta}{2}\right)$ is of diameter $< \delta$ but still is not contained in any single open set G_n. Thus $\mathcal{G}$ does not admit a Lebesgue number.

The relevant result in this regard is:

Lemma 6.31 (Lebesgue Covering Lemma) *In a compact metric space, every open cover admits a Lebesgue number.*

Proof Let X be a compact metric space and $\mathcal{G}$ be an open cover of X. Then each point $x \in X$ is contained in some open set G_x from $\mathcal{G}$. Thus for some $\delta_x > 0$, the open ball $B(x; 2\delta_x)$ is also contained in G_x. Consider the collection of open balls

$$\mathcal{B} := \{B(x; \delta_x) : x \in X\}.$$

$\mathcal{B}$ is also a cover of the compact space X and therefore admits a finite subcover, say $\mathcal{B}' = \{B(x_k; \delta_k) : k = 1, \ldots, n\}$. Choose $\delta = \min\{\delta_1, \ldots, \delta_n\} > 0$.

Let A be any subset of X with $\operatorname{diam} A < \delta$. Then for every point $a \in A$ fixed arbitrarily, $a \in B(x_k; \delta_k)$ for some k. For any other point $a' \in A$,

$$d(a', x_k) \leq d(a', a) + d(a, x_k) < \delta + \delta_k < 2\delta_k.$$

Hence $a' \in B(x_k; 2\delta_k)$ showing that $A \subset B(x_k; 2\delta_k) \subset G_{x_k}$. Thus δ is a Lebesgue number for $\mathcal{G}$. ∎

Exercises: 6.2

6.14. Provide a fourth proof of Heine-Borel theorem in $\mathbb{R}^n$ by showing that a closed and bounded subset of $\mathbb{R}^n$ is sequentially compact.

6.15. Let X be a metric space and $f: X \to \mathbb{R}$. f is said to be *locally bounded* on X if for each point $x \in X$, there exist $\delta_x > 0$ and $M_x > 0$ such that $\forall x' \in B(x; \delta_x)$, $|f(x')| \leq M_x$. Show that a locally bounded function defined over a compact domain is bounded.

6.16. Show that any open cover of the unit circle in $\mathbb{R}^2$ is also a cover of an annulus, i.e. if $\mathcal{G}$ is an open cover of $\{(x, y) \in \mathbb{R}^2 : x^2 + y^2 = 1\}$, then also covers an annulus $\{(x, y) \in \mathbb{R}^2 : 1 - \delta < x^2 + y^2 < 1 + \delta\}$ for some $\delta > 0$.

6.3 Continuous Functions on Compact Spaces

In an introductory real analysis course, we observed that a continuous function defined on a closed and bounded interval enjoys several desirable properties. A closed and bounded interval serves as a simple example of a compact and connected subset of the real line.

In this section, we explore how these properties extend to compact metric spaces, demonstrating their broader validity. We begin with a result of great importance.

Theorem 6.32 *Continuous image of a compact space is compact.*

Proof Let X, Y be metric spaces and $f: X \to Y$ be continuous. We will show that when X is compact, then $f(X)$ is a compact subset of Y.

Let $\mathcal{G}$ be an open cover of $f(X)$. As f is continuous, for every open set $G \in \mathcal{G}$, $f^{-1}(G)$ is open in X (Theorem 5.4.(*iv*)). Thus $\mathcal{G}' = \{f^{-1}(G) : G \in \mathcal{G}\}$ becomes an open cover for X. As X is compact, $\mathcal{G}'$ admits a finite subcover, say

$\mathcal{G}_0 = \{f^{-1}(G_k) : k = 1, \ldots, n\}$. Therefore,

$$X = \bigcup_{k=1}^{n} f^{-1}(G_k) = f^{-1}\left(\bigcup_{k=1}^{n} G_k\right)$$
$$\implies f(X) \subset \bigcup_{k=1}^{n} G_k.$$

Thus $\{G_1, \ldots, G_n\}$ is a finite subcover of $f(X)$ in Y. Hence $f(X)$ is compact. ■

Corollary 6.32.1 *Let f be a* homeomorphism *between two metric spaces X and Y. Then X is compact if and only if Y is compact.*

Corollary 6.32.2 *Let X be a compact metric space and $f: X \to \mathbb{R}$ be continuous. Then f is bounded and also attains its bounds.*

Proof Since X is compact, $f(X)$ is a compact subset of $\mathbb{R}$, thus bounded, i.e. f is bounded on X. Using Exercise 6.17, we conclude that f also attains its bounds. ■

Remarks

(a) This property shows why people like to define continuous functions on compact spaces.
(b) This property fails if X is not compact. For example $f: (0, 1) \to \mathbb{R}$ given by $x \mapsto \frac{1}{x}$ is not bounded. On the other hand, $f: (1, \infty) \to \mathbb{R}$ given by $x \mapsto \frac{1}{x}$ is bounded but does not attain its bounds.

Using the first corollary of the previous theorem, we now present another proof of the Heine-Borel theorem in $\mathbb{R}^n$. This proof is neither shorter nor less intricate than the previous ones—readers may even choose to skip it in the first read and proceed with the section.

Instead of brevity, this proof offers an insightful approach. It reduces the compactness of any closed and bounded subset of $\mathbb{R}^n$ to the compactness of the simplest closed and bounded interval, $[0, 1]$, in $\mathbb{R}$. Furthermore, the compactness of $[0, 1]$ is not derived from the Heine-Borel theorem in $\mathbb{R}$. Instead, it is established by leveraging the intrinsic rectilinear properties of the real number line.

(Fifth) Proof of *Heine-Borel Theorem in $\mathbb{R}^n$*

Step 1: Since $K \subset \mathbb{R}^n$ is bounded, for some $M > 0$, $K \subset [-M, M]^n$. As K is closed, it is enough to show $[-M, M]^n$ is compact (Exercise 6.9).

Step 2: Define $f: [0, 1] \to [-M, M]$ by $t \mapsto (2t - 1)M$. Then f is a homeomorphism. Therefore by Exercise 5.25, $[0, 1]^n$ is homeomorphic to $[-M, M]^n$ as well. Thus it is enough to show $[0, 1]^n$ is compact.

Step 3: We prove the hypercube $[0, 1]^n$ is compact by induction on n. Here we will use the d_∞ metric; by Example 3.19.1, it is equivalent to the usual Euclidean metric. Recall that in this metric, for a point $\mathbf{x} = (x_1, \ldots, x_n) \in \mathbb{R}^n$ and any

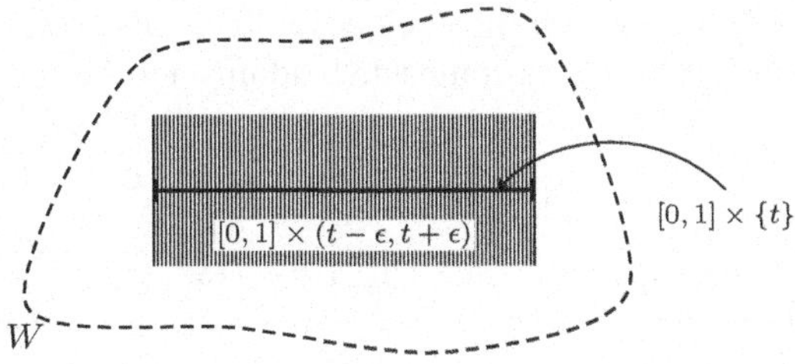

Fig. 6.10 Illustration of Tube lemma in $\mathbb{R}^2$

radius $r > 0$,

$$B(\mathbf{x}; r) = \prod_{k=1}^{n} B(x_k; r) = \prod_{k=1}^{n} (x_k - r, x_k + r).$$

We will first execute the inductive step. Let us assume $[0, 1]^{n-1}$ is compact. We will show $[0, 1]^n$ is also compact.

Let $\mathcal{G}$ be an open cover of $[0, 1]^n$. Now, for any point $t \in [0, 1]$ fixed arbitrarily, the map $\mathbf{x} \mapsto (\mathbf{x}, t)$ is a homeomorphism from $[0, 1]^{n-1}$ to $[0, 1]^{n-1} \times \{t\}$. Since by our assumption, $[0, 1]^{n-1}$ is compact, so is $[0, 1]^{n-1} \times \{t\}$. As $\mathcal{G}$ is a cover of $[0, 1]^n$, it also covers $[0, 1]^{n-1} \times \{t\}$, and the latter being compact admits a finite subcover from $\mathcal{G}$, say, $\mathcal{G}_t = \{G_t^{(1)}, \ldots, G_t^{(m_t)}\}$.

At this point, we require a result otherwise known as the *Tube lemma*, recorded in the following:

Claim Fix $t \in \mathbb{R}$. For an open subset W of $\mathbb{R}^n$ containing the slice $[0, 1]^{n-1} \times \{t\}$, there exists an $\epsilon > 0$ such that the tube

$$[0, 1]^{n-1} \times (t - \epsilon, t + \epsilon)$$

is also contained in W (Fig. 6.10).

Proof of Claim Fix $\mathbf{x} = (x_1, \ldots, x_{n-1}) \in \mathbb{R}^{n-1}$ arbitrarily. Let us agree to denote the point $(\mathbf{x}, t) = (x_1, \ldots, x_{n-1}, t)$ in $\mathbb{R}^n$ by $\mathbf{x}_t$. As W is open, for every point $\mathbf{x}_t \in W$, there will exist some positive $\epsilon = \epsilon(\mathbf{x})$ so that $B(\mathbf{x}_t; \epsilon) \subset W$. However, as mentioned earlier, in d_∞ metric,

$$\underbrace{B(\mathbf{x}_t; \epsilon)}_{\subset \mathbb{R}^n} = \underbrace{B(\mathbf{x}; \epsilon)}_{\subset \mathbb{R}^{n-1}} \times \underbrace{(t - \epsilon, t + \epsilon)}_{\subset \mathbb{R}}.$$

For clarity, let us agree to add the dimension of the balls into their notations. Thus, $B_n(\mathbf{x}_t; \epsilon)$ will denote the ball $B(\mathbf{x}_t; \epsilon)$ in $\mathbb{R}^n$, whereas $B_{n-1}(\mathbf{x}; \epsilon)$ will denote the ball $B(\mathbf{x}; \epsilon)$ in $\mathbb{R}^{n-1}$. With these notations, we have

$$B_n(\mathbf{x}_t; \epsilon(\mathbf{x})) \subset W. \tag{$*$}$$

Now consider the open cover $\mathcal{B} = \{B_{n-1}(\mathbf{x}; \epsilon(\mathbf{x})) : \mathbf{x} \in [0, 1]^{n-1}\}$ of $[0, 1]^{n-1}$. Since $[0, 1]^{n-1}$ is compact, $\mathcal{B}$ admits a finite subcover, say,

$$\mathcal{B}' = \{B_{n-1}(\mathbf{x}^{(1)}; \epsilon_1), \ldots, B_{n-1}(\mathbf{x}^{(p)}; \epsilon_p)\}.$$

Therefore, $[0, 1]^{n-1} \subset \bigcup_{k=1}^{p} B_{n-1}(\mathbf{x}^{(k)}; \epsilon_k)$. Call $\epsilon_0 = \min\{\epsilon_1, \ldots, \epsilon_p\} > 0$. Then for each k, $(t - \epsilon_0, t + \epsilon_0) \subset (t - \epsilon_k, t + \epsilon_k)$. Consequently,

$$\begin{aligned} B_{n-1}(\mathbf{x}^{(k)}; \epsilon_k) \times (t - \epsilon_0, t + \epsilon_0) \subset B_{n-1}(\mathbf{x}^{(k)}; \epsilon_k) &\times (t - \epsilon_k, t + \epsilon_k) \\ &= B_n(\mathbf{x}_t^{(k)}; \epsilon_k) \\ &= B_n(\mathbf{x}_t^{(k)}; \epsilon(\mathbf{x}_k)). \end{aligned}$$

Taking union over k, we obtain

$$\begin{aligned} \left[\bigcup_{k=1}^{p} B_{n-1}(\mathbf{x}^{(k)}; \epsilon_k)\right] \times (t - \epsilon_0, t + \epsilon_0) &\subset \bigcup_{k=1}^{n} B(\mathbf{x}_t^{(k)}; \epsilon(\mathbf{x}_k)) \\ \Longrightarrow \qquad [0, 1]^{n-1} \times (t - \epsilon_0, t + \epsilon_0) &\subset W. \qquad [\text{as } (*) \text{ is true } \forall \mathbf{x}_t \in W.] \end{aligned}$$

This proves the claim.

Now let us come back to our earlier concern, where $\mathcal{G}_t = \{G_t^{(1)}, \ldots, G_t^{(m_t)}\}$ is a finite subcover of $[0, 1]^{n-1} \times \{t\}$, i.e.

$$[0, 1]^{n-1} \times \{t\} \subset \bigcup_{k=1}^{m_t} G_t^{(k)}.$$

Write $G(t) = \bigcup_{k=1}^{m_t} G_t^{(k)} = \bigcup_{G \in \mathcal{G}_t} G$. Thus $G(t)$ is an open set in $\mathbb{R}^n$ itself. Therefore, using Tube lemma, we obtain $\epsilon_t > 0$ such that

$$[0, 1]^{n-1} \times (t - \epsilon_t, t + \epsilon_t) \subset G(t). \qquad (**)$$

As ϵ_t exists for every point $t \in [0, 1]$, the collection

$$\mathcal{I} = \{(t - \epsilon_t, t + \epsilon_t) : t \in [0, 1]\}$$

becomes an open cover for $[0, 1]$. Assuming $[0, 1]$ is compact (proof of this assumption is the base step of the induction process, which will follow next), $\mathcal{I}$

admits a finite subcover, say $\mathcal{T}' = \{(t_i - \epsilon_i, t_i + \epsilon_i) : i = 1, \ldots, q\}$. Therefore,

$$
\begin{aligned}
[0,1]^n &= [0,1]^{n-1} \times [0,1] \\
&\subset [0,1]^{n-1} \times \left[\bigcup_{i=1}^{q} (t_i - \epsilon_i, t_i + \epsilon_i)\right] \\
&= \bigcup_{i=1}^{q} \left[[0,1]^{n-1} \times (t_i - \epsilon_i, t_i + \epsilon_i)\right] \\
&\subset \bigcup_{i=1}^{q} G(t_i). \qquad \text{[as (**) is true } \forall t \in [0,1]] \\
&= \bigcup_{i=1}^{q} \left[\bigcup_{G \in \mathcal{G}_{t_i}} G\right].
\end{aligned}
$$

As each $\mathcal{G}_{t_i}$ is a collection of finitely many open sets from $\mathcal{G}$, finite union of $\mathcal{G}_{t_i}$'s is also a finite collection of open sets from $\mathcal{G}$. Consequently, $[0,1]^n$ is contained in finitely many open sets from $\mathcal{G}$, showing that $[0,1]^n$ is compact.

Step 4: Our whole proof has now been reduced to establishing the base step of the induction process, i.e. to show $[0,1]$ is compact. Let us now complete that. Let $\mathcal{U}$ be an open cover of $[0,1]$. Construct the set

$$S = \{x \in [0,1] : [0,x] \text{ is covered by finitely many open sets from } \mathcal{U}\}.$$

Then, $0 \in S$, and S is bounded above by 1. Thus $\sup S$ exists. Call $s = \sup S$. Thus $0 \leq s \leq 1$. At this point, we can achieve our goal by establishing the following:

Claim

(a) $s \in S$.
(b) $s = 1$.

Proof of Claim

(a) Assume $s \notin S$. Then s becomes a limit point of S.[4] Also, $0 \leq s < 1$. Since $\mathcal{U}$ is an open cover of $[0,1]$, s is contained in some open set U from $\mathcal{U}$. As U

[4] Recall the fact from real analysis: If $(\emptyset \neq) A \subset \mathbb{R}$ is bounded above and $\sup A \notin A$, then $\sup A$ is a limit point of A.

Fig. 6.11 Construction of S and $\sup S$

is open, for some $\delta > 0$, the neighbourhood $N(s;\delta) = (s-\delta, s+\delta)$ is also contained in U (Fig. 6.11). Again, as s is a limit point of S, we will find a point x_δ from S with $x_\delta \in N(s;\delta) \setminus \{s\}$. But $x_\delta \in S$ implies from definition that the interval $[0, x_\delta]$ is covered by finitely many open sets from $\mathfrak{U}$. If $x_\delta > s$, then $[0, s]$ is also covered by finitely many open sets from $\mathfrak{U}$, i.e. $s \in S$; otherwise, if $x_\delta < s$, then

$$[0, s] \subset [0, s+\delta) = [0, x_\delta] \cup N(s;\delta)$$

is also covered by finitely many open sets from $\mathfrak{U}$, a contradiction. Consequently, $s \in S$.

(b) Assume $s < 1$. Then by the last argument in (a), $\left[0, s+\frac{\delta}{2}\right]$ is also covered by finitely many open sets from $\mathfrak{U}$. Choose $0 < \epsilon < \frac{\delta}{2}$ such that $s+\epsilon < 1$. Then $s+\epsilon \in S$, contradicting the fact that $s = \sup S$. Hence $s = 1$.

This completes the proof.

Theorem 6.33 *Let X, Y be metric spaces and $f: X \to Y$ be bijective and continuous. If X is compact, then $f^{-1}: Y \to X$ is also continuous, i.e. $f: X \to Y$ is a homeomorphism.*

Proof Since $f: X \to Y$ is a bijection, $f^{-1}: Y \to X$ is well-defined. Let F be a closed subset of X. Since X is compact and $F \subset X$ is closed, F is also compact (Exercise 6.9). Therefore by Theorem 6.32, $f(F) \subset Y$ is also compact. In that case, Theorem 6.9 guarantees that $f(F)$ is closed in Y. But $f(F) = (f^{-1})^{-1}(F)$. Consequently, by Theorem 5.4.(*iii*), f^{-1} is continuous. ∎

Remark Let us pause for a while to appreciate this theorem. In case of an isomorphism (structure-preserving bijection between two algebraic systems), we need not worry about the structure-preserving property (viz., the homomorphism) of the inverse map. But in case of topological spaces, a continuous bijection need not have a continuous inverse in general. For instance, the map $f: [0, 1) \to S^1$[5] given by $f(t) = (\cos 2\pi t, \sin 2\pi t)$ is a continuous bijection with a discontinuous inverse (Fig. 6.12). Thus, to guarantee the equivalence of the topological structures between two spaces, we require to add the extra condition of continuity of the inverse function as well. However, for a compact space, this extra condition becomes redundant.

[5] Recall that S^1 is the circle $x^2 + y^2 = 1$ in $\mathbb{R}^2$.

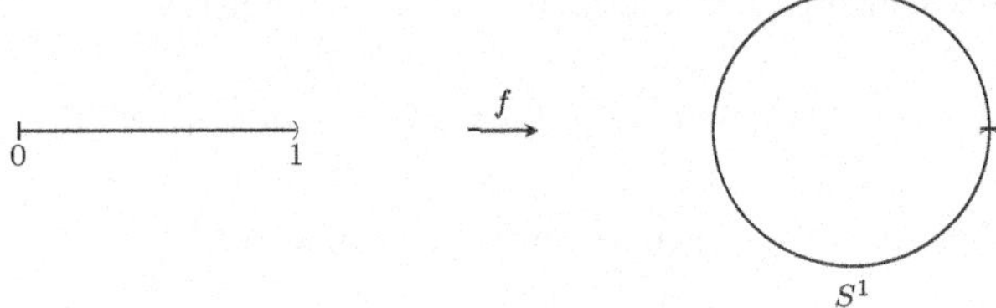

Fig. 6.12 $[0, 1)$ is not homeomorphic to S^1

Theorem 6.34 *A continuous function defined on a compact set is uniformly continuous.*

Proof Let (X, d_X) and (Y, d_Y) be metric spaces and $f: X \to Y$ be continuous. If X is compact, we want to show f is uniformly continuous on X.

Fix $\epsilon > 0$ arbitrarily. As f is continuous on X, for every point $x \in X$, we find some $\delta_x > 0$ so that the entire ball $B(x; \delta_x)$ in X gets mapped inside the ball $B(f(x); \frac{\epsilon}{2})$ in Y, i.e. for each point $x' \in B(x; \delta_x)$,

$$d_Y(f(x), f(x')) < \frac{\epsilon}{2}. \qquad (*)$$

Since δ_x exists for every $x \in X$, the collection $\mathcal{G} = \{B(x; \frac{1}{2}\delta_x) : x \in X\}$ forms an open cover of X. As X is compact, $\mathcal{G}$ admits a finite subcover, say,

$$\mathcal{G}' = \left\{B(x_1; \tfrac{1}{2}\delta_1), \ldots, B(x_n; \tfrac{1}{2}\delta_n)\right\}.$$

Call $\delta = \min\{\frac{1}{2}\delta_1, \ldots, \frac{1}{2}\delta_n\} > 0$.

Choose $x, x' \in X$ arbitrarily with $d_X(x, x') < \delta$. As $\mathcal{G}'$ covers X, let $x \in B(x_k; \frac{1}{2}\delta_k)$ for some $k \in \{1, \ldots, n\}$. Then

$$d_X(x', x_k) \leq d_X(x', x) + d_X(x; x_k) < \delta + \frac{1}{2}\delta_k \leq \delta_k. \qquad \text{[by definition of } \delta]$$

Therefore, $x' \in B(x_k; \delta_k)$. Consequently, by $(*)$,

$$d_Y(f(x), f(x')) \leq d_Y(f(x), f(x_k)) + d_Y(f(x_k), f(x')) < \epsilon.$$

As $x, x' \in X$ are arbitrary, f is uniformly continuous on X. ■

Another proof of this theorem, given by Lebesgue covering lemma (Lemma 6.31), goes as follows:

Another Proof The starting arguments remain the same, i.e. for any $\epsilon > 0$ chosen arbitrarily, and for every point $x \in X$, we find a $\delta_x > 0$ such that whenever $x' \in B(x; \delta_x)$, we have $d_Y(f(x), f(x')) < \frac{\epsilon}{2}$. Now consider the cover $\{B(x; \delta_x) : x \in X\}$ of X. By the Lebesgue covering lemma, this covering admits a Lebesgue number $\delta > 0$. Choose $x, x' \in X$ arbitrarily with $d_X(x, x') < \delta$. Thus both x and x' belong to a same open set $B(x_0; \delta_0)$ from $\mathcal{G}$. Therefore,

$$d_Y(f(x), f(x')) \leq d_Y(f(x), f(x_0)) + d_Y(f(x_0), f(x')) < \frac{\epsilon}{2} + \frac{\epsilon}{2} = \epsilon.$$

Consequently, f is uniformly continuous on X. ■

Miscellaneous Results

6.1. Show that in the space $\mathbb{C}$ of complex numbers, the open unit ball $B(0; 1) = \{z \in \mathbb{C} : |z| < 1\}$ and the closed unit ball $B[0; 1] = \{z \in \mathbb{C} : |z| \leq 1\}$ are not homeomorphic.

Proof. Easy! $B(0; 1)$ is not compact, whereas $B[0; 1]$ is compact. A homeomorphic image of a compact set must be compact. □

6.2. Let F be closed and K be compact in $\mathbb{R}^n$. If $F \cap K = \varnothing$, show that $d(F, K) > 0$, where $d(F, K)$ is the distance between the sets F and K (Fig. 6.13).

Proof. Define $f : K \to \mathbb{R}$ by

$$f(\mathbf{x}) = d(\mathbf{x}, F) := \inf\{d(\mathbf{x}, \mathbf{y}) : \mathbf{y} \in F\}.$$

Note from Definition 2.6 that

$$d(F, K) = \inf\{f(\mathbf{x}) : \mathbf{x} \in K\}.$$

Now, by miscellaneous result 5.6, f is continuous on K. As K is compact, by Corollary 6.32.2, f is bounded and attains its bounds. Thus for some $\mathbf{x}_0 \in X$,

$$f(\mathbf{x}_0) = \inf\{f(\mathbf{x}) : \mathbf{x} \in X\} = d(F, K).$$

Now, for every point $\mathbf{x} \in K$, $f(\mathbf{x}) \geq 0$. But,

$$f(\mathbf{x}_0) = d(\mathbf{x}_0, F) = 0 \implies \mathbf{x}_0 \in \overline{F}. \qquad \text{(see Miscellaneous result 3.3)}$$

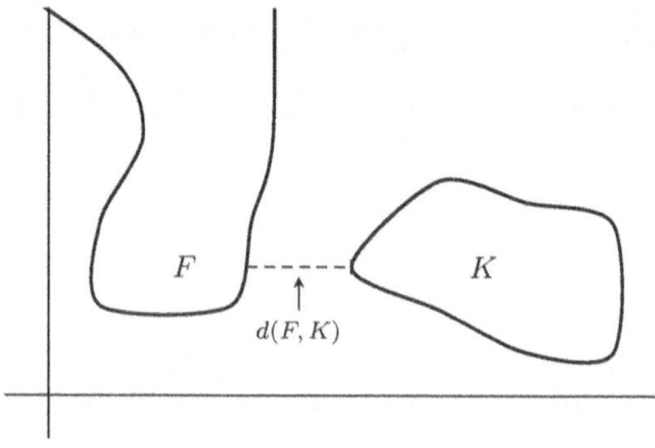

Fig. 6.13 Positive distance between disjoint compact and closed sets

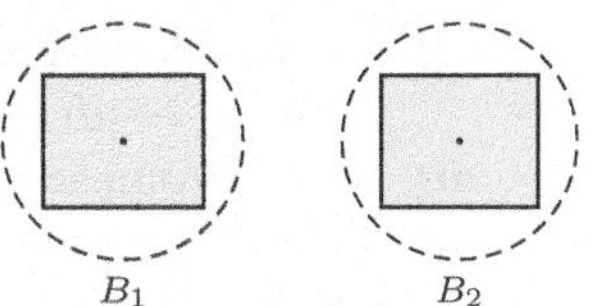

Fig. 6.14 X is union of two disjoint compact sets

As F is closed, $\overline{F} = F$; this concludes $\mathbf{x}_0 \in F \cap K$. However, it is a contradiction to the hypothesis $F \cap K = \varnothing$. Hence $d(F, K) = f(\mathbf{x}_0) > 0$. This completes the proof. □

6.3. Let (X, d_X) and (Y, d_Y) be metric spaces and $f : X \to Y$. f is called *locally Lipschitz* if for every point $x \in X$, we find a ball $B_x \subset X$ containing x and a constant $L_x > 0$ so that for any two points $x_1, x_2 \in B_x$,

$$d_Y(f(x_1), f(x_2)) \leq L_x d_X(x_1, x_2).$$

Show that if f is locally Lipschitz and X is compact, then f is Lipschitz.

Remark A common technique, as seen in the proof of Theorem 6.34, is to cover X with $\mathcal{G} = \{B_x : x \in X\}$ and obtain a finite subcover $\mathcal{G}'$ using the compactness of X. The uniform Lipschitz constant L is then chosen as the minimum of the finitely many L_x values corresponding to each covering ball in $\mathcal{G}'$. The final step is usually to verify that this choice of L is valid.

However, in this case, this approach fails. Suppose X consists of two disjoint closed rectangular regions, each contained in a separate open ball. See Fig. 6.14. Let $\mathcal{G}' = \{B_1, B_2\}$ and choose points $x_1 \in B_1$ and $x_2 \in B_2$. Since there is no transition point $x_0 \in B_1 \cap B_2$, we cannot estimate $d_Y(f(x_1), f(x_2))$ using L. (If such a point existed, we could write $d_Y(f(x_1), f(x_2)) \leq d_Y(f(x_1), f(x_0)) + d_Y(f(x_0), f(x_2))$.)

Thus, we take a different approach to solve the problem.

Proof. We start by showing that f is continuous at every point $x \in X$. Fix $\epsilon > 0$ arbitrarily. Choose $0 < \delta_x < \frac{\epsilon}{L_x}$ so small that $B(x; \delta_x) \subset B_x$. Then for any point $x' \in B(x; \delta_x)$,

$$d_Y(f(x), f(x')) \leq L_x d_X(x, x') < \epsilon.$$

Thus f is continuous at x.

Now if possible, assume that f is not Lipschitz (see Example 5.8.3). Then for every value of $L > 0$, we will find points x_L and y_L from X so that $d_Y(f(x_L), f(y_L)) > L d_X(x_L, y_L)$. Choose $L = 1, 2, 3, \ldots$ successively. This way we obtain two sequences $\{x_n\}$ and $\{y_n\}$ from X so that

$$\text{for each positive integer } n, \quad \frac{d_Y(f(x_n), f(y_n))}{d_X(x_n, y_n)} > n. \qquad (*)$$

The following arguments are in order:

i) In any metric space (M, d), the distance function $d: M \times M \to \mathbb{R}$ given by $(x, y) \mapsto d(x, y)$ is continuous (miscellaneous result 5.5)
ii) As X is compact, so is $X \times X$. (Theorem 6.27)
iii) Since X is compact, and f is continuous, $f(X)$ is also compact (Theorem 6.32).
iv) Thus $f(X) \times f(X)$ is also compact.
v) Both the sets

$$\{d_Y(f(x_n), f(y_n)) : n \in \mathbb{N}\} \quad \text{and} \quad \{d_X(x_n, y_n) : n \in \mathbb{N}\}$$

are bounded (Corollary 6.32.2).

From $(*)$, we conclude that $\frac{d_Y(f(x_n), f(y_n))}{d_X(x_n, y_n)}$ diverges to ∞, and from the above arguments, the only possibility that guarantees such is $d_X(x_n, y_n) \to 0$.

Now, as X is compact, it is sequentially compact (Theorem 6.23). Hence $\{x_n\}$ admits a convergent subsequence, say $\{x_{n_k}\}$. Then $d_X(x_n, y_n) \to 0$ implies for the same subsequence $\{y_{n_k}\}$ of $\{y_n\}$,

$$\lim y_{n_k} = \lim x_{n_k} = \xi, \text{ say.}$$

Therefore, f is locally Lipschitz at ξ. Thus we have a $\delta > 0$ and a constant $L_\xi > 0$ such that for any two points $x, y \in B(\xi; \delta)$,

$$d_Y(f(x), f(y)) \leq L_\xi \, d_X(x, y). \tag{$**$}$$

But $x_{n_k} \to \xi$ and $y_{n_k} \to \xi$ imply that there exists some $k_1 \in \mathbb{N}$ such that for any index $k \geq k_1$, the points $x_{n_k}, y_{n_k} \in B(\xi; \delta)$. Choose $k_0 > k_1$ big enough so that $n_{k_0} > L_\xi$. Write $x_0 = x_{n_{k_0}}$ and $y_0 = y_{n_{k_0}}$. Hence $x_0, y_0 \in B(\xi; \delta)$. However, from $(*)$, we get

$$\frac{d_Y(f(x_0), f(y_0))}{d_X(x_0, y_0)} > n_{k_0} > L_\xi, \quad \text{i.e.} \quad d_Y(f(x_0), f(y_0)) > L_\xi d_X(x_0, y_0),$$

a contradiction to $(**)$. This completes the proof. □

6.4. Let X be a compact metric space and $f: X \to X$ be continuous. Show that there exists a non-empty subset $A \subset X$ such that $f(A) = A$.

Proof. Call $A_1 = f(X)$ and construct $A_{n+1} = f(A_n)$ inductively. Since f is continuous on the compact space X, $f(X)$ is also compact and hence a closed subset of X. Again, as $A_1 = f(X)$ is compact, $A_2 = f(A_1)$ is also compact hence closed and so on. Also $A_1 = f(X) \subset X$, $A_2 = f(A_1) = f(f(X)) \subset f(X) = A_1$, and so on. Thus the sequence $\{A_n\}$ is such that:

i) Each A_n is non-empty and closed.
ii) For every n, $A_{n+1} \subset A_n$.

Therefore for every n, $\bigcap_{k=1}^{n} A_k = A_n \neq \emptyset$, i.e. $\{A_n\}$ satisfies finite intersection property. Hence by Theorem 6.7, $\bigcap_n A_n$ is also non-empty. Call $A = \bigcap_{n\in\mathbb{N}} A_n$. We will show $f(A) = A$.
Let $y \in f(A)$ be arbitrary. Therefore, for some point $x \in A$, $f(x) = y$. But $x \in A$ implies that for each n, $x \in A_n$. Thus $y = f(x) \in A_{n+1}$ holds for every n. Therefore, $y \in A$ as well, i.e. $f(A) \subset A$.
Conversely, choose $y \in A$ arbitrarily. Then for each $n \in \mathbb{N}$, $y \in A_{n+1}$. Therefore, there exist points $x_n \in A_n$ so that $f(x_n) = y$. Consider the sequence $\{x_n\}$ in the compact space X. As X is also sequentially compact (Theorem 6.23), $\{x_n\}$ will admit a subsequence $\{x_{n_k}\}$ converging to a point x, say. Now, for each index k, the tail

$$\{x_{n_p} : p \geq k\}$$

of the sequence $\{x_{n_k}\}$ is a subsequence converging to x in the closed set A_{n_k}. Consequently, $x \in A_{n_k}$ for each k. However, $A_{n_k} \subset A_k$ for every k. Thus $x \in A_k$ for each k, i.e. $x \in A$.
Now, as f is continuous,

$$f(x) = f\left(\lim_{k\to\infty} x_{n_k}\right) = \lim_{k\to\infty} f(x_{n_k}) = y.$$

Thus $y \in f(A)$, i.e. $A \subset f(A)$. This completes the proof. □

6.5. Let X be a compact metric space and $f: X \to X$ be such that for any pair of distinct points $x, y \in X$, $d(f(x), f(y)) < d(x, y)$. Show that f has a fixed point. Is this fixed point unique?
Proof. For $x \in X$, consider the function $\varphi: X \to \mathbb{R}$ given by $x \mapsto d(x, f(x))$. A fixed point of f is given by the zero of φ and vice versa, i.e. $\xi \in X$ will be a fixed point of f iff $\varphi(\xi) = 0$.
Now, for $x, x' \in X$,

$$\begin{aligned} & d(x, f(x)) \leq d(x, x') + d(x', f(x')) + d(f(x'), f(x)) \\ \Longrightarrow\ & d(x, f(x)) - d(x', f(x')) \leq d(x, x') + d(f(x), f(x')) < 2d(x, x') \\ \Longrightarrow\ & \varphi(x) - \varphi(x') < 2d(x, x'). \end{aligned}$$

Interchanging x and x', we get $|\varphi(x) - \varphi(x')| < 2d(x, x')$, showing that φ is uniformly continuous, and hence, continuous. Since X is compact, φ is bounded and attains its bounds. Let $\varphi(\xi) = \inf\{\varphi(x) : x \in X\}$. Clearly, $\varphi(\xi) \geq 0$. Assume, if possible, that $\varphi(\xi) > 0$, i.e. ξ and $f(\xi)$ are distinct points in X. Then by the hypothesis, we obtain

$$\varphi(\xi) = d(\xi, f(\xi)) > d(f(\xi), f(f(\xi))) = \varphi(f(\xi)),$$

contradicting that $\varphi(\xi)$ is the smallest of all the φ values. Therefore $\varphi(\xi) = 0$, and hence $f(\xi) = \xi$.

This fixed point ξ is unique. Because otherwise, if for $\eta \in X$, $f(\eta) = \eta$ and $\xi \neq \eta$, then

$$d(\xi, \eta) = d(f(\xi), f(\eta)) < d(\xi, \eta),$$

a contradiction. □

Remark Continuity of φ can also be proved in the following way:

For any two points $x, y \in X$, $d(f(x), f(y)) \leq d(x, y)$ (equality occurs only for $x = y$) implies that f is uniformly continuous, hence continuous. Thus the map $x \mapsto (x, f(x))$ is continuous (miscellaneous result 5.2). Again as d is continuous (miscellaneous result 5.5), so is φ.

6.6. Show that a metric space X is compact if and only if every real-valued continuous function on X is bounded.

Proof. *Necessary part:* Let X be compact. As continuous image of a compact set is compact, for any continuous $f: X \to \mathbb{R}$, $f(X)$ will be compact in $\mathbb{R}$. Hence, by Heine-Borel theorem, $f(X)$ is bounded.

Sufficient part: Let every continuous real-valued function on (X, d) be bounded. Assume if possible, X is not compact. We will produce a counter-example of a continuous unbounded function $f: X \to \mathbb{R}$.

As X is not compact, it is not sequentially compact. WLOG, let $\{x_n\} \subset X$ be a sequence of *distinct* terms having no convergent subsequence.[6] For each $k \in \mathbb{N}$, construct the set

$$A_k = \left\{\{x_n\}_{n=1}^{\infty} : n \neq k\right\}$$

by removing the kth term from the sequence $\{x_n\}$. Each term x_k stays at a positive distance apart from A_k, i.e. there exists $\delta_k > 0$ such that $B(x_k; \delta_k) \cap A_k = \varnothing$. Because otherwise, if for every $\delta > 0$, $B(x_k; \delta) \cap A \neq \varnothing$, then x_k becomes a limit point of A_k (as $x_k \notin A_k$). Hence a sequence from A_k, a.k.a. a subsequence of $\{x_n\}$ (possibly rearranged) will converge to x_k.

Observe that for any $m \neq n$, $d(x_m, x_n) > \max\{\delta_m, \delta_n\}$ (see Fig. 6.15). Now, for each $k \in \mathbb{N}$, call $r_k = \frac{1}{3}\delta_k$. We refine our observation by claiming that for

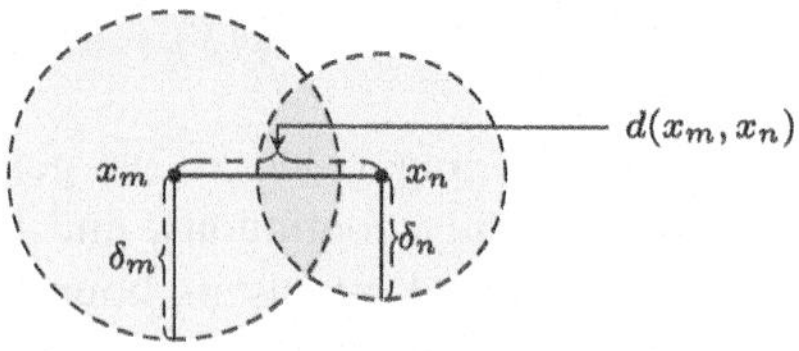

Fig. 6.15 $d(x_m, x_n) > \max\{\delta_m, \delta_n\}$

[6] As X is not sequentially compact, it admits a sequence $\{p_n\}$ with no convergent subsequence. Therefore, no term of $\{p_n\}$ can appear infinitely often. Thus we can always extract a subsequence with no repeating term. We work with this subsequence, and for brevity of notations, denote by $\{x_n\}$.

any $m \neq n$,

$$\overline{B(x_m; r_m)} \cap \overline{B(x_n; r_n)} = \varnothing.$$

To prove the claim, fix $m \neq n$ arbitrarily. Let some point $x \in \overline{B(x_m; r_m)} \cap \overline{B(x_n; r_n)}$. Then for any pre-fixed $\epsilon > 0$, we will find points $x' \in B(x_m; r_m)$ and $x'' \in B(x_n; r_n)$ with $d(x, x') < \epsilon$ and $d(x, x'') < \epsilon$ (miscellaneous result 3.3). Then

$$\begin{aligned} d(x_m, x_n) &\leq d(x_m, x') + d(x', x) + d(x, x'') + d(x'', x_n) \\ &< r_m + 2\epsilon + r_n \\ &\leq 2\max\{r_m, r_n\} + 2\epsilon = \frac{2}{3}\max\{\delta_m, \delta_n\} + 2\epsilon. \end{aligned} \tag{6.1}$$

Choosing ϵ small enough so that $2\epsilon < \frac{1}{3}\max\{\delta_m, \delta_n\}$, we obtain

$$d(x_m, x_n) < \max\{\delta_m, \delta_n\},$$

a contradiction to our primary observation. Thus $\overline{B(x_m; r_m)} \cap \overline{B(x_n; r_n)} = \varnothing$. This proves the claim.

For $n \in \mathbb{N}$, denote by $B_n = \overline{B(x_n; r_n)}$. Thus each B_n is closed and disjoint from each other. Therefore it makes sense to define $f : X \to \mathbb{R}$ by

$$f(x) = \begin{cases} n\left(1 - \frac{d(x, x_n)}{r_n}\right), & \text{if } x \in B_n \text{ for some } n \\ 0, & \text{otherwise} \end{cases}.$$

We will show that f is continuous and unbounded on X. Figure 6.16 contains an illustration of f with the choices $X = \mathbb{R}$, $x_n = n$, and $r_n = \frac{1}{3}$.

To show f is unbounded is easy; as for each n, $f(x_n) = n$. We will use the sequential criterion (Theorem 5.4.(*ii*)) to establish the continuity of f on X. This part will be a little tricky, as we need to consider a few cases and subcases. Fix $x \in X$ arbitrarily and let $\{a_k\} \subset X$ converge to x. We will show $\{f(a_k)\}$ converges to $f(x)$. Let us agree to write $B = \bigcup_n B_n$.

Case 1: $x \in B_n$ for some n. As each B_n is closed and disjoint from each other, $\{a_k\}$ cannot have a subsequence in another B_m. Keeping this observation in mind, the following subcases arise:

Subcase (i): $\{a_k\}$ has no subsequence from outside B_n, i.e. for large enough values of k, $a_k \in B_n$. As $y \mapsto d(y, x_n)$ is a continuous function (Example 5.4.5), $d(a_k, x_n) \xrightarrow{k} d(x, x_n)$. Therefore,

$$\lim_{k\to\infty} f(a_k) = \lim_{k\to\infty} n\left(1 - \frac{d(a_k, x_n)}{r_n}\right) = n\left(1 - \frac{d(x, x_n)}{r_n}\right) = f(x),$$

i.e. f is continuous at x.

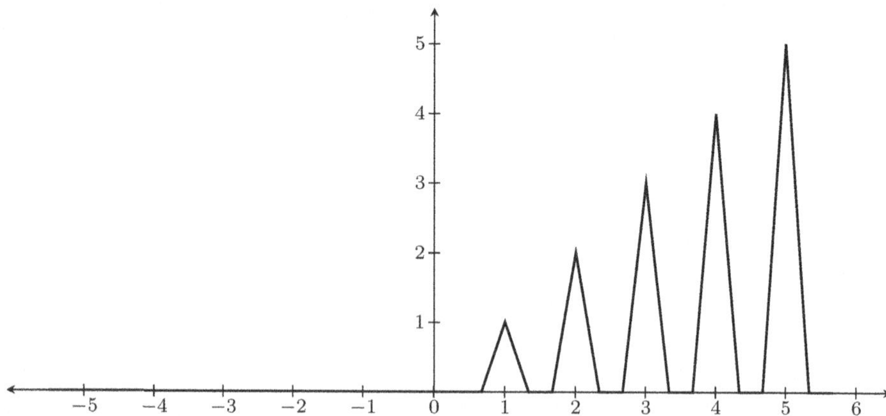

Fig. 6.16 Illustration of $f:\mathbb{R} \to \mathbb{R}$

Subcase (ii): $\{a_k\}$ has a subsequence outside B_n. WLOG, we may assume that $\{a_k\}$ has no subsequence in B_n, as that case is already dealt above. Therefore, $a_k \in X \setminus B$ for large enough values of k. Moreover, $a_k \to x$ and $x \in B_n$. Thus x is a limit point of $X \setminus B$, and consequently, a limit point of $X \setminus B_n$. Therefore

$$x \in \overline{B_n} \cap \overline{X \setminus B_n} = \partial B_n.$$

(See Definition 3.30 and Exercise 3.30(a).) However, using Exercise 3.13, this fact concludes that $d(x, x_n) = r_n$. In that case, $f(a_k) = 0$ for large enough values of k, and $f(x) = 0$ as well. Thus $f(a_k) \to f(x)$, showing f is continuous at x.

Case 2: x is an element of no B_n, i.e. $x \in X \setminus B$. Then $f(x) = 0$. We establish the following claim first:

Claim: $X \setminus B$ is an open set, i.e. there is some $\delta > 0$ so that $B(x; \delta)$ does not intersect any B_n.

Proof of Claim Assume otherwise; i.e. for each $\delta > 0$, the ball $B(x; \delta)$ intersects some B_n. For every choice of $\delta = \frac{1}{i}$, $i \in \mathbb{N}$, choose $y_i \in B(x; \frac{1}{i}) \cap B_{n_i}$. Then $y_i \to x$. WLOG, we may assume that B_{n_i}'s are all distinct. Because, one particular B_{n_i} containing infinitely many y_i's will have a limit point outside of it. (Recall $x \notin B_{n_i}$). As each B_{n_i} is closed, this leads to a contradiction.

Now, as $\{y_i\}$ is convergent, it is Cauchy. Fix $\epsilon > 0$ arbitrarily. Then for some large enough i and j, $d(y_i, y_j) < \epsilon$. Let x_i and x_j be the centres of the balls B_{n_i} and B_{n_j}. Using similar arguments as shown in Eq. (6.1), we then arrive at same contradiction about $d(x_i, x_j)$ for some choice of ϵ. This proves the claim.

Now, as $a_k \to x$, for some $k_0 \in \mathbb{N}$, $a_k \in B(x;\delta) \subset X \setminus B$ holds for each $k \geq k_0$. Thus, for each $k \geq k_0$, $f(a_k) = 0$. Therefore, $f(a_k) \to 0 = f(x)$. Consequently, f is continuous at x.
With all cases combined, we can say that f is continuous at each point $x \in X$. This concludes the proof. □

Remark If X happens to be a subset of $\mathbb{R}^n$, compactness of X can be proved much easily:
The fact that X is bounded can be easily seen by considering the projection maps (Example 5.4.4) in each coordinate. Since these maps are continuous, boundedness of each of them will guarantee the boundedness of X. To see that X is closed in $\mathbb{R}^n$, consider a limit point $\mathbf{a}$ of X. (If X does not have a limit point, we are already through.) If $\mathbf{a} \notin X$, then the function $\mathbf{x} \mapsto \frac{1}{\|\mathbf{x}-\mathbf{a}\|}$ is well-defined, and by Example 5.4.5, continuous as well. However, as $\mathbf{a}$ is a limit point of X, $\inf\{\|\mathbf{x}-\mathbf{a}\| : \mathbf{x} \in K\} = 0$, showing that this continuous function $\mathbf{x} \mapsto \frac{1}{\|\mathbf{x}-\mathbf{a}\|}$ is unbounded. Hence $\mathbf{a} \in X$, i.e. X is closed. Consequently, by Heine-Borel theorem, X is compact.

6.7. Let $\{x_n\}$ be a sequence in a compact metric space X. If every convergent subsequence of $\{x_n\}$ has the same limit x, show that $\{x_n\}$ also converges to x. Give an example to show that this may not happen in a non-compact metric space.
Proof. Assume the contrary. Then $\{x_n\}$ has some subsequence $\{x_{n_k}\}$ which does not converge to x, i.e. for some $\epsilon_0 > 0$, the inequality $d(x_{n_k}, x) \geq \epsilon_0$ holds for each index $k \in \mathbb{N}$. Now, as X is compact, it is sequentially compact. Thus the sequence $\{x_{n_k}\}$ must admit a convergent subsequence, say $\{x_p\}_p$. However, $\{x_p\}$ is a subsequence of $\{x_n\}$ as well. Therefore by the hypothesis, $x_p \to x$. On the other hand, for each p, the constraint $d(x_p, x) \geq \epsilon_0$ poses a contradiction to this result. Therefore $x_n \to x$.
Consider the sequence $\left\{n^{(-1)^n}\right\}$ in the non-compact space $[0, \infty)$. The only convergent subsequence $\left\{\frac{1}{2n-1}\right\}$ converges to 0, but definitely $\{x_n\}$ is not convergent. □

6.8. Let X, Y be metric spaces and Y be compact. For a function $f: X \to Y$, the graph of f is the subset

$$\Gamma_f = \{(x, f(x)) : x \in X\} \subset X \times Y.$$

Show that f is continuous if and only if Γ_f is a closed subset of $X \times Y$.
Proof. Let f be continuous on X and $(x, y) \in X \times Y$ be a limit point of Γ_f. We want to show $(x, y) \in \Gamma_f$, i.e. $y = f(x)$.
As (x, y) is a limit point of Γ_f, we find a sequence $\{(x_n, f(x_n))\}$ from Γ_f converging to (x, y). Then $x_n \to x$ and $f(x_n) \to y$ individually (Example 3.39.6). However, as f is continuous at x, by sequential criterion,

$$y = \lim f(x_n) = f(x).$$

Conversely, assume Γ_f is closed. Choose $x \in X$ arbitrarily and consider a sequence $\{x_n\} \subset X$ converging to x. We will show $\{f(x_n)\}$ converges to $f(x)$ in Y.

As Y is (sequentially) compact, $\{f(x_n)\}$ admits a convergent subsequence. Let $\{f(x_p)\}$ be *any* convergent subsequence of $\{f(x_n)\}$. Now, the subsequence $\{x_p\}$ of $\{x_n\}$ converges to x. Therefore, the sequence $\{(x_p, f(x_p))\}$ from Γ_f converges, and as Γ_f is closed, converges to a point $(a, b) \in \Gamma_f$. Therefore, $x_p \to a$ and $f(x_p) \to b$. Thus, $a = x$ and $b = f(x)$.

The above arguments show that every convergent subsequence of $\{f(x_n)\}$ converges to $f(x)$. Therefore by the previous example (miscellaneous result 6.7), $f(x_n) \to f(x)$. This completes the proof. □

Remark Observe that we did not use the compactness of Y to prove Γ_f is closed. Therefore, the graph of a continuous function $f: X \to Y$ is always closed.

Exercises: 6.3

6.17. If K is a compact subset of $\mathbb{R}$, show that both $\inf K$ and $\sup K$ belong to K.

6.18. Let X be a compact metric space and $f: X \to (0, \infty)$ be continuous on X. Show that $f(X) \subset [a, \infty)$ for some $a > 0$.

6.19. Show that no bijection from $[0, 1]$ to $\mathbb{R}$ can be continuous.

6.20. Let f be a continuous mapping of a compact metric space X into itself such that $f(X)$ is everywhere dense in X. Prove that $f(X) = X$.

6.21. If E is a non-empty compact subset of a metric space X, then show that there exist points x and y in E such that $\operatorname{diam} E = d(x, y)$.

6.22. Let $\{x_n\}$ be a sequence converging to x. Prove that the union $\{x\} \cup \{x_n : n \in \mathbb{N}\}$ is a compact in the underlying space. Use this to show that if f is a mapping of a metric space X into a metric space Y and if $f|_A$ is continuous on A for all compact subsets A of X, then f is continuous on X.

6.23. Let f be a continuous mapping of a compact metric space X into a metric space. Prove that for all subsets A of X, $f(\overline{A}) = \overline{f(A)}$.

6.24. A metric space X is called *locally compact* if each point x of X has a neighbourhood $B(x; r)$ with compact closure. Let X be locally compact and $f: X \to Y$ be a continuous, open map.[7] Show that $f(X)$ is also locally compact.

[7] A map $f: X \to Y$ is called an *open map* if f maps every open set of X to an open set in Y.

6.4 Compact Sets in Particular Spaces

In this final section of this chapter, we furnish examples and conditions of compact sets in the sequence spaces ℓ_p (as introduced in Example 2.2.6) and in the function space $\mathcal{C}[a, b]$ with the uniform metric (introduced in Example 2.2.9). These two spaces carry their fair shares of importance in the literature.

6.4.1 Compact Sets in Sequence Spaces (ℓ_p Spaces)

The ℓ_p space, as a whole, is not compact. Although it is complete (Example 4.5.5), it fails to be bounded. For any $M > 0$, however big, the element

$$\{2M, 0, 0, \dots\}$$

of ℓ_p stays at a $2M$ ($> M$) distance from the origin $\mathbf{0} = \{0, 0, \dots\}$. The following theorem provides an NASC for a subset of ℓ_p to be compact.

Theorem 6.35 *A non-empty subset $X \subset \ell_p$ is compact if and only if:*

(i) X is bounded.
(ii) For any $\epsilon > 0$ arbitrary, there exists a positive integer N such that, for every element $\mathbf{x} = \{x_k\} \in X$,

$$\left[\sum_{k=N}^{\infty} |x_k|^p\right]^{1/p} < \epsilon,$$

i.e. contributions from the tails of the members of X vanish uniformly.

Proof *Necessary Part*: Let $X \subset \ell_p$ be compact:

(i) X is compact implies it is totally bounded and hence bounded.
(ii) As X is totally bounded, for any $\epsilon > 0$ chosen arbitrarily, we get a finite $\frac{\epsilon}{2}$-net, say A, where each member $\mathbf{a} \in A$ denotes the element

$$\mathbf{a} = \{a_k\}_k = \{a_1, a_2, a_3, \dots\}$$

in ℓ_p. However, $\mathbf{a} \in \ell_p$ implies that $\sum_k |a_k|^p < \infty$. Therefore for every $\mathbf{a} \in A$, we obtain positive integers $n_{\mathbf{a}}$ such that $\left[\sum_{k=n_{\mathbf{a}}}^{\infty} |a_k|^p\right]^{1/p} < \frac{\epsilon}{2}$. Call $N = \max\{n_{\mathbf{a}} : \mathbf{a} \in A\}$. Thus for every $\mathbf{a} \in A$,

$$\left[\sum_{k=N}^{\infty} |a_k|^p\right]^{1/p} < \frac{\epsilon}{2}. \tag{6.2}$$

This is our required N. We now proceed to prove that this choice of N is such that (6.2) is satisfied by every member $\mathbf{x}$ from X.
Choose $\mathbf{x} = \{x_k\} \in X$ arbitrarily. As A is an $\frac{\epsilon}{2}$-net of X, we can find some $\mathbf{a}$ from A with $d_p(\mathbf{x}, \mathbf{a}) < \frac{\epsilon}{2}$. Hence

$$\left[\sum_{k=1}^{\infty}|x_k - a_k|^p\right]^{1/p} < \frac{\epsilon}{2}, \quad \text{and this implies} \quad \left[\sum_{k=N}^{\infty}|x_k - a_k|^p\right]^{1/p} < \frac{\epsilon}{2}.$$

Therefore, by Minkowski's inequality (item 4),

$$\begin{aligned}
\left[\sum_{k=N}^{\infty}|x_k|^p\right]^{1/p} &= \left[\lim_{B\to\infty}\sum_{k=N}^{B}|x_k|^p\right]^{1/p} \\
&= \lim_{B\to\infty}\left[\sum_{k=N}^{B}|x_k|^p\right]^{1/p} \\
&\qquad\qquad \text{[as } x \mapsto x^{1/p} \text{ is a continuous function]} \\
&\le \lim_{B\to\infty}\left[\left[\sum_{k=N}^{B}|x_k - a_k|^p\right]^{1/p} + \left[\sum_{k=N}^{B}|a_k|^p\right]^{1/p}\right] \\
&= \lim_{B\to\infty}\left[\sum_{k=N}^{B}|x_k - a_k|^p\right]^{1/p} + \lim_{B\to\infty}\left[\sum_{k=N}^{B}|a_k|^p\right]^{1/p} \\
&\qquad\qquad \text{[as both limits exist seperately]} \\
&= \left[\sum_{k=N}^{\infty}|x_k - a_k|^p\right]^{1/p} + \left[\sum_{k=N}^{\infty}|a_k|^p\right]^{1/p} \\
&< \frac{\epsilon}{2} + \frac{\epsilon}{2} = \epsilon. \qquad \text{[using (6.2)]}
\end{aligned}$$

This proves the necessary part of the theorem.

Sufficient Part Let $X \subset \ell_p$ satisfy the hypotheses *(i)* and *(ii)*. We will prove X is totally bounded and complete.

First we show X is totally bounded. Choose $\epsilon > 0$ arbitrarily. Then from *(ii)*, we obtain $N \in \mathbb{N}$ such that

$$\text{for every } \mathbf{x} = \{x_k\} \in X, \quad \sum_{k=N}^{\infty}|x_k|^p < \frac{\epsilon^p}{2}. \tag{6.3}$$

Associate with every

$$\mathbf{x} = \{x_1, x_2, \ldots, x_k, \ldots\} \in X,$$

the truncated element

$$\widetilde{\mathbf{x}} = (x_1, \ldots, x_N) \in \mathbb{R}^N.$$

Call the collection $\widetilde{X}_\epsilon = \{\widetilde{\mathbf{x}} : \mathbf{x} \in X\} \subset \mathbb{R}^N$ thus generated. As X is bounded, $\widetilde{X}_\epsilon$ is also so in the metric space $(\mathbb{R}^N, d_p)$. Therefore, by Exercise 6.25, $\widetilde{X}_\epsilon$ is totally bounded.

Let $\widetilde{A}$ be a finite $\frac{\epsilon}{2}$-net of $\widetilde{X}_\epsilon$, where an element $\mathbf{a}$ of $\widetilde{A}$ is of the form $\widetilde{\mathbf{a}} = (a_1, \ldots, a_N)$. Then for each $\widetilde{\mathbf{x}} \in \widetilde{X}_\epsilon$, we find some $\widetilde{\mathbf{a}}$ such that $d_p(\widetilde{\mathbf{a}}, \widetilde{\mathbf{x}}) < \sqrt{\frac{\epsilon^p}{2}}$, i.e.

$$\sum_{k=1}^{N} |x_k - a_k|^p < \frac{\epsilon^p}{2}. \tag{6.4}$$

Associate each term $\widetilde{\mathbf{a}}$ from $\widetilde{A}$ with $\mathbf{a} = \{a_1, \ldots, a_N, 0, 0, \ldots\}$ in X; and call

$$A = \{\mathbf{a} : \widetilde{\mathbf{a}} \in \widetilde{A}\} \subset X.$$

We will show A is a finite ϵ-net for X.

For any chosen $\mathbf{x} \in X$, consider the corresponding element $\widetilde{\mathbf{x}} \in \widetilde{X}_\epsilon$ and hence find $\widetilde{\mathbf{a}} \in \widetilde{A}$ satisfying (6.4). Therefore for the corresponding $\mathbf{a}$ from A, and for the chosen $\mathbf{x} \in X$, we have

$$\sum_{k=1}^{\infty} |x_k - a_k|^p = \sum_{k=1}^{N} |x_k - a_k|^p + \sum_{k=N+1}^{\infty} |x_k - a_k|^p < \epsilon^p, \qquad \text{by (6.3) and (6.4).}$$

Thus $d_p(\mathbf{x}, \mathbf{a}) < \epsilon$. Consequently, A is a finite ϵ-net of X, i.e. X is totally bounded.

Next we will show X is complete. Let $\{\mathbf{x}^{(n)}\}_n$ be any Cauchy sequence in X. Since ℓ_p is complete, (Example 4.5.5), $\{\mathbf{x}^{(n)}\}$ admits a limit $\mathbf{x}$ in ℓ_p. We only need to show $\mathbf{x} \in X$.

Let $\mathbf{x}^{(n)} = \{x_1^{(n)}, x_2^{(n)}, \ldots\}$ and $\mathbf{x} = \{x_1, x_2, \ldots\}$. Now, $\mathbf{x}^{(n)} \to \mathbf{x}$ is defined by $d_p(\mathbf{x}^{(n)}, \mathbf{x}) \to 0$, and this implies that for any chosen $\epsilon > 0$, we get $n_0 \in \mathbb{N}$ such that

$$\text{for each index } n \geq n_0, \quad \sum_{k=1}^{\infty} |x_k^{(n)} - x_k|^p < \epsilon^p.$$

In particular,

$$\sum_{k=1}^{\infty} |x_k^{(n_0)} - x_k|^p < \epsilon^p. \tag{6.5}$$

Hence by Minkowski's inequality, and the corresponding manipulation exhibited in the proof of necessary part of this theorem,

$$\begin{aligned} \left[\sum_{k=N}^{\infty} |x_k|^p\right]^{1/p} &\le \left[\sum_{k=N}^{\infty} |x_k - x_k^{(n_0)}|^p\right]^{1/p} + \left[\sum_{k=N}^{\infty} |x_k^{(n_0)}|^p\right]^{1/p} \\ &< \epsilon + \epsilon \qquad \text{[by (6.5) and hypothesis (ii) , as } \mathbf{x}^{(n_0)} \in X] \\ &= \epsilon. \end{aligned}$$

Thus $\mathbf{x} \in X$ and X is complete. Consequently, X is compact. ■

Remark The same result holds for compact subsets of the metric space ℓ_∞ (Example 2.2.7) as well. The proof is similar and is left for the reader as an exercise.

Examples

6.35.1. The *Hilbert cube* is the subset of ℓ_2 containing the elements $\mathbf{x} = \{x_k\}_k$ such that $|x_k| \le \frac{1}{k}$. We usually denote it by I^ω or I^∞. I^ω is a compact subset of ℓ_2.

Proof. We will apply the last theorem. Recall the fact that the infinite series $\sum_k \frac{1}{k^2}$ converges to $\frac{\pi^2}{6}$.

For any element $\mathbf{x}$ from I^ω, the distance of $\mathbf{x}$ from the origin $\mathbf{0} = \{0, 0, \dots\}$ can be estimated as

$$\|\mathbf{x} - \mathbf{0}\| = \sqrt{\sum_k x_k^2} \le \sqrt{\sum_k \frac{1}{k^2}} = \frac{\pi}{\sqrt{6}}.$$

Thus X is bounded. Also, as the series $\sum_k \frac{1}{k^2}$ is convergent, for any $\epsilon > 0$ chosen arbitrarily, there exists $N \in \mathbb{N}$ such that $\sum_{k\ge N} \frac{1}{k^2} < \epsilon^2$. Therefore, for any element $\mathbf{x} \in I^\omega$,

$$\sqrt{\sum_{k\ge N} |x_k|^2} \le \sqrt{\sum_{k\ge N} \frac{1}{k^2}} < \epsilon.$$

Consequently, by the last theorem, I^ω is a compact subset of ℓ_2. □

6.4.2 Compact Sets in Function Spaces ($\mathcal{C}(X)$ Space)

Like the ℓ_p spaces, the space $\mathcal{C}_{\text{sup}}[a, b]$ (see Example 2.2.9) of continuous functions defined over the compact interval domain $[a, b]$ is also not compact. In this subsection, we will study conditions for a subset of $\mathcal{C}_{\text{sup}}[a, b]$ to be compact. An NASC in this regard is given by the Arzelà-Ascoli theorem. The framework for this theorem generally replaces $[a, b]$ with an arbitrary compact metric space.

Definition 6.36 Let (X, d_X) and (Y, d_Y) be metric spaces and x be a point in X. A family $\mathcal{A}$ of functions from X to Y is said to be *equicontinuous* at x if for any $\epsilon > 0$ chosen arbitrarily, we find a $\delta > 0$ such that for each function $f \in \mathcal{A}$ and for points $x' \in X$,

$$d_X(x, x') < \delta \text{ implies } d_Y(f(x), f(x')) < \epsilon.$$

We say $\mathcal{A}$ is *equicontinuous* on X if $\mathcal{A}$ is equicontinuous at every point x of X.

Here δ depends on both ϵ and x. However, if we get a $\delta > 0$ which works for every point $x \in X$, we get:

Definition 6.37 Let (X, d_X) and (Y, d_Y) be metric spaces. A family $\mathcal{A}$ of functions from X to Y is said to be *uniformly equicontinuous* if for any $\epsilon > 0$ chosen arbitrarily, we find a $\delta > 0$ such that for each function $f \in \mathcal{A}$ and for each pair of points x, x' from X,

$$d_X(x, x') < \delta \text{ implies } d_Y(f(x), f(x')) < \epsilon.$$

Remark It is evident from the above definition that every member of a uniformly equicontinuous family is also uniformly continuous.

Proposition 6.38 *Let X be a compact metric space and $\mathcal{A}$ be a family of equicontinuous functions from X to another metric space Y. Then $\mathcal{A}$ is a uniformly equicontinuous family.*

The proof is left for the reader as an exercise.

Remark For our discussion in this subsection, the underlying metric space X will be compact. Thus, supported by the above proposition, we will use the term "equicontinuous family" to denote a uniformly equicontinuous family of functions.

Examples

6.38.1. Each finite subset of $\mathcal{C}[0, 1]$ is equicontinuous on $[0, 1]$.

Proof. Let $\mathcal{A}$ be a finite collection of continuous functions on $[0, 1]$. Fix $x \in [0, 1]$ and $\epsilon > 0$ arbitrarily. Then for each function $f \in \mathcal{A}$, we find $\delta_f > 0$ so that for a point $x' \in [0, 1]$ with $|x - x'| < \delta_f$, one will have $|f(x) - f(x')| < \epsilon$. Choose $\delta = \min\{\delta_f : f \in \mathcal{A}\} > 0$. Thus for each function $f \in \mathcal{A}$ and for points $x' \in [0, 1]$ with $|x - x'| < \delta$, we will have $|f(x) - f(x')| < \epsilon$. Consequently, $\mathcal{A}$ is equicontinuous. □

6.38.2. Let $B[\mathbf{0}; 1]$ be the closed unit ball in $\mathcal{C}[0, 1]$, i.e.

$$B[\mathbf{0}; 1] = \left\{ f \in \mathcal{C}[0, 1] : \sup_{0 \le x \le 1} |f(x)| \le 1 \right\}.$$

For $f \in B[\mathbf{0}; 1]$, define

$$g(x) = \int_0^x f(t)\,\mathrm{d}t, \quad 0 \le x \le 1$$

and call the set of all g's as $\mathcal{B}$, i.e.

$$\mathcal{B} = \left\{ g(x) = \int_0^x f(t)\,\mathrm{d}t : f \in \mathcal{C}[0, 1] \right\}.$$

Then $\mathcal{B}$ is an equicontinuous family of functions from $\mathcal{C}[0, 1]$.

Proof. For each function $g \in \mathcal{B}$ and for each pair of points $x, x' \in [0, 1]$, we have

$$\begin{aligned} |g(x) - g(x')| &= \left| \int_0^x f(t)\,\mathrm{d}t - \int_0^{x'} f(t)\,\mathrm{d}t \right| \\ &= \left| \int_x^{x'} f(t)\,\mathrm{d}t \right| \\ &\le \sup_{0 \le t \le 1} |f(t)| \times \left| \int_{x'}^x \mathrm{d}t \right| \\ &\le |x - x'|. \qquad \text{[as f} \in B[\mathbf{0}; 1], \sup f \le 1] \end{aligned}$$

Therefore, for any $\epsilon > 0$, $|x - x'| < \epsilon$ implies $|g(x) - g(x')| < \epsilon$. Thus $\mathcal{B}$ is equicontinuous. □

6.38.3. Let X be a compact metric space and $F \colon X \times X \to Z$ be continuous. For any fixed $y_0 \in X$, let $f_{y_0}(x) := F(x, y_0)$. Call $\mathcal{A} := \{f_y : y \in X\}$. Then $\mathcal{A}$ is an equicontinuous family.

Proof. Since X is compact, $X \times X$ is also so (Theorem 6.27) and hence F is uniformly continuous. Therefore for any chosen $\epsilon > 0$, we find a $\delta > 0$ such that for any pair of points (x, y) and (x', y') from $X \times X$,

$$d_{X \times X}((x, y), (x', y')) < \delta \text{ implies } d_Z(F(x, y), F(x', y')) < \epsilon.$$

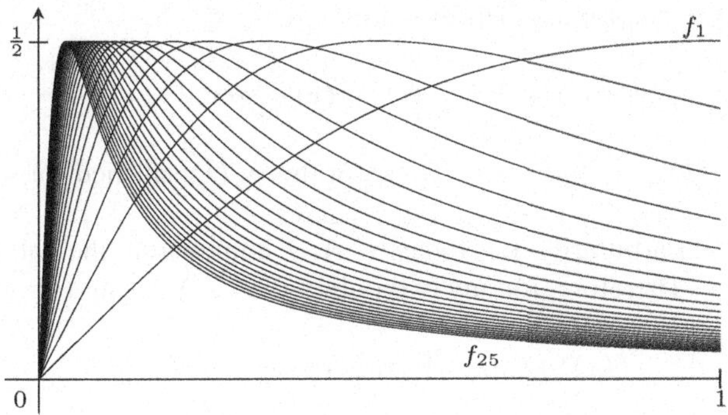

Fig. 6.17 $f_n(x) = \frac{nx}{1+n^2x^2}$

Choose $y' = y$. Therefore, for the chosen $\epsilon > 0$ and for any $y \in X$, $d_X(x, x') < \delta$ implies

$$d_Z(F(x, y), F(x', y)) < \epsilon$$
$$\text{i.e.} \quad d_Z(f_y(x), f_y(x')) < \epsilon.$$

Consequently, $\mathcal{A}$ is an equicontinuous family. □

6.38.4. For each $n \in \mathbb{N}$, define

$$f_n(x) = \frac{nx}{1+n^2x^2}, \quad x \in [0, 1].$$

Then $\{f_n\}_{n\in\mathbb{N}}$ is *not* equicontinuous (Fig. 6.17).
Proof. Assume the family $\{f_n\}$ is equicontinuous. Then for any chosen $\epsilon > 0$, around the point $x = 0$, we obtain a $\delta > 0$ so that for each point $x' \in [0, \delta)$, and each $n \in \mathbb{N}$, $|f_n(x') - f_n(0)| = |f_n(x')| < \epsilon$. Now, observe that each f_n assumes its maximum value $\frac{1}{2}$ at $x = \frac{1}{n}$, i.e. $f_n(\frac{1}{n}) = \frac{1}{2}$. Choose n large enough so that $\frac{1}{n} < \delta$. However, if we start by choosing $0 < \epsilon < \frac{1}{2}$, this leads to a contradiction. □

We now see the culminating result of this subsection.

Theorem 6.39 (Arzelà-Ascoli Theorem) *Let (X, d) be a compact metric space and $\mathbb{K}$ be $\mathbb{R}$ or $\mathbb{C}$ with usual Euclidean metric. Let $\mathcal{C}(X, \mathbb{K})$ denote the metric space of all continuous functions from X to $\mathbb{K}$ in sup norm metric. A subset $\mathcal{B}$ of $\mathcal{C}(X, \mathbb{K})$ is compact if and only if $\mathcal{B}$ is closed, bounded and equicontinuous.*

Proof *Necessary Part*: Let $\mathcal{B}$ be compact. Then it is closed and totally bounded. Hence it is bounded. Thus we are only left to show that $\mathcal{B}$ is equicontinuous. Fix $\epsilon > 0$ arbitrarily. As $\mathcal{B}$ is totally bounded, we obtain a finite subset A of $\mathcal{B}$ so that $\mathcal{B} \subset \bigcup_{g\in A} B(g; \epsilon)$. Now, as X is compact, every function $f \in \mathcal{C}(X; \mathbb{K})$ is uniformly continuous, hence so is every function $g \in A$. Therefore, for each $g \in A$, we find $\delta_g > 0$ so that for any pair of points $x, x' \in X$,

$$d(x, x') < \delta_g \implies |g(x) - g(x')| < \epsilon. \tag{$*$}$$

Choose $\delta = \min\{\delta_g : g \in A\} > 0$. Thus, for this δ, $(*)$ satisfied for every function $g \in A$.

Choose $f \in \mathcal{B}$ arbitrarily. Let for $g_0 \in A$, f comes from the ball $B(g_0; \epsilon)$, i.e. $\|f - g_0\|_\infty < \epsilon$. Therefore, for any two points $x, x' \in X$, using $(*)$, we obtain

$$|f(x) - f(x')| \leq |f(x) - g_0(x)| + |g_0(x) - g_0(x')| + |g_0(x') - f(x')| < 3\epsilon.$$

As $\epsilon > 0$ is arbitrary, this concludes that $\mathcal{B}$ is equicontinuous on X.

Sufficient Part: Let $\mathcal{B}$ be closed, bounded and equicontinuous. As $\mathcal{C}(X, \mathbb{K})$ is complete w.r.t. the sup metric (Exercise 4.10), the closed subset $\mathcal{B}$ is also complete (Theorem 4.6). Hence we only need to show that $\mathcal{B}$ is totally bounded.
Choose $\epsilon > 0$ arbitrarily. Since $\mathcal{B}$ is bounded, there will exist some $M > 0$ such that each function f from $\mathcal{B}$ will satisfy the bound $|f(x)| < M$ for each point $x \in X$. Again, as $\mathcal{B}$ is an equicontinuous family as well, we will find $\delta > 0$ such that for every member $f \in \mathcal{B}$ and for any pair of points $x, x' \in X$, $d(x, x') < \delta$ will imply $|f(x) - f(x')| < \epsilon$. For referring purpose, let us record the above fact:

$$\forall x, x' \in X, \quad d(x, x') < \delta \implies |f(x) - f(x')| < \epsilon. \tag{$*$}$$

Now, as X is compact, it is totally bounded. Therefore we can find a set of finitely many points A from X such that $X = \bigcup_{a \in A} B_X(a; \delta)$. On the other hand, the boundedness of $\mathcal{B}$ guarantees that the range of all functions from $\mathcal{B}$ is contained in the bounded ball $B_{\mathbb{K}}(0; M)$ in $\mathbb{K}$. But by Proposition 6.25, $B_{\mathbb{K}}(0; M)$ is totally bounded. Thus we can again find a set of finitely many points B from $B_{\mathbb{K}}(0; M)$ such that $B_{\mathbb{K}}(0; M) \subset \bigcup_{b \in B} B_{\mathbb{K}}(b; \epsilon)$. Consider the set

$$\mathcal{A} := \{\alpha \mid \alpha\colon A \to B\}.$$

Essentially, $\mathcal{A}$ can be thought as a set of line segments joining the grid points on the mesh $A \times B$. Observe that the number of elements in $\mathcal{A}$ is n^m, i.e. $\mathcal{A}$ is finite. For each $\alpha \in \mathcal{A}$, collect all the functions f from $\mathcal{B}$ which stay in the close vicinity to α at the grid points. Formally, define

$$U_\alpha = \{f \in \mathcal{B} : |f(a) - \alpha(a)| < \epsilon,\ a \in A\}. \tag{$**$}$$

An illustrative figure is given in Fig. 6.18. We prove the following claim regarding U_α's.

Claim For each $\alpha \in \mathcal{A}$, $\operatorname{diam} U_\alpha < 4\epsilon$.

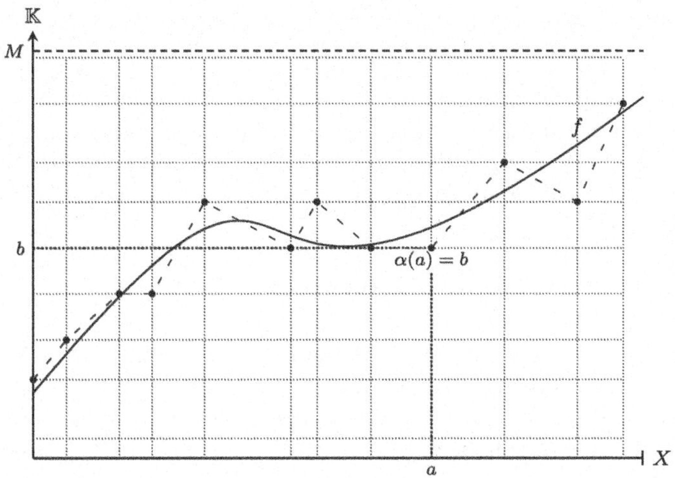

Fig. 6.18 Construction of U_α

Proof of Claim Choose $f, g \in U_\alpha$ arbitrarily. We want to prove

$$d(f, g) = \sup\{|f(x) - g(x)| : x \in X\} < 4\epsilon$$
$$\iff \text{for each point } x \in X, \ |f(x) - g(x)| < 4\epsilon.$$

Choose $x \in X$ arbitrarily. Find a point $a \in A$ so that $x \in B_X(a; \delta)$. Therefore from $(*)$, we conclude that $|f(x) - f(a)| < \epsilon$ and $|g(x) - g(a)| < \epsilon$. Hence using $(**)$, we finally obtain that for the chosen point $x \in X$

$$\begin{aligned} |f(x) - g(x)| &\le |f(x) - f(a)| + |f(a) - \alpha(a)| + |\alpha(a) - g(a)| + |g(a) - g(x)| \\ &< \epsilon + \epsilon + \epsilon + \epsilon = 4\epsilon. \end{aligned}$$

As $x \in X$ is arbitrary, the above concludes $\operatorname{diam} U_\alpha < 4\epsilon$, and thus the claim is proved.

We now show that $\bigcup_{\alpha \in \mathcal{A}} U_\alpha = \mathcal{B}$. Choose $f \in \mathcal{B}$. For each $a \in A$, choose $b_a \in B$ such that $f(a) \in B_{\mathbb{K}}(b_a; \epsilon)$. Let α be that member from $\mathcal{A}$ so that for every a, $\alpha(a) = b_a$. Then $|f(a) - \alpha(a)| < \epsilon$ holds for all $a \in A$, i.e. $f \in U_\alpha$. Consequently, $\bigcup_{\alpha \in \mathcal{A}} U_\alpha = \mathcal{B}$.

As $\mathcal{A}$ is finite and for each $\alpha \in \mathcal{A}$ and $\operatorname{diam} U_\alpha < 4\epsilon$, there exists a finite 4ϵ-net for $\mathcal{B}$. Since $\epsilon > 0$ is arbitrary, $\mathcal{B}$ is totally bounded.

Therefore $\mathcal{B}$ is complete and totally bounded and hence compact. ∎

Remark The statement of the last theorem can be strengthened a little further. The following lemma helps in that regard.

Lemma 6.40 *Let X be a metric space and $\mathcal{B}$ be a non-empty subset of $\mathcal{C}(X,\mathbb{K})$. If $\mathcal{B}$ is bounded and equicontinuous, so is the closure of $\mathcal{B}$ in $\mathcal{C}(X,\mathbb{K})$.*

Proof We prove the boundedness of $\overline{\mathcal{B}}$ first. Since $\mathcal{B}$ is bounded, there will exist some $M > 0$ such that each function f from $\mathcal{B}$ will satisfy the bound $|f(x)| < M$ for each point $x \in X$. Let $g \in \overline{\mathcal{B}}$ be arbitrary. Then we will find a function $f \in \mathcal{B}$ so that $\|f - g\|_\infty < 1$, i.e. for each point $x \in X$, $|f(x) - g(x)| < 1$. Thus, for each point $x \in X$,

$$|g(x)| \leq |g(x) - f(x)| + |f(x)| < 1 + M,$$

i.e. $g \in B(\mathbf{0}; M + 1)$. Since $g \in \overline{\mathcal{B}}$ is arbitrary, $\overline{\mathcal{B}}$ is bounded.

To prove the equicontinuity of $\overline{\mathcal{B}}$, fix $\epsilon > 0$ and $x \in X$ arbitrarily. As $\mathcal{B}$ is equicontinuous, there will exist a $\delta > 0$ so that for every function $f \in \mathcal{B}$ and for points $x' \in B(x;\delta)$, we will have $|f(x) - f(x')| < \frac{\epsilon}{3}$. Let $g \in \overline{\mathcal{B}}$ be arbitrary. Find $f \in \mathcal{B}$ so that for any point $x'' \in X$, $|g(x'') - f(x'')| < \frac{\epsilon}{3}$. Thus, for any point $x' \in B(x;\delta)$, we will have

$$|g(x)-g(x')| \leq |g(x)-f(x)|+|f(x)-f(x')|+|f(x')-g(x')| < \frac{\epsilon}{3}+\frac{\epsilon}{3}+\frac{\epsilon}{3} = \epsilon.$$

As $g \in \overline{\mathcal{B}}$ is arbitrary, $\overline{\mathcal{B}}$ is equicontinuous at x. This completes the proof. ■

In the light of the above lemma and Exercise 6.30, Theorem 6.39 takes the following form:

Theorem 6.41 (Arzelà-Ascoli Theorem) *Let (X, d) be a compact metric space and $\mathbb{K}$ be $\mathbb{R}$ or $\mathbb{C}$ with usual Euclidean metric. Let $\mathcal{C}(X,\mathbb{K})$ denote the metric space of all continuous functions from X to $\mathbb{K}$ in sup norm metric. A subset $\mathcal{B}$ of $\mathcal{C}(X,\mathbb{K})$ has compact closure if and only if $\mathcal{B}$ is pointwise bounded and equicontinuous.*

Miscellaneous Results

6.9. Let X be a compact metric space and $\{f_n\}$ be a sequence of continuous functions from X to $\mathbb{R}$ or $\mathbb{C}$. Let f_n converge to f pointwise on X. Show that if $\{f_n\}$ is an equicontinuous family, then the convergence $f_n \to f$ is uniform.

Proof. We need to show that for every $\epsilon > 0$ fixed arbitrarily, there exists a positive integer k so that for every $n \geq k$ and for every point $x \in X$, $|f_n(x) - f(x)| < \epsilon$.

Fix $\epsilon > 0$ arbitrarily. As $\{f_n\}$ is equicontinuous on X, there exists $\delta > 0$ so that

$$\forall n \in \mathbb{N} \text{ and } \forall x, x' \in X \text{ with } d(x, x') < \delta,\ |f_n(x) - f_n(x')| < \frac{\epsilon}{3}. \quad (*)$$

As the above inequality is true for every value of n, letting $n \to \infty$, we obtain

$$\forall x, x' \in X \text{ with } d(x, x') < \delta,\ |f(x) - f(x')| \leq \frac{\epsilon}{3}. \quad (**)$$

Fix a point $x \in X$ and another point $x' \in B(x; \delta)$ arbitrarily. As $f_n \to f$ pointwise in X, there will exist some $k \in \mathbb{N}$ so that

$$\forall n \geq k, \ |f_n(x') - f(x')| < \frac{\epsilon}{3}. \qquad (***)$$

Therefore, we finally get that for any positive integer $n \geq k$,

$$\begin{aligned} |f_n(x) - f(x)| &\leq |f_n(x) - f_n(x')| + |f_n(x') - f(x')| + |f(x') - f(x)| \\ &< \frac{\epsilon}{3} + \frac{\epsilon}{3} + \frac{\epsilon}{3} \qquad \text{[using (*), (**) and (***)]} \\ &= \epsilon. \end{aligned}$$

As $x \in X$ and $\epsilon > 0$ are arbitrary, this completes the proof. □

Remark Using the stronger version of the Arzelà-Ascoli theorem (Theorem 6.41), the above proof can be argued in the following way as well:

Proof. Since $f_n \to f$ pointwise on X, a uniformly convergent subsequence of $\{f_n\}$ (if any) must converge to f as well.

Consider the collection $\mathcal{B} = \{f_n : n \in \mathbb{N}\}$. By the hypothesis, $\mathcal{B}$ is equicontinuous. Also, as $\{f_n\}$ is pointwise convergent, it is pointwise bounded. Thus by Arzelà-Ascoli theorem, $\overline{\mathcal{B}}$ is compact in $\mathcal{C}(X, \mathbb{K})$. Therefore, any sequence in $\overline{\mathcal{B}}$ has a convergent subsequence. Recall that convergence in $\mathcal{C}(X, \mathbb{K})$ guarantees uniform convergence for a sequence of functions. Therefore any sequence in $\mathcal{B}$, which is in turn a subsequence of $\{f_n\}$, admits a convergent subsequence. But as stated earlier, any uniformly convergent subsequence of $\{f_n\}$ must converge to f. Consequently, by Exercise 3.19, f_n converges to f uniformly. □

6.10. Prove that the family $\{\sin nx\}_{n\in\mathbb{N}}$ is not compact in $\mathcal{C}[-\pi, \pi]$.

Proof. This family is not equicontinuous. Because for any $\delta > 0$, we can get $k \in \mathbb{N}$ such that $\frac{\pi}{k} < \delta$. Then for $x = \frac{\pi}{2k}$ and $y = -\frac{\pi}{2k}$, $|x - y| < \delta$, but $|f_k(x) - f_k(y)| = 2$. Hence by Arzelà-Ascoli theorem, $\{\sin nx\}_n$ is not compact in $\mathcal{C}[-\pi.\pi]$. □

Exercises: 6.4

6.25. Let for a fixed value of $p \geq 1$, S be a bounded subset of $\mathbb{R}^n$ in the metric d_p. Show that S is totally bounded. Also show that this result is true in d_∞ metric as well.

6.26. Consider the space ℓ_∞ as described in Example 2.2.7. Let X be the subset of ℓ_∞ defined as

$$X = \{\mathbf{x} = \{x_k\} \in \ell_\infty : |x_k| \leq a_k\},$$

where $\{a_k\}$ is a sequence of non-negative real numbers. Show that X is compact if and only if $\lim a_k = 0$.

6.27. Let X be a compact metric space and $\mathcal{A}$ be a family of equicontinuous functions from X to another metric space Y. Show that $\mathcal{A}$ is a uniformly equicontinuous family.

6.28. Let $\mathcal{B}$ be a subset of $\mathcal{C}[0, 1]$ such that every $f \in \mathcal{B}$ is differentiable and that there exists an $M > 0$ such that for all $f \in \mathcal{B}$, $\sup_{x\in X}|f'(x)| \leq M$. Prove that $\mathcal{B}$ is equicontinuous.

6.29. Let X be a compact metric space and $\{f_n\}$ be a sequence in $\mathcal{C}(X, \mathbb{K})$. Show that $\{f_n\}$ is convergent in $\mathcal{C}(X, \mathbb{K})$ iff $\{f_n\}$ is equicontinuous and pointwise convergent on X. [*Hint:* The "if" part is already done in miscellaneous result 6.9.] Give counter-examples to show that none of the above implications is necessarily true if X is not compact.

6.30. Let X be a metric space and let $\mathcal{B} \subset \mathcal{C}(X, \mathbb{K})$. We say that $\mathcal{B}$ is *pointwise bounded* on X if for each point $x \in X$, there exists a positive number M_x so that for each function $f \in \mathcal{B}$, $|f(x)| \leq M_x$. Show that if $\mathcal{B}$ is pointwise bounded and equicontinuous, then $\mathcal{B}$ is also bounded in the sup norm metric in $\mathcal{C}(X, \mathbb{K})$.

6.31. Use Arzelà-Ascoli theorem to show that the set $\{f_n(x) : x \in [0, 1], n \in \mathbb{N}\}$, where

$$f_n(x) = \frac{x^2}{x^2 + (1 - nx)^2},$$

is not compact in $\mathcal{C}[0, 1]$.

Chapter 7
Connectedness

The concept of connectedness in a metric space is more intuitive than compactness, as it determines whether a space is a single, unified whole or fragmented into multiple disconnected components. While our initial intuition about connected sets stems from intervals on the real line, the richness of the study of connectedness becomes more apparent in higher dimensional spaces.

Consider a real-valued continuous function defined on an interval. It exhibits a highly desirable property known as the intermediate value property: To attain every value between any two of its distinct values. This ensures that the graph of such a function forms an unbroken curve in the plane. Such behaviour of a continuous function serves as an anchor when characterising connectedness in broader settings.

Despite the appeal of connected spaces, disconnected spaces are equally significant. They reveal how much a given space can be "broken"—analogous to a shattered vase, where the fragments range from large pieces to nearly indistinguishable granules. A striking example of a completely disconnected space is the Cantor set.

This chapter formalises these intuitive ideas into precise mathematical definitions and principles. The first section introduces the definition of connectedness and explores the role of continuous functions in identifying connected spaces. The second section examines disconnected spaces and related properties. Finally, the last section delves into the stronger notion of path connectedness.

7.1 Definition and Examples

Consider the following subsets of $\mathbb{R}$ as shown in Fig. 7.1. A set is connected if it is in one piece. With this understanding, X_1 and X_2 are connected, while X_3, X_4, and X_5 are disconnected.

S. Paul, *Metric Spaces*, University Texts in the Mathematical Sciences,
https://doi.org/10.1007/978-981-96-9259-0_7

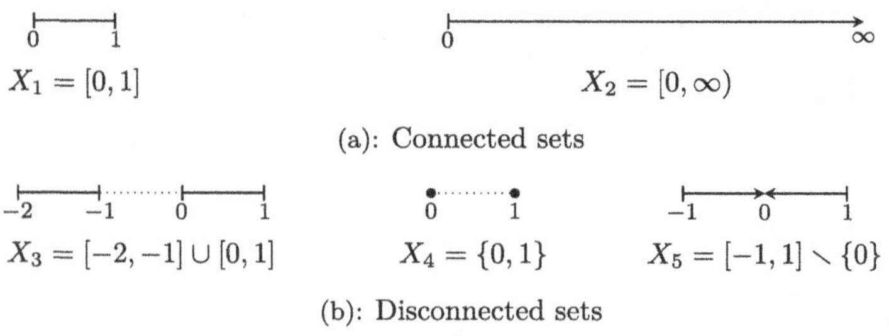

$X_1 = [0, 1]$ $X_2 = [0, \infty)$

(a): Connected sets

$X_3 = [-2, -1] \cup [0, 1]$ $X_4 = \{0, 1\}$ $X_5 = [-1, 1] \setminus \{0\}$

(b): Disconnected sets

Fig. 7.1 Connected and disconnected sets

Now, we attempt to formally define connectedness through a series of refinements, improving upon initial shortcomings until we arrive at a precise definition.

Let X be a metric space. A naive approach to defining disconnectedness is to say that X is disconnected if it can be written as the union of two non-empty, disjoint subsets: $X = A \cup B$, with $A \cap B = \varnothing$. While this works for the sets in Fig. 7.1b, this fails for those in Fig. 7.1a. For, under this definition, we can write $X_1 = [0, \frac{1}{2}] \cup (\frac{1}{2}, 1]$ or $X_2 = [0, 1] \cup (1, \infty)$, making them appear disconnected. More generally, this approach incorrectly classifies any space with at least two elements as disconnected.

To improve this, let us ask for some positive distance between A and B. This refinement correctly classifies the connected sets X_1 and X_2 and also works for two of the disconnected ones, viz., X_3 and X_4. However, it still misclassifies $X_5 = [-1, 0) \cup (0, 1]$ as connected.

To resolve this issue, we refine the definition further: A space X is connected if it *cannot* be written as the union of two non-empty, disjoint subsets A and B such that neither set contains a limit point of the other, i.e. $A' \cap B = \varnothing$ and $A \cap B' = \varnothing$. This formulation aligns well with intuition and provides a solid mathematical criterion for connectedness. However, it is still somewhat descriptive rather than fully concise. We now refine it further into a more compact and precise definition:

- $X = A \cup B$ and $A \cap B = \varnothing$ together imply that $B = A^c$ and $A = B^c$.
- $A' \cap B = \varnothing \implies A' \cap A^c = \varnothing \implies A' \subset A$. Hence A is closed. Therefore A^c or B is open.
- $A \cap B' = \varnothing \implies B^c \cap B' = \varnothing \implies B' \subset B$. Hence B is also closed. Therefore B^c or A is open.
- Combining everything together, we conclude that A is a proper non-empty subset of X that is both closed and open. A rather fancy term to denote such a set is *clopen*.

Thus we obtain the concise and formal definition of a connected space as

Definition 7.1 A metric space X is said to be *connected* if there is no proper non-empty clopen subset of X.

Equivalently, a metric space X is said to be *connected* if the only clopen subsets of X are $\varnothing$ and X itself.

A subset $A \subset X$ is said to be *connected* if A as a subspace of X is connected, i.e. the only clopen subsets of A *in* A are $\varnothing$ and A.

Definition 7.2 A metric space is called *disconnected* if it is not connected.

Remark Like compactness, connectedness is also a topological property of the space as it is defined only in terms of its open sets.

Examples

7.2.1. Both $\varnothing$ and any singleton set are always connected.

7.2.2. $\mathbb{R}$ with respect to the usual Euclidean metric is connected.

Proof. Assume the contrary, i.e. there is a non-empty proper subset A of $\mathbb{R}$ which is both open and closed in $\mathbb{R}$. Now, choose $a \in A$ and $b \in A^c$. WLOG, assume $a < b$. Consider the set $A_0 = \{x \in \mathbb{R} : [a, x] \subset A\}$. Clearly, A_0 is a non-empty subset of A as $a \in A_0$, and also A_0 is bounded above by b. Hence $a_0 = \sup A_0$ exists. Where does this a_0 belong to? Let us consider cases:

Case 1: If $a_0 \in A$, since A is open, for some $\delta > 0$, $(a_0 - \delta, a_0 + \delta) \subset A$. But then $\left[a, a_0 + \frac{\delta}{2}\right] = [a, a_0] \cup \left[a_0, a_0 + \frac{\delta}{2}\right] \subset A$, showing that $a_0 + \frac{\delta}{2} \in A_0$, a contradiction to the fact that $a_0 = \sup A_0$.

Case 2: If $a_0 \notin A$, then as $a_0 = \sup A_0$, for every $\epsilon > 0$, there exists some $a_\epsilon \in A_0$ such that $a_0 - \epsilon < a_\epsilon < a_0$. As $A_0 \subset A$, this says that a_0 is a limit point of A. Since A is closed, $a_0 \in A$. Hence $a_0 \notin A$ implies $a_0 \in A$, a contradiction.

Thus our assumption was not tenable, i.e. $\mathbb{R}$ w.r.t. the usual metric is connected. □

7.2.3. Let X be a set with more than one element equipped with the discrete metric. Then X is disconnected.

Proof. Recall that every subset in a discrete space is both open and closed. Thus, if $a \in X$, then $\{a\}$ is a non-empty proper clopen subset of X, making X disconnected. □

Remark As a matter of fact, any finite space with more than one element is disconnected. All metrics on a finite space are equivalent to the discrete metric (Exercise 3.7).

Before seeing more examples, let us have two immediate results recorded.

Theorem 7.3 *Continuous image of a connected set is connected.*

Proof Let X be a connected metric space and $f: X \to Y$ be continuous. Since the map $f: X \to f(X)$ is also continuous (Exercise 5.2(a)), we may assume, WLOG, that $f: X \to Y$ is continuous and onto. Assume, if possible, that Y is disconnected. Let V be a non-empty proper clopen subset of Y. As f is continuous and onto, $f^{-1}(V)$ is also a non-empty proper clopen subset of X, concluding X is disconnected as well. Thus, our assumption is not tenable, and Y must be connected. This completes the proof. ■

Corollary 7.3.1 *A metric space X is connected if and only if every continuous function $f: X \to \{0, 1\}$ is a constant function.*

Proof By Example 7.2.3, we know that the discrete space $\{0, 1\}$ is disconnected, and by Example 7.2.1, we know only connected subsets of $\{0, 1\}$ are $\varnothing$, $\{0\}$, or $\{1\}$. Thus, by the last theorem, as X is connected, and f is continuous, $f(X)$ must either be $\{0\}$ or $\{1\}$. Consequently, f is a constant function.

Conversely, assume that X is not connected. Let $X = A \cup B$ where both A and B are proper non-empty open disjoint sets. Define $f: X \to \{0, 1\}$ by

$$f(x) = \begin{cases} 0, & \text{if } x \in A \\ 1, & \text{if } x \in B \end{cases}.$$

Then as $\varnothing$, $\{0\}$, $\{1\}$, and $\{0, 1\}$ are the only open sets in $\{0, 1\}$, we see that the inverse images of all of them under f are open in X. Hence f is continuous. Therefore, f is a non-constant continuous function defined on X, a contradiction to the hypothesis. Hence the proof. ■

Remark The set $\{0, 1\}$ serves as a prototype for the discrete two-point space, which is the simplest example of a disconnected space. In fact, as all two-point discrete spaces are homeomorphic, there is essentially only one such space—differences arise only in notation. We choose to label the elements as 0 and 1, thinking of them as real numbers.

By the previous theorem, the continuous image of a connected set is connected. Consequently, any continuous function from a connected space into the discrete two-point space must be constant, as a single point is the only possible connected subset in that space.

Examples

7.3.1. *Connected subsets of $\mathbb{R}$: A non-empty set $I \subset \mathbb{R}$ is connected if and only if I is an* interval.

Proof. An approach similar to the proof of Example 7.2.2 shows that any interval $I \subset \mathbb{R}$ is connected. The details of the proof are left for the reader as an exercise.

Conversely, assume I is a connected subset of $\mathbb{R}$. If I is a singleton, we are done. If not, choose $x < y$ from I. Let $x < z < y$. Assume $z \notin I$. Consider the map $f: \mathbb{R} \setminus \{z\} \to \{0, 1\}$ defined by

$$f(t) = \begin{cases} 0, & \text{if } t < z \\ 1, & \text{if } t > z \end{cases}.$$

By the same arguments laid in the converse part of the proof of Corollary 7.3.1, f is a continuous function on $\mathbb{R} \setminus \{z\}$, and hence on I as well (Exercise 5.2(b)). Hence I is not connected, a contradiction. Therefore, $z \in I$, and thus I is an interval. □

Remark A real-valued continuous function defined on a connected space satisfies the intermediate value property (IVP). Theorem (iv) is a special case when $X = [a, b]$. The general statement is recorded as follows:

Theorem 7.4 (Intermediate Value Theorem) *Let (X, d) be a connected metric space and $f: X \to \mathbb{R}$ be continuous. Let y_1 and y_2 be two distinct real numbers attained by f. If y is any real number lying between y_1 and y_2, then f attains y as well, i.e. there exists some point $x \in X$ so that $f(x) = y$.*

The nature of connected subsets of $\mathbb{R}$, as proved in the last result, along with Theorem 7.3, proves the IVT. The converse of this result is also true and is recorded in miscellaneous result 7.1.

7.4.2. The circle $S^1 = \{(x, y) \in \mathbb{R}^2 : x^2 + y^2 = 1\}$ is connected.
Proof. By the last result, the interval $[0, 2\pi)$ is connected. Define $\phi: [0, 2\pi) \to S^1$ by $\varphi(t) = (\cos t, \sin t)$. Clearly, ϕ is continuous, and hence by Theorem 7.3, S^1 is also connected. □

7.4.3. The rectangular hyperbola $xy = 1$ is disconnected.
Proof. Denote by

$$H = \{(x, y) \in \mathbb{R}^2 : xy = 1\}.$$

Call

$$H_1 = \{(x, y) \in H : x > 0\},$$

and

$$H_2 = \{(x, y) \in H : x < 0\}.$$

Then $H = H_1 \cup H_2$, $H_1 \cap H_2 = \varnothing$, and both H_1 and H_2 are closed in H. Consequently, H is disconnected. □

Miscellaneous Results

7.1 Let (X, d) be a metric space. If every continuous function $f: X \to \mathbb{R}$ satisfies the intermediate value property (i.e. if y_1 and y_2 are two distinct values from the range of f, and y is any value lying between y_1 and y_2, then for some $x \in X$, $f(x) = y$), then X is a connected metric space.
Proof. Assume if possible that X is disconnected. Then by Corollary 7.3.1, there exists a continuous onto map $g: X \to D$, where $D = \{0, 1\}$ is the two-point discrete space. Define $h: D \to \mathbb{R}$ by $h(0) = 0$ and $h(1) = 1$. As every map defined over a discrete space is continuous, h is also so. Hence the composite map $h \circ g: X \to \mathbb{R}$ is also continuous. However, $h \circ g$ does not satisfy the intermediate value property, as there is no point $x \in X$ so that $(h \circ g)(x) = \frac{1}{2}$. Thus, our assumption is not tenable. Consequently, X must be connected. □

7.2 Let $f: S^1 \to \mathbb{R}$ be continuous. Show that there exists a point $\boldsymbol{\xi} \in S^1$ such that $f(\boldsymbol{\xi}) = f(-\boldsymbol{\xi})$.
Proof. Consider the function $g: S^1 \to \mathbb{R}$ given by

$$g(\mathbf{x}) = f(\mathbf{x}) - f(-\mathbf{x}).$$

Observe that g is continuous, and for every point $\mathbf{x} \in S^1$, $g(-\mathbf{x}) = -g(\mathbf{x})$. If g is identically 0 on S^1, we are done. Otherwise, there exists some point $\boldsymbol{\alpha} \in S^1$ such that $g(\boldsymbol{\alpha}) \neq 0$. WLOG, assume $g(\boldsymbol{\alpha}) > 0$. Then $g(-\boldsymbol{\alpha}) < 0$. Hence by the intermediate value theorem (Theorem 7.4), there must exist $\boldsymbol{\xi} \in S^1$ such that $g(\boldsymbol{\xi}) = 0$, i.e. $f(\boldsymbol{\xi}) = f(-\boldsymbol{\xi})$. □

Exercise: 7.1

7.1. Two non-empty subsets A and B of a metric space are called *separated* if $\overline{A} \cap B = \varnothing$ and $A \cap \overline{B} = \varnothing$:

(a) Show that if the subsets A, B are separated, also B, C are separated, then the subsets B and $A \cup C$ are also separated.
(b) If A and B are closed sets in a metric space X, show that the sets $A \cap B^c$ and $A^c \cap B$ are separated sets.
(c) For two subsets A and B of a metric space, if the distance $d(A, B) > 0$, show that A and B are separated. Also show by an example that this condition is not necessary.
(d) If A and B are both open (or closed), then A and B are separated iff $A \cap B = \varnothing$.
(e) Let Y be a subspace of a metric space X. The pair of subsets (A, B) of X is called a *separation* of Y if A and B are separated sets and $A \cup B = Y$. Show that Y is a disconnected space iff there exists a separation of Y. [*Hint*: Use Exercise 3.15.]
(f) Let Y be a connected subspace of a connected metric space X. Let (A, B) be a separation of $X \setminus Y$. Show that $Y \cup A$ and $Y \cup B$ are connected.

7.2. Use Corollary 7.3.1 to prove Theorem 7.3.
7.3. Show that an interval $I \subset \mathbb{R}$ is a connected subset of $\mathbb{R}$.

7.2 Disconnected Spaces

In this section, we explore the common structure of disconnected spaces and examine some of their key properties. We begin by paraphrasing the definition of a disconnected space from the previous section.

Definition 7.5 A metric space X is *disconnected* if there exist two non-empty, disjoint subsets A and B that are both open and closed (clopen) and satisfy $X = A \cup B$. The pair (A, B) is called a *disconnection* or *separation* of X.

With this definition, we can establish several important and useful results about disconnected spaces. These results can be derived in two ways—either directly from the definition or using the previously established fact in Corollary 7.3.1. We present both approaches.

Proposition 7.6 *Let X be a disconnected metric space where (A, B) is a disconnection of X. If Y is a connected subset of X, then Y is either contained entirely in A or is contained entirely in B.*

Proof As A and B are clopen sets in X, the sets $A_1 = A \cap Y$ and $B_1 = B \cap Y$ are clopen in Y. Also, $A_1 \cap B_1 = \varnothing$ and $A_1 \cup B_1 = Y$. If both A_1 and B_1 are non-empty, then (A_1, B_1) will constitute a disconnection of Y. Hence, at least one of A_1 or B_1 must be empty. This completes the proof. ■

Proposition 7.7 *Let A be a connected set in a metric space and let $A \subset B \subset \overline{A}$. Then B is connected.*

Proof Let B be disconnected and (C, D) be a disconnection of B. Then by the previous result, A must entirely be contained either in C, or in D. WLOG, assume $A \subset C$. In that case, $\overline{A} \subset \overline{C} = C$, as C is closed. However, as $B \subset \overline{A}$, this implies $B \subset C$, leaving $D = \varnothing$, a contradiction. Consequently, B must be connected. ■

Another Proof Let $f: B \to \{0, 1\}$ be continuous. Then f is continuous on A as well (Exercise 5.2(b)). Since A is connected, f is constant on A, say 0. Choose $b \in B$ arbitrarily. Since $B \subset \overline{A}$, we find a sequence $\{a_n\}_n \subset A$ such that $a_n \to b$. But for every n, $a_n \in A$ implies $f(a_n) = 0$. Hence by continuity of f, $f(b) = 0$ as well. Since $b \in B$ is arbitrary, f is constant on B. Hence B is connected. ■

Corollary 7.7.1 *If A is a connected set in a metric space, then $\overline{A}$ is also connected.*

Proposition 7.8 *Let $\{A_\lambda : \lambda \in \Lambda\}$ be a collection of connected sets in a metric space with the property that for each $\alpha, \beta \in \Lambda$, $A_\alpha \cap A_\beta \neq \varnothing$. Then $\bigcup_{\lambda \in \Lambda} A_\lambda$ is connected.*

Proof Call $A = \bigcup_{\lambda \in \Lambda} A_\lambda$. Assume A is disconnected and (B, C) be a disconnection of A. As each A_λ is connected, each of them must be contained entirely either in B or in C. Let $A_\beta \subset B$ and $A_\gamma \subset C$. However, as $A_\beta \cap A_\gamma \neq \varnothing$, this will imply $B \cap C \neq \varnothing$ as well, a contradiction. Hence A must be connected. ■

Another Proof Call $A = \bigcup_{\lambda \in \Lambda} A_\lambda$. Assume A is disconnected. Then we have a continuous function f from A onto $\{0, 1\}$. Then for every index $\lambda \in \Lambda$, f is continuous on A_λ. Let for $a, b \in A$, $f(a) = 0$ and $f(b) = 1$. Assume for $\alpha, \beta \in \Lambda$, $a \in A_\alpha$ and $b \in A_\beta$. Since A_α is connected and $f(a) = 0$, f is identically 0 on A_α. By the same argument, A_β is connected and $f(b) = 1$ together implies f is identically 1 on A_β. But as $A_\alpha \cap A_\beta \neq \varnothing$, for $c \in A_\alpha \cap A_\beta$, $f(c)$ is 0 as well as 1, a contradiction. Consequently, A must be connected. ■

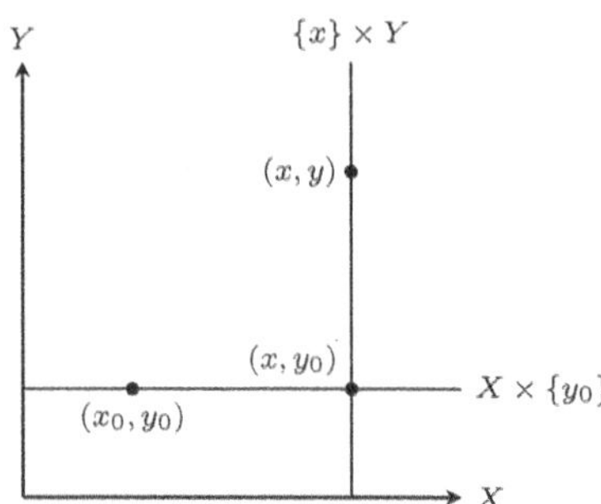

Fig. 7.2 Product of two connected spaces

Theorem 7.9 *Product of two connected spaces is a connected space.*

Proof Let X and Y be two connected metric spaces. Fix a point $(x_0, y_0) \in X \times Y$. See Fig. 7.2. The respective horizontal and vertical lines $X \times \{y_0\}$ and $\{x\} \times Y$ are connected, as they are homeomorphic images of the connected spaces X and Y. Thus, for each point $x \in X$, the T-shaped space

$$T_x = \big(X \times \{y_0\}\big) \cup \big(\{x\} \times Y\big)$$

is also connected, as they intersect at (x, y_0). However, observe that $X \times Y = \bigcup_x T_x$. But, for any two $x, x' \in X$, T_x and $T_{x'}$ intersect along the horizontal line $X \times \{y_0\}$. Hence, by the previous result, $\bigcup_x T_x$ is connected. Consequently, $X \times Y$ is connected. ■

Another Proof Let X and Y be two connected metric spaces. Fix a point $(x_0, y_0) \in X \times Y$. See Fig. 7.2. We start from the point (x_0, y_0) and go to (x, y_0) via the connected set $X \times \{y_0\}$. So the f values at both the points (x_0, y_0) and (x, y_0) remain same. Then from (x, y_0), we reach the point (x, y) via the connected set $\{x\} \times Y$. Hence $f(x, y) = f(x, y_0)$. Therefore, finally, $f(x, y) = f(x_0, y_0)$. We make this geometric intuition precise in the following paragraph.

Recall that for every point $x \in X$ and $y \in Y$, the inclusion maps (Exercise 5.3) $i_y : X \to X \times Y$ given by $i_y(x) := (x, y)$ and $i_x : Y \to X \times Y$ given by $i_x(y) := (x, y)$ are continuous. Since X is connected, by Theorem 7.3, $i_{y_0}(X) = X \times \{y_0\}$ is connected. Hence, f, being continuous on the connected subset $X \times \{y_0\}$, is constant. Hence, $f(x, y_0) = f(x_0, y_0)$. Again, as i_x is continuous and Y is connected, $i_x(Y) = \{x\} \times Y$ is connected. Moreover, f being continuous on the connected subset $\{x\} \times Y$ is constant. Hence, $f(x, y) = f(x, y_0)$. Therefore, for any arbitrary point $(x, y) \in X \times Y$, $f(x, y) = f(x_0, y_0)$, i.e. f is a constant. Since f is arbitrary, $X \times Y$ is connected. ■

Remarks

(a) This theorem can be extended to any finite product of connected spaces. In particular, it shows that $\mathbb{R}^n$ is connected.

(b) The converse of this theorem is also true, i.e. if $X \times Y$ is a connected space, then both X and Y must also be connected. The continuity of the projection maps (Example 5.4.4) in both coordinates guarantees that.

A disconnected space is composed of multiple "pieces." Our main objective is to formally identify and study these pieces, known as *connected components*. To approach this systematically, we use an algebraic framework to address the abstract nature of the space.

Let X be a metric space. Define a relation $\sim$ on X by, for $x, x' \in X$,

$$x \sim x' \text{ if and only if } x \text{ and } x' \text{ belong to some connected subset of } X.$$

Now, this relation $\sim$ is an equivalence relation on X. Because:

- Every point $x \in X$ belongs to the connected subset $\{x\}$, so $x \sim x$.
- If $x \sim x'$, then x' and x are in the same connected subset, i.e. $x' \sim x$.
- Finally, for $x, x', x'' \in X$, assume $x \sim x'$ and $x' \sim x''$. Thus we have two connected subsets C_1, C_2 of X such that $x, x' \in C_1$ and $x', x'' \in C_2$. Then $x' \in C_1 \cap C_2$ showing that $C_1 \cap C_2 \neq \varnothing$. Hence, by Proposition 7.8, $C_1 \cup C_2$ is connected containing all three points x, x', and x''. Hence, $x \sim x''$.

Definition 7.10 The equivalence classes of X by the equivalence relation $\sim$ are called the *connected components* or simply *components* of X.

Remarks

(a) Since the equivalence classes create a partition of X, the connected components are pairwise disjoint and their union is the whole space X.
(b) For $x \in X$, let $C[x]$ denote the equivalence class containing x, defined as

$$C[x] = \{x' \in X : x \text{ and } x' \text{ belong to some connected subset of } X\}.$$

Let C be any connected subset of X that contains x. Then, for any $x' \in C$, it follows trivially that $x' \in C[x]$. Hence, $C \subset C[x]$.

Now, let $\mathcal{C}(x)$ be the collection of all connected subsets of X that contain x. Taking the union over all such subsets, we get

$$\bigcup_{C \in \mathcal{C}(x)} C \subset C[x].$$

Conversely, if $x' \in C[x]$, then x' belongs to some connected subset $C' \in \mathcal{C}(x)$, implying

$$C[x] \subset \bigcup_{C \in \mathcal{C}(x)} C.$$

Thus, we conclude that

$$C[x] = \bigcup_{C \in \mathcal{C}(x)} C.$$

Since the subsets in $\mathcal{C}(x)$ are not disjoint (they all contain x), their union remains connected by Proposition 7.8. Consequently, $C[x]$ is a connected subset of X. Moreover, $C[x]$ is the largest connected subset containing x, as any connected subset containing x must be included within $C[x]$. Therefore, $C[x]$ is called the connected component of x.

(c) The above two remarks justify the definition to call the $\sim$ equivalence classes as "connected components" of X. They decompose X into maximal connected subsets of X.

(d) If X is connected, the only component X has is X itself.

Examples

7.10.1. The connected components of the space of all non-zero real numbers are the set of all positive reals $\mathbb{R}_+$ and the set of all negative reals $\mathbb{R}_-$.

7.10.2. Let $X = \mathbb{Q}$. The connected component for $x \in \mathbb{Q}$ is $\{x\}$. In other words, any subset A of $\mathbb{Q}$ containing more than one point is disconnected. Because, if $A \subset \mathbb{Q}$ is connected and $x, y \in A$ with $x < y$, then for any irrational α between x and y, $(-\infty, \alpha) \cap A$ and $(\alpha, \infty) \cap A$ are two disjoint open sets in A whose union is A. Hence A is disconnected.

7.10.3. Let $Y \subset \mathbb{R}^2$ be the subspace consisting of the segments joining the origin to the points $\left\{(1, \frac{1}{n}) : n \in \mathbb{N}\right\}$ together with the segment $[\frac{1}{2}, 1]$. See Fig. 7.3. The line joining $(0, 0)$ and $\left(1, \frac{1}{n}\right)$ is the image of the connected set $[0, 1]$ under the continuous map $y = \frac{x}{n}$ and, therefore, connected. If X denotes the union of these lines, then X is connected since the origin is common to all the line segments. Finally, Y is such that $X \subset Y \subset \overline{X}$, where $\overline{X} = X \cup (0, 1]$,

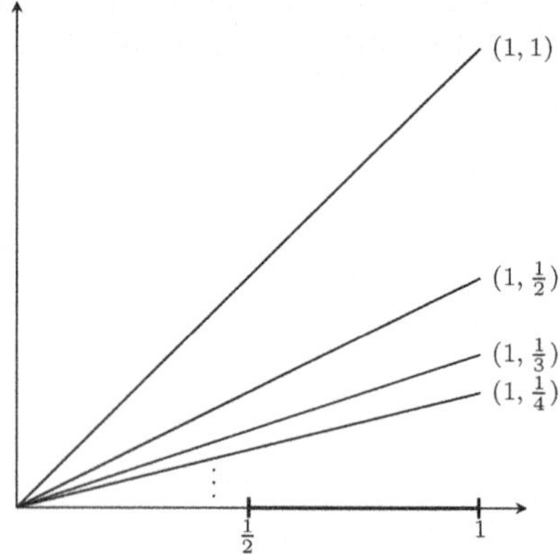

Fig. 7.3 X and Y

and so Y is connected, by Proposition 7.7. However, $Y \setminus \{(0,0)\}$ is not connected. In $Y \setminus \{(0,0)\}$, the component of each point is the segment containing it.

Theorem 7.11 *Let X be a metric space. Then:*

i) Each point $x \in X$ is contained in exactly one component of X.
ii) Each connected subset of X is contained in exactly one component of X.
iii) Each component of X is closed.
iv) Each non-empty connected subset of X that is both open and closed is a component of X.

Proof

i) Follows directly from the definition of components.
ii) Follows from Remark (b) on page 271.
iii) If C is a connected component of X, by Proposition 7.7, $\overline{C}$ is also connected. For $x \in C$, C is the largest connected subset of X containing x. But then the fact $\overline{C}$ is also connected must imply $\overline{C} = C$. Hence C is closed.
iv) Let A be a non-empty connected subset of X which is both open and closed. Then by (*ii*), A is contained in some component C of X. Now, if $A \subsetneq C$, then as A is both open and closed,

$$C = (C \cap A) \cup (C \setminus A)$$

is a disconnection of C, a contradiction as C must be connected. Hence, $A = C$.

■

Remark A natural question arises: "Are the components always open?" The answer is "No". Consider the space $X = \left\{\frac{1}{n} : n \in \mathbb{N}\right\} \cup \{0\}$ in the metric induced from $\mathbb{R}$. Here, $\{0\}$ is a component. Because if $C[0]$, the component containing 0, has another point, say, $\frac{1}{m}$ from X, then for some irrational number α with $0 < \alpha < \frac{1}{m}$,

$$C[0] = \Big((-\infty, \alpha) \cap C[0]\Big) \cup \Big((\alpha, \infty) \cap C[0]\Big)$$

is a disconnection of $C[0]$. Hence $C[0] = \{0\}$. But clearly $\{0\}$ is not open in X as its complement $\left\{\frac{1}{n} : n \in \mathbb{N}\right\}$ is not closed in X.

However, it is important to note that if X has only finitely many components, then each component is both open and closed.

Definition 7.12 A metric space is called *totally disconnected* if the only components of X are singleton sets, i.e. every subset of X with more than one point is disconnected.

Examples

7.12.1. Both the set of all rational numbers and the set of all irrational numbers are totally disconnected.

7.12.2. A discrete space with more than one element is totally disconnected.

7.12.3. Cantor set is totally disconnected. See Appendix D for more details.

Miscellaneous Results

7.3 Refer to Example 6.29.3. Consider the function $\varphi: \mathcal{O}_n(\mathbb{R}) \to \mathbb{R}$ given by $\varphi(A) = \det A$, $A \in \mathcal{O}_n(\mathbb{R})$. Now, determinant is a continuous map (Example 5.6.2) which maps entire $\mathcal{O}_n(\mathbb{R})$ to $\{-1, 1\}$ which is disconnected. Hence as a contrapositive argument to Theorem 7.3, $\mathcal{O}_n(\mathbb{R})$ must be disconnected. On the other hand, $\mathcal{SO}_2(\mathbb{R})$ is given by

$$\mathcal{SO}_2(\mathbb{R}) = \left\{ \begin{pmatrix} \cos t & \sin t \\ -\sin t & \cos t \end{pmatrix} : t \in [0, 2\pi) \right\}.$$

Then, $\mathcal{SO}_2(\mathbb{R})$ is the image of the connected set $[0, 2\pi)$ by the continuous map $t \mapsto \begin{pmatrix} \cos t & \sin t \\ -\sin t & \cos t \end{pmatrix}$ (prove this!). Hence it is connected.

Remark Both $\mathcal{SO}_n(\mathbb{R})$ and $SL_n(\mathbb{R})$ are also connected spaces. But the proofs of them are beyond the scope of this book.

7.4 Refer to Example 6.29.4. Here also the determinant function maps $GL_n(\mathbb{R})$ to the disconnected set $\mathbb{R} \setminus \{0\}$. Hence it is disconnected.

7.5 Show that $\mathbb{R}^2 \setminus \{\mathbf{0}\}$ is connected.

Proof. For two points $\mathbf{x}$ and $\mathbf{y}$ in $\mathbb{R}^2 \setminus \{(0, 0)\}$, denote the line segment joining $\mathbf{x}, \mathbf{y}$ as

$$[\mathbf{x}, \mathbf{y}] = \{t\mathbf{x} + (1 - t)\mathbf{y} : t \in [0, 1]\},$$

see Fig. 7.4. If the line segment $[\mathbf{x}, \mathbf{y}]$ does not contain the origin, then it is the continuous image of the connected set $[0, 1]$ via the map $\varphi: [0, 1] \to \mathbb{R}^2 \setminus \{\mathbf{0}\}$ given by $\varphi(t) = t\mathbf{x} + (1 - t)\mathbf{y}$. Hence is connected.

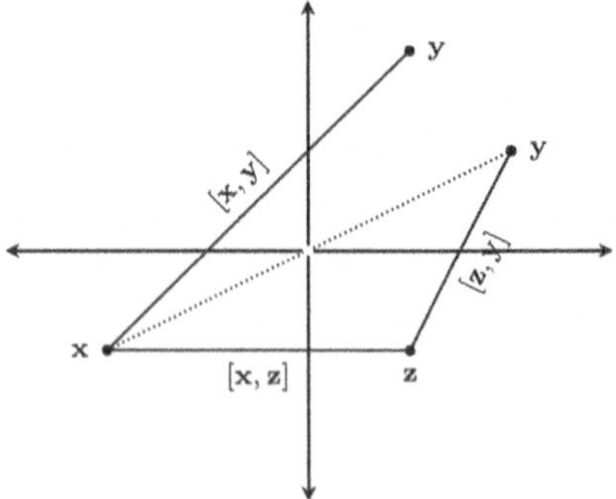

Fig. 7.4 $\mathbb{R}^2 \setminus \{\mathbf{0}\}$ is connected

Fig. 7.5 Hawaiian earring

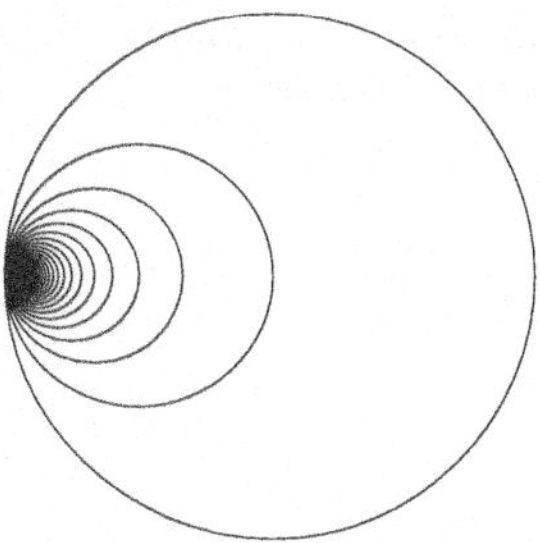

If the line segment contains the origin, choose a point $\mathbf{z}$ outside the segment. Then by the above argument, both the line segments $[\mathbf{x}, \mathbf{z}]$ and $[\mathbf{z}, \mathbf{y}]$ are connected, which intersect at the point $\mathbf{z}$. Hence $[\mathbf{x}, \mathbf{z}] \cup [\mathbf{z}, \mathbf{y}]$ is also connected by Proposition 7.8.
Hence for any two points $\mathbf{x}, \mathbf{y} \in \mathbb{R}^2 \setminus \{\mathbf{0}\}$, there is a connected set containing them. Therefore, by Exercise 7.4, $\mathbb{R}^2 \setminus \{\mathbf{0}\}$ is connected. □

Remarks

(a) In exactly same way, we can prove that for $n > 1$, $\mathbb{R}^n \setminus \{\mathbf{0}\}$ is connected.
(b) Another way to see the result for $\mathbb{R}^2$ is to realise that

$$\begin{aligned}\mathbb{R}^2 \setminus \{(0,0)\} &= \{(r\cos\theta, r\sin\theta) : \theta \in [0, 2\pi),\ r > 0\} \\ &\cong \{(\cos\theta, \sin\theta, r) : \theta \in [0, 2\pi),\ r > 0\} \\ &= \left\{(\cos\theta, \sin\theta) : \theta \in [0, 2\pi)\right\} \times \{r : r > 0\} \\ &= S^1 \times (0, \infty).\end{aligned}$$

Since both S^1 and $(0, \infty)$ are connected (Example 7.4.2), so is $\mathbb{R}^2 \setminus \{\mathbf{0}\}$, by Theorem 7.9.

7.6 The unit sphere $S^n = \{\mathbf{x} \in \mathbb{R}^{n+1} : \|\mathbf{x}\| = 1\}$ is the image of $\mathbb{R}^{n+1} \setminus \{\mathbf{0}\}$ by the continuous map $\mathbf{x} \mapsto \frac{\mathbf{x}}{\|\mathbf{x}\|}$. Hence is connected.

7.7 A beautiful example of a connected space is the *Hawaiian earring*. For every $n \in \mathbb{N}$, let C_n be the circle in $\mathbb{R}^2$ with centre at $\left(\frac{1}{n}, 0\right)$ and radius $\frac{1}{n}$. The Hawaiian earring H is the union of all such C_n's. H is a connected space as every circle C_n is connected, and they all intersect at $(0, 0)$ (Fig. 7.5).

7.8 Let X, Y be metric spaces. A map $f : X \to Y$ is called *locally constant* if for every point $x \in X$, there exists some open set U_x in X containing x such that f is constant on U_x. Show that if X is connected, then a locally constant map is always constant.
Proof. Fix $p \in X$. Construct a set

$$E = \{x \in X : f(x) = f(p)\}.$$

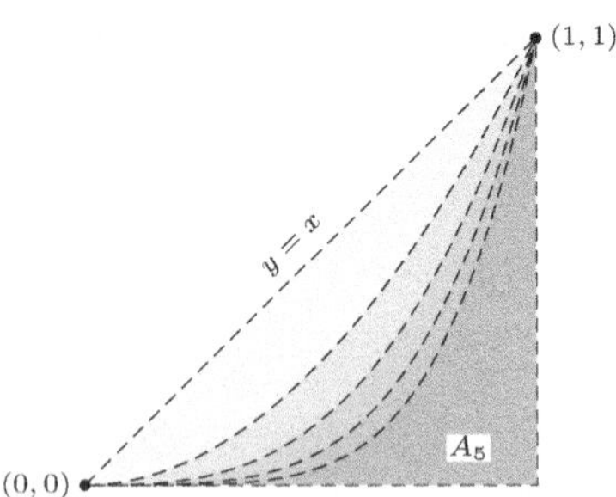

Fig. 7.6 Sets A_1 to A_5

We will show that $E = X$. As $p \in E$, $E \neq \varnothing$. As f is locally constant, for each point $x \in E$, there is some open $U_x \subset X$ so that for any point $x' \in U_x$, $f(x') = f(x) = f(p)$. Thus $U_x \subset E$. As $x \in E$ is arbitrary, this implies that E is open.
If $E \neq X$, choose $y \in X \setminus E$. Then $f(y) \neq f(p)$. Again, as f is locally constant, there is some open $U_y \subset X$, such that for any point $y' \in U_y$, $f(y') = f(y) \neq f(p)$. Hence $U_y \subset X \setminus E$. Thus, $X \setminus E$ is open as well.
Consequently, $(E, X \setminus E)$ becomes a disconnection of X, a contradiction. Therefore, $E = X$, i.e. f is constant on X. □

7.9 Let X be an unbounded connected metric space. For any fixed $x \in X$ and $r > 0$, show that there always exists some point $y \in X$ such that $d(x, y) = r$.
Proof. For $y \in X$, consider the map $\varphi: X \to \mathbb{R}$ given by $y \mapsto d(x, y)$. This is a continuous map by miscellaneous result 5.6. Since X is connected, $\varphi(X)$ is a connected subset of $\mathbb{R}$. Hence $\varphi(X)$ must be an interval. Again, as X is unbounded, $\varphi(X) = [0, \infty)$. Hence for any arbitrary $r > 0$, we find a $y \in X$ such that $d(x, y) = r$. □

7.10 Show that $\mathbb{R}$ is not homeomorphic to $\mathbb{R}^2$.
Proof. Refer to the fact recorded in Exercise 5.27. If $\mathbb{R}^2$ is homeomorphic to $\mathbb{R}$, then $\mathbb{R}^2 \setminus \{(0, 0)\}$ should also be homeomorphic to $\mathbb{R}$ with a single point excluded. But that leads to a contradiction as $\mathbb{R}^2 \setminus \{(0, 0)\}$ is connected and $\mathbb{R}$ with a single point excluded is not. □

7.11 Give an example of a sequence $\{A_n\}_n$ of connected subsets of $\mathbb{R}^2$ such that for each $n \in \mathbb{N}$, $A_{n+1} \subset A_n$, but $\bigcap_{n \in \mathbb{N}} A_n$ is not connected.
Proof. For $n \in \mathbb{N}$, choose

$$A_n = \{(x, y) \in \mathbb{R}^2 : 0 < y < x^n,\ 0 < x < 1\} \cup \{(0, 0), (1, 1)\}.$$

Then A_n's are all connected (prove this!), but $\bigcap_{n \in \mathbb{N}} A_n = \{(0, 0), (1, 1)\}$ is disconnected (Fig. 7.6). □

7.12 Consider the metric space of $\mathbb{R}$ with the usual metric. Show that a subset $A \subset \mathbb{R}$ is totally disconnected if and only if A has empty interior.
Proof. Let A be totally disconnected and if possible, let $\operatorname{int} A \neq \varnothing$. Then for $a \in \operatorname{int} A$, there exists some $\delta > 0$ such that the neighbourhood interval $B(a; \delta) = (a - \delta, a + \delta) \subset A$. But $B(a; \delta)$ is definitely a connected subset

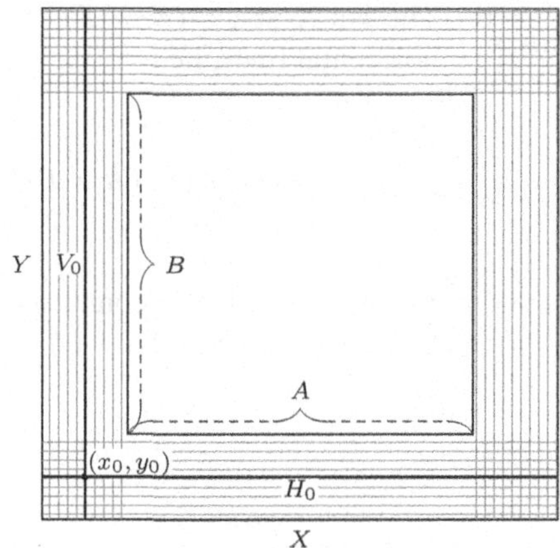

Fig. 7.7 $(X \times Y) \setminus (A \times B)$ is connected

of A which is not singleton, a contradiction to the hypothesis that A is totally disconnected.

Conversely, let $\operatorname{int} A = \varnothing$ and A be not totally disconnected. Then there is a connected subset C of A containing at least two points $a, b \in A$. But as the only connected subsets of $\mathbb{R}$ are the intervals, $a, b \in C$ and C is connected must together imply $[a, b] \subset C$. Hence the open interval $(a, b) \subset C \subset A$ is a non-empty open set in A, contradicting the hypothesis that A has empty interior. □

7.13 A counter-intuitive result: We have seen that $\mathbb{R} \setminus \mathbb{Q}$ is totally disconnected. However, the set $\mathbb{R}^2 \setminus \mathbb{Q}^2$ is connected! There are two more general statements in this regard:

(a) If X and Y are connected metric spaces and A and B are proper subsets of X and Y, respectively, then $(X \times Y) \setminus (A \times B)$ is connected in $X \times Y$.
(b) For $n \geq 2$, $\mathbb{R}^n \setminus \mathbb{Q}^n$ is connected.

Here we prove (a) to establish $\mathbb{R}^2 \setminus \mathbb{Q}^2$ is connected, and leave (b) in Exercise 7.9.

Proof. We imitate the approach in Theorem 7.9. Fix a point $(x_0, y_0) \in (X \times Y) \setminus (A \times B)$. Look at Fig. 7.7. We conclude the proof with the following observations and arguments:

(a) For each point $x \in X \setminus A$, the vertical line $V_x = \{x\} \times Y$ is connected, as it is a homeomorphic image of the connected space Y.
(b) For each point $y \in Y \setminus B$, the horizontal line $H_y = X \times \{y\}$ is connected, as it is a homeomorphic image of the connected space X.
(c) The connected sets $V_0 = \{x_0\} \times Y$ and $H_0 = X \times \{y_0\}$ intersect at (x_0, y_0). Hence the T-shaped set $T_0 = V_0 \cup H_0$ is connected.
(d) For any $x \in X \setminus A$, and any $y \in Y \setminus B$, the vertical line V_x and the horizontal line H_y intersect T_0.

(e) The entire space $(X \times Y) \setminus (A \times B)$ is the union of such vertical and horizontal lines, i.e.

$$(X \times Y) \setminus (A \times B) = \left[\bigcup_{x \in X \setminus A} V_x \right] \cup \left[\bigcup_{y \in Y \setminus B} H_y \right].$$

Since they all are connected and intersect the connected set T_0, by Exercise 7.6, $(X \times Y) \setminus (A \times B)$ is connected.

□

Exercise: 7.2

7.4. Let X be a metric space such that for any two points $x, x' \in X$, there is a connected set $A \subset X$ containing x and x'. Show that X is connected.

7.5. Let $\{A_n\}$ be a sequence of connected subsets of a metric space X. If for each n, $A_n \cap A_{n+1} \neq \emptyset$, show that $\bigcup_n A_n$ is connected.

7.6. Let $\{A_\lambda : \lambda \in \Lambda\}$ be a collection of connected subsets of X; let A be a connected subset of X. Show that if for each λ, $A \cap A_\lambda \neq \emptyset$, then $A \cup (\bigcup_\lambda A_\lambda)$ is connected.

7.7. Let X, Y be metric spaces and $f: X \to Y$ be continuous. Show that if X is connected, then the graph of f

$$\Gamma_f = \{(x, f(x)) : x \in X\}$$

is a connected subset of $X \times Y$. [*Hint*: Use Theorem 7.9]. Also show by an example that a function with a connected graph need not be continuous. [*Hint*: Topologist's sine curve].

7.8. Let $U \subset \mathbb{R}^n$ be open and connected and $f: U \to \mathbb{R}^n$ be differentiable. If for every point $\mathbf{x} \in U$, $Df(\mathbf{x}) = 0$, show that f is locally constant on U, and hence using miscellaneous result 7.8, show that f is constant on U.

7.9. Show that for $n \geq 2$, $\mathbb{R}^n \setminus \mathbb{Q}^n$ is connected.

7.10. If A is a connected subspace of a metric space X, what can you conclude about connectedness of int A and ∂A? (∂A denotes the boundary of A.)

7.11. Let X be a metric space and $A \subset X$. A connected subset $C \subset X$ be such that C intersects both A and $X \setminus A$. Show that C intersects ∂A.

7.12. Show that the set of irrationals is totally disconnected in $\mathbb{R}$.

7.13. Show that the letters X and Y (considered as metric spaces) are not homeomorphic. [*Hint*: Remove a special point from X and apply Exercise 5.27]

Fig. 7.8 A path

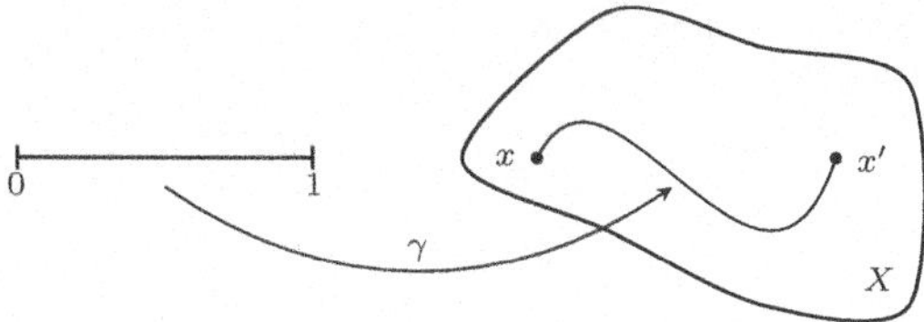

7.3 Path Connected Spaces

Path connectedness offers a more intuitive and geometric perspective on connectedness. It is a stronger condition than connectedness and provides a clearer, more visual understanding of a space. This property is particularly significant in complex analysis. As we will demonstrate later in this section, any open and connected region of the complex plane is necessarily path connected, a fact upon which many proofs in the subject rely.

The primary goals of this section are to establish three key results. First, as mentioned, we will prove that every open, connected region of the complex plane is path connected. Additionally, we will show that path connectedness is a topological property and that every path connected space is necessarily connected. Finally, we will explore a classic yet intriguing example of a connected space that is not path connected.

Definition 7.13 A *path* is a continuous map $\gamma\colon [0, 1] \to X$. If for two points x and x' from X, $\gamma(0) = x$, and $\gamma(1) = x'$, then γ is a path which joins x to x'. Also we say that x *is path connected to* x'.

Remarks

(a) A path is a continuous image of the connected set $[0, 1]$ in $\mathbb{R}$ (Fig. 7.8). Hence every path in X is a connected subset of X.
(b) To refer the following remark on a later occasion, we record it in:

Proposition 7.14 *If for $x, x' \in X$, x is path connected to x', then x' is also path connected to x.*

Proof Since x is path connected to x', we have a continuous map $\gamma\colon [0, 1] \to X$ such that $\gamma(0) = x$ and $\gamma(1) = x'$. Then the map $\sigma\colon [0, 1] \to X$ given by $\sigma(t) = \gamma(1-t)$ is also continuous and $\sigma(0) = x'$ and $\sigma(1) = x$. Hence x' is path connected to x. ∎

Definition 7.15 A metric space X is said to be *path connected* if for any pair of points $x, x' \in X$, there exists a path $\gamma\colon [0, 1] \to X$ such that $\gamma(0) = x$ and $\gamma(1) = x'$.

Examples

7.15.1. Any interval in $\mathbb{R}$ is path connected.

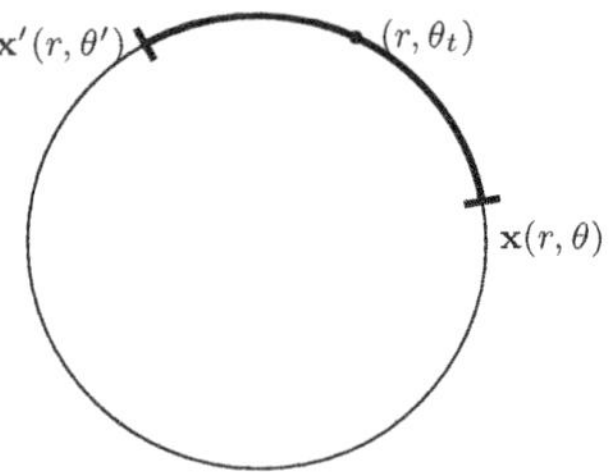

Fig. 7.9 A circle is path connected

Proof. Let $I \subset \mathbb{R}$ be any interval and $x, x' \in I$. Then $\gamma: [0, 1] \to I$ given by

$$\gamma(t) = tx + (1 - t)x'$$

is a path in I between x and x'. □

7.15.2. For every $n \in \mathbb{N}$, $\mathbb{R}^n$ is path connected.
Proof. For $\mathbf{x}, \mathbf{y} \in \mathbb{R}^n$, $\gamma: [0, 1] \to \mathbb{R}^n$ given by

$$\gamma(t) = t\mathbf{x} + (1 - t)\mathbf{y}$$

is a path between $\mathbf{x}$ and $\mathbf{y}$. □

7.15.3. The circle $C_r = \{(x, y) \in \mathbb{R}^2 : x^2 + y^2 = r^2\}$ centred at origin and radius $r > 0$ is path connected in $\mathbb{R}^2$ (Fig. 7.9).
Proof. Observe that for any given $r > 0$, we may write

$$C_r = \{(r \cos\theta, r \sin\theta) : \theta \in [0, 2\pi)\}.$$

Choose any two points $\mathbf{x} = (r \cos\theta, r \sin\theta)$ and $\mathbf{x}' = (r \cos\theta', r \sin\theta')$ from C_r arbitrarily. WLOG, assume $0 \leq \theta \leq \theta' < 2\pi$. Then

$$\gamma(t) := (r \cos\theta_t, r \sin\theta_t), \quad \text{where } \theta_t = t\theta + (1 - t)\theta',$$

is a continuous map from $[0, 1]$ to C_r carrying $\gamma(0) = \mathbf{x}$ to $\gamma(1) = \mathbf{x}'$. Consequently, C_r is path connected. □

7.15.4. Similarly, a parabola, an ellipse, one branch of a hyperbola are all path connected spaces, whereas an entire hyperbola with both branches is not path connected. Proofs of these facts are easy and left for the reader as exercises.

7.15.5. The union of the parabolas $y = x^2$ and $y^2 = x$ is path connected in $\mathbb{R}^2$.
Proof. Denote the parabolas by $P_1 : y = x^2$ and $P_2 : y^2 = x$. Choose arbitrary points $\mathbf{x}_1, \mathbf{x}_2$ from $P_1 \cup P_2$. If both $\mathbf{x}_1$ and $\mathbf{x}_2$ lie on the same parabola, then by the previous example, they are path connected. If $\mathbf{x}_1$ and $\mathbf{x}_2$ come from different parabolas (say $\mathbf{x}_1 \in P_1$ and $\mathbf{x}_2 \in P_2$), we proceed in the following way:

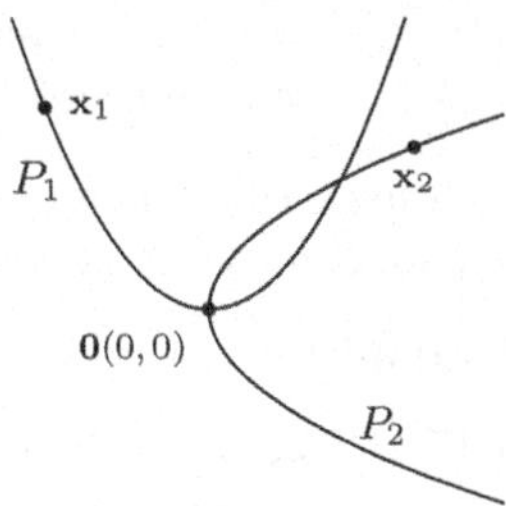

Fig. 7.10 Intersecting parabolas

Look at Fig. 7.10. Find and fix a point of intersection of P_1 and P_2, say $\mathbf{0}(0, 0)$. As P_1 and P_2 are individually path connected spaces, we can obtain paths γ_1 and γ_2 joining $\mathbf{x}_1$ to $\mathbf{0}$ in P_1, and $\mathbf{0}$ to $\mathbf{x}_2$ in P_2, respectively. Then the conjoined path γ_1 followed by γ_2 will join $\mathbf{x}_1$ and $\mathbf{x}_2$. Formally, the conjoined path $\gamma: [0, 1] \to P_1 \cup P_2$ will look like

$$\gamma(t) = \begin{cases} \gamma_1(2t), \text{ when } 0 \leq t < \frac{1}{2} \\ \gamma_2(2t-1), \text{ when } \frac{1}{2} \leq t \leq 1 \end{cases},$$

where for $i = 1, 2$, $\gamma_i: [0, 1] \to P_i$ are continuous maps so that

$$\gamma_1(0) = \mathbf{x}_1, \ \gamma_1(1) = \mathbf{0}, \quad \text{and} \quad \gamma_2(0) = \mathbf{0}, \ \gamma_2(1) = \mathbf{x}_2.$$

(To find explicit expressions for γ_1 and γ_2 remains a part of Exercise 7.14 mentioned in the previous example.) □

Remark The above proof provides a method to demonstrate that in a metric space, if a point x is path connected to x' and x' is path connected to x'', then x' is path connected to x''. A formal proof of this result is left as an exercise for the reader.

The above remark also gives the following fact:

Proposition 7.16 *A metric space is path connected if and only if there exists a point $a \in X$ which is path connected to every $x \in X$.*

A formal proof is left for the reader in Exercise 7.16(b).

Now, we channelise our efforts to prove the following major facts about path connectedness.

Theorem 7.17 *Continuous image of a path connected space is path connected.*

Proof Let X, Y be metric spaces, X is path connected, and $f: X \to Y$ be continuous. For any two points $y, y' \in f(X)$, there will exist points $x, x' \in X$ so that $y = f(x)$ and $y' = f(x')$. Now, as X is path connected, we have a path $\gamma: [0, 1] \to X$ satisfying $\gamma(0) = x$ and $\gamma(1) = x'$. Therefore, by Proposition 5.6, $f \circ \gamma$ is a continuous map from $[0, 1]$ to $f(X)$ satisfying $f \circ \gamma(0) = y$ and

$f \circ \gamma(1) = y'$, i.e. y and y' are path connected. Since $y, y' \in f(X)$ are arbitrary, $f(X)$ is path connected. ■

Remark This theorem shows that homeomorphic image of a path connected space is path connected, i.e. path connectedness is a topological property.

Examples

7.17.1. The unit sphere $S^n = \{\mathbf{x} \in \mathbb{R}^{n+1} : \|\mathbf{x}\| = 1\}$ is the image of $\mathbb{R}^{n+1} \setminus \{\mathbf{0}\}$ by the continuous map $\mathbf{x} \mapsto \frac{\mathbf{x}}{\|\mathbf{x}\|}$. Hence, by Exercise 7.17, it is path connected.

Theorem 7.18 *A path connected metric space is connected.*

Proof Let X be a path connected metric space. Fix a point $a \in X$. As each path is a connected subset, for any point $x \in X$, the path $C_a(x)$ joining a to x is a connected subset of X. As $x \in X$ is arbitrary, we may write

$$X = \bigcup_x C_a(x).$$

Again, as $\bigcap_x C_a(x) = \{a\}$ is non-empty, by Proposition 7.8, X is connected. ■

Examples

7.18.1. The metric space $\mathcal{C}[a, b]$ with sup metric is path connected. If f, g are two continuous functions, then for every $\lambda \in [0, 1]$, $\lambda f + (1 - \lambda)g$ is also a continuous function on $[a, b]$. Hence, $\mathcal{C}[a, b]$ is path connected and by the above theorem is connected as well.

Topologist's Sine Curve
Now, we will see a classic example of a connected space which is not path connected. Consider the set

$$A = \left\{ \left(x, \sin \frac{1}{x} \right) \in \mathbb{R}^2 : 0 < x \leq 1 \right\}.$$

A is the graph of the continuous function $x \mapsto \sin \frac{1}{x}$ over the connected set $(0, 1] \subset \mathbb{R}$ (Fig. 7.11). Therefore A is connected (Exercise 7.7). Consider the vertical line segment

$$B = \{0\} \times [-1, 1] = \{(0, y) : -1 \leq y \leq 1\}.$$

Choose $y \in [-1, 1]$ arbitrarily. Let $\theta_y \in [-\frac{\pi}{2}, \frac{\pi}{2}]$ be the principal angle for which $\sin \theta_y = y$. Then for the choice of $x_n = \frac{1}{2n\pi + \theta_y}$, $\left\{(x_n, \sin \frac{1}{x_n})\right\}_n$ is a sequence in A converging to $(0, y) \in B$. Hence, $B \subset \overline{A}$. It only remains for a routine check to observe that $\overline{A} = A \cup B$. Since A is connected, by Proposition 7.7, $\overline{A}$ is also connected.

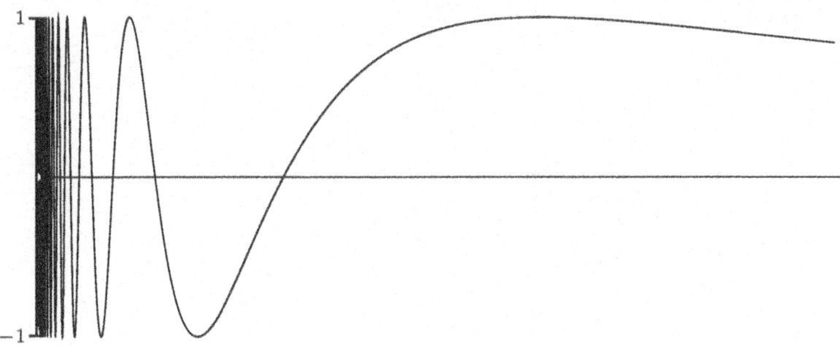

Fig. 7.11 Topologist's sine curve

Now to see $\overline{A}$ is not path connected, let us assume the contrary. Keep in mind the facts mentioned in the following items from earlier chapters: Example 3.39.3, Exercise 3.20, and Exercise 5.5.

According to our assumption, there is a continuous path $\gamma\colon [0, 1] \to \overline{A}$ such that $\gamma(0) = (0, 0)$ and $\gamma(1) = (\frac{1}{\pi}, 0)$. Let $\gamma(t) = (x(t), y(t))$ be the coordinate functions of γ. As γ is continuous, both $x(t)$ and $y(t)$ are continuous for every point $t \in [0, 1]$. Now, as $x(0) = 0$ and $x(1) = \frac{1}{\pi}$, by the intermediate value property (IVP), there exists a point $t_1 \in (0, 1)$ such that $x(t_1) = \frac{1}{3\pi/2}$. Applying IVP again, we can find another point t_2 between 0 and t_1 so that $x(t_2) = \frac{1}{5\pi/2}$. Continuing thus, we get a sequence $\{t_n\}$ of points from [0, 1] such that:

(a) $0 < \cdots < t_n < \cdots < t_2 < t_1$, i.e. $\{t_n\}$ is a decreasing sequence bounded below by 0 and hence convergent.
(b) $x(t_n) = \frac{1}{(2n+1)\frac{\pi}{2}}$.

Since for every n, $x(t_n) > 0$, $y(t_n)$ is given by $y(t_n) = \frac{1}{\sin(1/x(t_n))} = (-1)^n$. Call $\tau = \lim t_n$. As [0, 1] is closed, $\tau \in [0, 1]$. Thus γ is continuous at τ. Hence, the sequence $\{\gamma(t_n)\}$ must converge to $\gamma(\tau)$. However, this is possible if and only if both the sequences $\{x(t_n)\}$ and $\{y(t_n)\}$ converge to $x(\tau)$ and $y(\tau)$, respectively. But $\{y_n = (-1)^n\}$ is not a convergent sequence. Hence the contradiction. Therefore, $\overline{A}$ is not path connected.

The set $\overline{A}$ is called the *topologist's sine curve.*

Theorem 7.19 *Let Ω be an open connected set in $\mathbb{R}^n$. Then Ω is path connected.*

Proof If $\Omega = \varnothing$, nothing to prove. So assume $\Omega \neq \varnothing$ and choose $\mathbf{a} \in \Omega$. Consider the set

$$A = \{\mathbf{x} \in \Omega : \mathbf{x} \text{ is path connected to } \mathbf{a}\}.$$

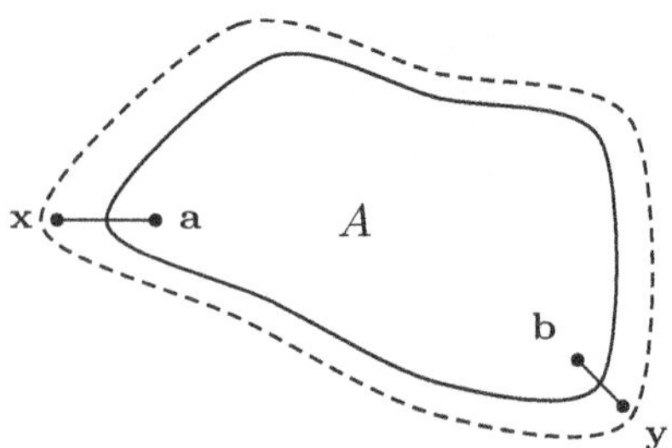

Fig. 7.12 ϵ-neighbourhood of A

As the point $\mathbf{a}$ is trivially path connected to itself, $A \neq \varnothing$. Thus A is a non-empty path connected subset of Ω. We will show that A is both open and closed in Ω. As Ω is connected, this will conclude that $A = \Omega$.

Let $\mathbf{x} \in A$ be arbitrary. Then $\mathbf{x} \in \Omega$, and as Ω is open, there is some $\delta > 0$ such that the neighbourhood $B(\mathbf{x}; \delta) \subset \Omega$. Now, by Exercise 7.20, $B(\mathbf{x}; \delta)$ is convex and hence path connected. Therefore, for any $\mathbf{y} \in B(\mathbf{x}; \delta)$, $\mathbf{y}$ is path connected to $\mathbf{x}$ and $\mathbf{x}$ is path connected to $\mathbf{a}$. Hence $\mathbf{y}$ is path connected to $\mathbf{a}$ (Exercise 7.16(a)). So $B(\mathbf{x}; \delta) \subset A$. Hence A is open.

To see A is closed in Ω, let $\mathbf{x} \in \Omega$ be any limit point of A. Let $\{\mathbf{a}_n\}$ be a sequence of points from A converging to $\mathbf{x}$. As Ω is open, for some $\delta > 0$ the neighbourhood $B(\mathbf{x}; \delta)$ of $\mathbf{x}$ is contained in Ω. Again, as $\mathbf{a}_n \to \mathbf{x}$, for some index k, $\mathbf{a}_k \in B(\mathbf{x}; \delta)$. Consequently, we have the following implications in order:

(a) $B(\mathbf{x}; \delta)$ is convex and hence path connected.
(b) $\mathbf{a}_k \in B(\mathbf{x}; \delta)$ implies $\mathbf{a}_k$ is path connected to $\mathbf{x}$.
(c) $\mathbf{a}_k \in A$ implies $\mathbf{a}_k$ is path connected to $\mathbf{a}$.
(d) $\mathbf{x}$ is path connected to $\mathbf{a}$, i.e. $\mathbf{x} \in A$.
(e) A is closed.

Thus A is both open and closed, and as Ω is connected, this implies $A = \Omega$. This completes the proof. ∎

Remark The requirement that Ω is open cannot be dropped as topologist's sine curve poses as a counter-example.

Miscellaneous Results

7.14 Let A be a connected subset in $\mathbb{R}^n$ and $\epsilon > 0$ be arbitrary. Show that the ϵ-neighbourhood of A given by

$$B(A; \epsilon) = \{\mathbf{x} \in \mathbb{R}^n : d(\mathbf{x}, A) < \epsilon\}$$

is path connected (Fig. 7.12).

Proof. We only need to show that $B(A; \epsilon)$ is open and connected. To see that $B(A; \epsilon)$ is open, recall that the map $f : X \to \mathbb{R}_{\geq 0}$ given by

$$f(x) = d(x, A)$$

is continuous (miscellaneous result 5.6), and $[0, \epsilon)$ is open in $\mathbb{R}_{\geq 0}$. Hence $f^{-1}[0, \epsilon)$ is open in X. However, observe that $B(A; \epsilon) = f^{-1}[0, \epsilon)$. Hence $B(A; \epsilon)$ is open.

To see $B(A; \epsilon)$ is connected, choose points $\mathbf{x} \in B(A; \epsilon)$ arbitrarily. Then we get a corresponding point $\mathbf{a_x} \in A$, such that $d(\mathbf{x}, \mathbf{a_x}) = \|\mathbf{x} - \mathbf{a_x}\| < \epsilon$. Now, the line segment $[\mathbf{x}, \mathbf{a_x}]$ is given by the set

$$[\mathbf{x}, \mathbf{a_x}] = \{t\mathbf{x} + (1-t)\mathbf{a_x} : t \in [0, 1]\}.$$

Also, $[\mathbf{x}, \mathbf{a_x}]$ is in fact a path, hence connected, and it intersects the connected set A as $\mathbf{a_x} \in [\mathbf{x}, \mathbf{a_x}] \cap A$. Therefore, $[\mathbf{x}, \mathbf{a_x}] \cup A$ is connected. Moreover, observe that for any value of $t \in [0, 1]$, the t-point $\mathbf{x}_t = t\mathbf{x} + (1-t)\mathbf{a_x}$ in the line segment $[\mathbf{x}, \mathbf{a_x}]$ satisfies

$$d(\mathbf{x}_t, \mathbf{a_x}) = \|t\mathbf{x} + (1-t)\mathbf{a_x} - \mathbf{a_x}\| = t\|\mathbf{x} - \mathbf{a_x}\| \leq \|\mathbf{x} - \mathbf{a_x}\| < \epsilon.$$

Thus, $[\mathbf{x}, \mathbf{a_x}] \subset B(A; \epsilon)$. Writing $C_\mathbf{x}$ for the connected set $[\mathbf{x}, \mathbf{a_x}] \cup A$, we can express

$$B(A; \epsilon) = \bigcup_{\mathbf{x} \in B(A;\epsilon)} C_\mathbf{x}.$$

As each $C_\mathbf{x}$ is connected, and for any $\mathbf{x}, \mathbf{y} \in B(A; \epsilon)$, $A \subset C_\mathbf{x} \cap C_\mathbf{y}$, by Proposition 7.8, $B(A; \epsilon)$ is connected.

Therefore, by the last theorem, $B(A; \epsilon)$ is path connected. □

7.15 Show that $\mathbb{R}^2 \setminus \mathbb{Q}^2$ is path connected.

Proof. $\mathbb{R}^2 \setminus \mathbb{Q}^2$ contains all the points in xy-plane with at least one irrational coordinate. Write X for $\mathbb{R}^2 \setminus \mathbb{Q}^2$. Let (a, b) and (c, d) be two arbitrary points from X. We will establish a path from (a, b) to (c, d) lying completely in X. Let a be irrational. The case when b is irrational can be dealt similarly. Now consider the following cases:

Case 1: c is irrational. See Fig. 7.13. Then the line segments l_1, l_2 and l_3, where

l_1 is the line segment joining (a, b) to (a, c),

l_2 is the line segment joining (a, c) to (c, c),

and

l_3 is the line segment joining (c, c) to (c, d),

lie completely in X. Now, a line segment in $\mathbb{R}^2$ is homeomorphic to a closed and bounded interval in $\mathbb{R}$, and every such interval in $\mathbb{R}$ is a homeomorphic copy of $[0, 1]$. Therefore, each l_i, $i = 1, 2, 3$, is a path,

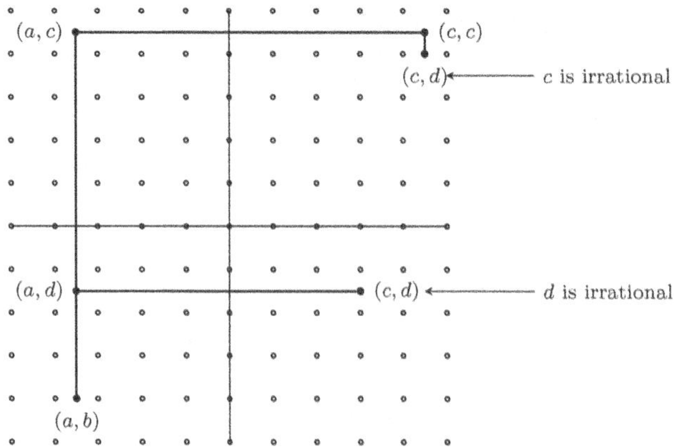

Fig. 7.13 Illustration for paths in $\mathbb{R}^2 \setminus \mathbb{Q}^2$

and hence, by Exercise 7.16(a), $l_1 \cup l_2 \cup l_3$ is also a path which joins (a, b) to (c, d).

Case 2: d is irrational. Following the same argument as in case 1, the union of the line segments l_1 joining (a, b) to (a, d), and l_2 joining (a, d) to (c, d), is a path joining (a, b) to (c, d) which lies completely in X.

This completes the proof. □

Exercise: 7.3

7.14. Show that a parabola, an ellipse, one branch of a hyperbola are all path connected spaces, whereas an entire hyperbola with both branches is not path connected:

7.15. (a) Refer to Example 7.15.3. Let C be any circle on the 2-dimensional plane. Establish a homeomorphism between C_1 (unit circle centred at origin) and C to conclude that C is path connected.

(b) Let E be any ellipse on the 2-dimensional plane. Establish a homeomorphism between E and C to conclude that E is path connected as well.

7.16. (a) In a metric space X, if x is path connected to x' and x' is path connected to x'', show that x is path connected to x''.

(b) Prove Proposition 7.16.

7.17. Show that for any $n \geq 1$, $\mathbb{R}^n \setminus \{\mathbf{0}\}$ is path connected.

7.18. Is the Hawaiian earring path connected?

7.19. If A and B are two path connected subsets of a metric space such that $A \cap B \neq \varnothing$, show that $A \cup B$ is path connected.

7.20. A subset $X \subset \mathbb{R}^n$ is called *convex* if for any pair of points $\mathbf{x}, \mathbf{x}' \in X$, the line segment

$$[\mathbf{x}, \mathbf{x}'] = \{t\mathbf{x} + (1-t)\mathbf{x}' : t \in [0, 1]\}$$

lies entirely on X.

(a) Show that for any $\mathbf{x} \in \mathbb{R}^n$ and any $r > 0$, the open r-neighbourhood $B(\mathbf{x}; r)$ of $\mathbf{x}$ is a convex set.
(b) Show that any convex set in $\mathbb{R}^n$ is path connected.
(c) Show that the unit sphere $S^n = \{\mathbf{x} = (x_1, \ldots, x_n, x_{n+1}) \in \mathbb{R}^{n+1} : \|\mathbf{x}\| = 1\}$ is path connected in the following way:

(*i*) Decompose S^n into two hemispheres as the following:

$$S^n_+ = \{\mathbf{x} = (x_1, \ldots, x_n, x_{n+1}) \in S^n : x_{n+1} \geq 0\},$$

$$S^n_- = \{\mathbf{x} = (x_1, \ldots, x_n, x_{n+1}) \in S^n : x_{n+1} \leq 0\}.$$

The hemispheres intersect at the equator given by $\{\mathbf{x} \in S^n : x_{n+1} = 0\}$.

(*ii*) Show that each hemisphere is homeomorphic to the unit disc

$$D_{n+1} = \{\mathbf{x} \in \mathbb{R}^{n+1} : \|\mathbf{x}\| \leq 1,\ x_{n+1} = 0\}.$$

(*iii*) Since D_{n+1} is convex, it is path connected; hence so are the two hemispheres.

(*iv*) Since the hemispheres intersect, by the last exercise, deduce that the entire sphere S^n is path connected.

7.21. Show that the *topologist's sine circle*, viz., the set $A \cup B \cup C$, where

$$A = \left\{\left(x, \sin\frac{1}{x}\right) : 0 < x \leq \frac{1}{\pi}\right\},$$

$$B = \{0\} \times [-1, 1] = \{(0, y) : -1 \leq y \leq 1\},$$

C is a circular arc joining$(0, 0)$ and $(\frac{1}{\pi}, 0)$,

is path connected. See Fig. 7.14.

7.22. Assume that a path $\gamma \colon [0, 1] \to \mathbb{R}^n$ connects a point $\mathbf{x} \in B(\mathbf{0}; 1) \subset \mathbb{R}^n$ to a point $\mathbf{y}$ with $\|\mathbf{y}\| > 1$. Show that there must be some $t_0 \in [0, 1]$ such that $\|\gamma(t_0)\| = 1$.

7.23. Show that a product of path connected spaces is path connected.

7.24. If A is a path connected subset of a metric space, is $\overline{A}$ necessarily path connected?

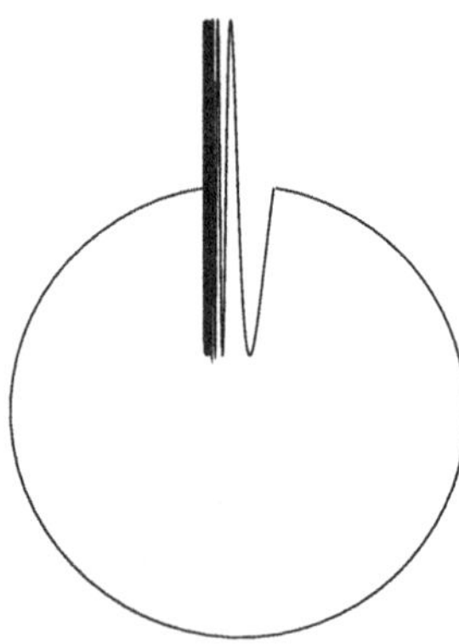

Fig. 7.14 Topologist's sine circle

7.25. If $\{A_\lambda : \lambda \in \Lambda\}$ is a collection of path connected subspaces of a metric space X with $\bigcap_{\lambda\in\Lambda} A_\lambda \neq \varnothing$, show that $\bigcup_{\lambda\in\Lambda} A_\lambda$ is path connected in X.

7.26. Let X be a metric space. Consider the relation $\sim$ on X given by

$$\text{for } x, x' \in X,\ x \sim x' \text{ iff } x \text{ is path connected to } x'.$$

(a) Show that $\sim$ is an equivalence relation. The equivalence classes generated by $\sim$ are called *path components* of X.
(b) Exhibit a counter-example to establish that a path component is not necessarily closed.
(c) Refer to the topologists sine curve as described on page 282. Create a set B_1 by removing all points with rational y-coordinates from the line segment B along the y-axis. Show that the set $A \cup B_1$ is connected, yet have uncountably many path components.

Appendix A
Hölder's and Minkowski's Inequalities

We will prove here the Hölder's and Minkowski's inequalities as mentioned in the subsubsection "Few useful inequalities" on page 53. We start with the following observation.

Consider the graph of the function $y = \log x$ for $x > 0$ (Fig. A.1). Let us, for $0 < a \leq b$, consider the line segment AB, where $A(a, \log a)$ and $B(b, \log b)$ are two points on the graph. For arbitrary $0 < \theta < 1$, let $x_\theta = \theta a + (1 - \theta)b$ be an intermediate value between a and b. Then the corresponding points to x_θ on the line AB, and on the graph $y = \log x$, are respectively given by

$$P \equiv (x_\theta, \theta \log a + (1 - \theta) \log b),$$

and

$$Q \equiv (x_\theta, \log x_\theta).$$

Now, observe the function $y = \log x$ is concave downward ($y'' < 0$) everywhere in its domain. Therefore, the point P on the line segment AB must lie below the point Q on the graph. Thus, comparing their ordinates, we obtain,

$$\theta \log a + (1 - \theta) \log b \leq \log x_\theta = \log(\theta a + (1 - \theta)b),$$

equality that occurs if and only if $a = b$. This produces

$$a^\theta b^{1-\theta} \leq \theta a + (1 - \theta)b. \tag{A.1}$$

Note that (A.1) also holds for any $a, b \geq 0$. Let for $k = 1, \ldots, n$, $\{a_k\}$, $\{b_k\}$ be two sets of positive reals. Call $A = \sum_{k=1}^n a_k$ and $B = \sum_{k=1}^n b_k$. Then for each k,

S. Paul, *Metric Spaces*, University Texts in the Mathematical Sciences,
https://doi.org/10.1007/978-981-96-9259-0

Fig. A.1 $y = \log x$

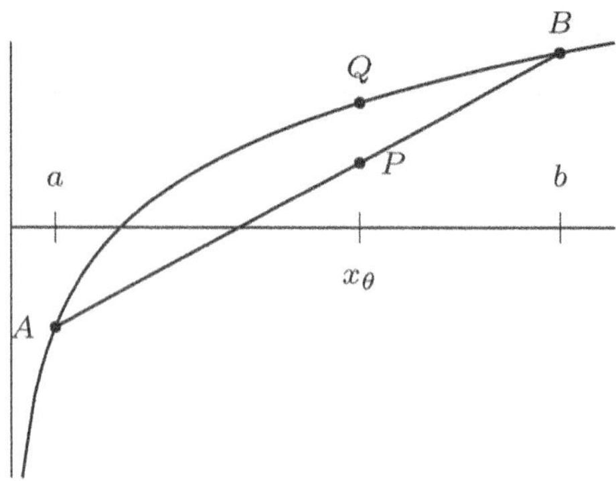

replacing a by a_k/A and b by b_k/B in (A.1), we obtain

$$
\begin{aligned}
& \left(\frac{a_k}{A}\right)^{\theta}\left(\frac{b_k}{B}\right)^{1-\theta} \leq \theta\left(\frac{a_k}{A}\right)+(1-\theta)\left(\frac{a_k}{A}\right) \\
\Longrightarrow \quad & \sum_{k=1}^{n}\left(\frac{a_k}{A}\right)^{\theta}\left(\frac{b_k}{B}\right)^{1-\theta} \leq \sum_{k=1}^{n}\left[\theta\left(\frac{a_k}{A}\right)+(1-\theta)\left(\frac{a_k}{A}\right)\right] \\
\Longrightarrow \quad & \frac{1}{A^{\theta}B^{1-\theta}}\sum_{k=1}^{n} a_k^{\theta} b_k^{1-\theta} \leq 1 \\
\Longrightarrow \quad & \sum_{k=1}^{n} a_k^{\theta} b_k^{1-\theta} \leq A^{\theta}B^{1-\theta} \\
\Longrightarrow \quad & \sum_{k=1}^{n} a_k^{\theta} b_k^{1-\theta} \leq \left(\sum_{k=1}^{n} a_k\right)^{\theta}\left(\sum_{k=1}^{n} b_k\right)^{1-\theta}.
\end{aligned} \tag{A.2}
$$

Equality occurs in (A.2) if and only if for every value of k, the relation $a_k/A = b_k/B$ holds, i.e., for some constant λ independent of k, $a_k = \lambda b_k$.

Now, write $a_k^{\theta} = c_k$, $b_k^{1-\theta} = d_k$, $\theta = \frac{1}{p}$, and $1-\theta = \frac{1}{q}$. Then c_k and d_k's are non-negative real numbers, p, q are greater than 1 satisfying the relation $\frac{1}{p}+\frac{1}{q} = 1$, and $a_k = c_k^p$, $b_k = d_k^q$. Thus (A.2) reduces to

$$
\boxed{\sum_{k=1}^{n} c_k d_k \leq \left(\sum_{k=1}^{n} c_k^p\right)^{1/p}\left(\sum_{k=1}^{n} d_k^q\right)^{1/q}.} \tag{A.3}
$$

Equation (A.3) holds for every set of non-negative reals $\{c_k\}$, $\{d_k\}$ with equality occurring if and only if for every k, each c_k is a constant multiple of d_k.

The inequality in (A.3) is famously known as the *Hölder's inequality*. The particular case when $p = q = 2$ is known as *Cauchy-Schwarz's inequality* as given in (A.4).

$$\boxed{\sum_{k=1}^{n} c_k d_k \le \sqrt{\left(\sum_{k=1}^{n} c_k^2\right)\left(\sum_{k=1}^{n} d_k^2\right)}.} \tag{A.4}$$

Manipulating further, observe that

$$\begin{aligned}
\sum_{k=1}^{n}(a_k+b_k)^p &= \sum_{k=1}^{n}(a_k+b_k)(a_k+b_k)^{p-1} \\
&= \sum_{k=1}^{n} a_k(a_k+b_k)^{p-1} + \sum_{k=1}^{n} b_k(a_k+b_k)^{p-1} \\
&\le \left(\sum_{k=1}^{n} a_k^p\right)^{1/p}\left(\sum_{k=1}^{n}(a_k+b_k)^{q(p-1)}\right)^{1/q} \\
&\qquad + \left(\sum_{k=1}^{n} b_k^p\right)^{1/p}\left(\sum_{k=1}^{n}(a_k+b_k)^{q(p-1)}\right)^{1/q} \\
&\qquad\qquad \text{[using (A.3)]} \\
&\le \left[\left(\sum_{k=1}^{n} a_k^p\right)^{1/p} + \left(\sum_{k=1}^{n} b_k^p\right)^{1/p}\right]\left(\sum_{k=1}^{n}(a_k+b_k)^{q(p-1)}\right)^{1/q}.
\end{aligned}$$

$$\sum_{k=1}^{n}(a_k+b_k)^p \le \left[\left(\sum_{k=1}^{n} a_k^p\right)^{1/p} + \left(\sum_{k=1}^{n} b_k^p\right)^{1/p}\right]\left(\sum_{k=1}^{n}(a_k+b_k)^{p}\right)^{1-\frac{1}{p}}. \tag{A.5}$$

Thus (A.5) finally reduces to

$$\boxed{\left(\sum_{k=1}^{n}(a_k+b_k)^p\right)^{1/p} \le \left(\sum_{k=1}^{n} a_k^p\right)^{1/p} + \left(\sum_{k=1}^{n} b_k^p\right)^{1/p}.} \tag{A.6}$$

(A.6) holds for every set of non-negative reals $\{a_k\}$, $\{b_k\}$ with equality occurring if and only if for every k, a_k is a constant multiple of b_k.

The inequality given in (A.6) is known as the *Minkowski's inequality*.

Appendix B
Stereographic Projection

B.1 Stereographic Projection on Complex Plane

The *extended set of complex numbers* comprises of all complex numbers with an ideal number ∞. Denoted by $\mathbb{C}^* = \mathbb{C} \cup \{\infty\}$, one representation of the extended system is given by the *stereographic projection* of the sphere $S^2 := x^2+y^2+z^2 = 1$ onto the plane $z = 0$, taking the point $(0, 0, 1)$ as the vertex of projection. The projection works in the following way: Let $z = x + iy$ be any complex number (Fig. B.1). We identify z as the point $P(x, y, 0)$ in the xy-plane in $\mathbb{R}^3$. Let $N(0, 0, 1)$ be the north pole of S^2. Let $Q(\xi, \eta, \zeta)$ be the point where NP intersects S^2. Then P is the stereographic projection of Q. This association $P \leftrightarrow Q$ is a continuous bijection between $\mathbb{C}$ and $S^2 \setminus \{(0, 0, 1)\}$, which we will explore in detail in a general setup in the next section. Now, to find the coordinates of Q, observe that the straight line joining P and N has all its points in the form $(rx, ry, 1 - r)$, where r is a real number. We find for which value of r does the point $(rx, ry, 1 - r)$ lie on S^2. This way, $r = \frac{2}{1+x^2+y^2}$, and hence $Q(\xi, \eta, \zeta)$ is given by

$$\xi = \frac{2x}{1 + x^2 + y^2}, \qquad \eta = \frac{2y}{1 + x^2 + y^2}, \qquad \zeta = \frac{x^2 + y^2 - 1}{x^2 + y^2 + 1}.$$

In terms of z, the coordinates of Q are given by

$$\xi = \frac{z + \overline{z}}{1 + |z|^2}, \qquad \eta = \frac{z - \overline{z}}{1 + |z|^2}, \qquad \zeta = \frac{|z|^2 - 1}{|z|^2 + 1}.$$

From here, we also conclude that

$$\xi + i\eta = \frac{2z}{1 + |z|^2} \qquad \text{and} \qquad \zeta = \frac{|z|^2 - 1}{|z|^2 + 1}.$$

S. Paul, *Metric Spaces*, University Texts in the Mathematical Sciences,
https://doi.org/10.1007/978-981-96-9259-0

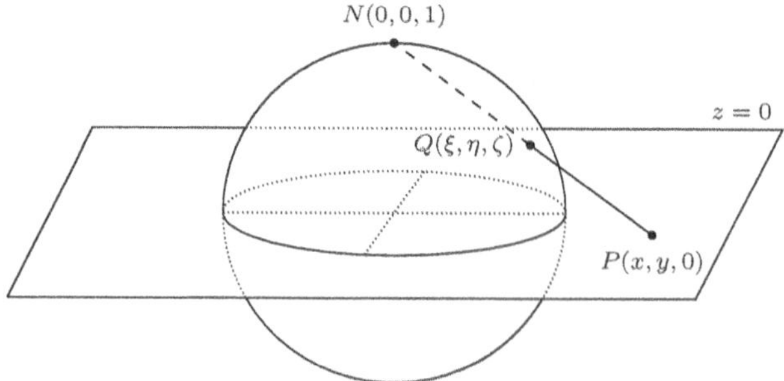

Fig. B.1 Stereographic projection for complex numbers

Conversely, z is given in terms of the coordinates of Q by the relation

$$z = \frac{\xi + i\eta}{1 - \zeta}.$$

Remarks

(a) The sphere $x^2 + y^2 + z^2 = 1$ in this context is known as the *Riemann sphere*.

(b) If you imagine the complex plane as an infinite field, then it deforms into the Riemann sphere through the inverse map of the stereographic projection. Think of a timelapse video of a bud blooming into a flower. If you play the video in reverse, it will bear much analogy to the above situation. The "infinity", situated at the horizon in every direction, merges into the single point $(0, 0, 1)$ at the north pole of the Riemann sphere.

If z_1, z_2 are two complex numbers with the corresponding points $Q_1(\xi_1, \eta_1, \zeta_1)$ and $Q_2(\xi_2, \eta_2, \zeta_2)$, respectively, on S^2, then a notion of distance between z_1 and z_2 can be defined by the chordal distance between Q_1 and Q_2. This way,

$$\begin{aligned}
[d(z_1, z_2)]^2 &= \|Q_1 - Q_2\|^2 \\
&= (\xi_1 - \xi_2)^2 + (\eta_1 - \eta_2)^2 + (\zeta_1 - \zeta_2)^2 \\
&= (\xi_1 - \xi_2)^2 - [i(\eta_1 - \eta_2)]^2 + (\zeta_1 - \zeta_2)^2 \\
&= (\xi_1 + i\eta_1 - \xi_2 - i\eta_2)(\xi_1 - i\eta_1 - \xi_2 + i\eta_2) + (\zeta_1 - \zeta_2)^2 \\
&= \left(\frac{2z_1}{1 + |z_1|^2} - \frac{2z_2}{1 + |z_2|^2}\right)\left(\frac{2\overline{z_1}}{1 + |z_1|^2} - \frac{2\overline{z_2}}{1 + |z_2|^2}\right) \\
&\qquad + \left(\frac{|z_1|^2 - 1}{|z_1|^2 + 1} - \frac{|z_2|^2 - 1}{|z_2|^2 + 1}\right)^2
\end{aligned}$$

$$= 4\left[\frac{z_1\overline{z}_1}{\left(1+|z_1|^2\right)^2} + \frac{z_2\overline{z}_2}{\left(1+|z_2|^2\right)^2} - \frac{z_1\overline{z}_2 + \overline{z}_1 z_2}{\left(1+|z_1|^2\right)\left(1+|z_2|^2\right)}\right]$$

$$+ \left[\frac{\left(1-|z_1|^2\right)\left(1+|z_2|^2\right) - \left(1+|z_1|^2\right)\left(1-|z_2|^2\right)}{\left(1+|z_1|^2\right)\left(1+|z_2|^2\right)}\right]^2$$

$$= 4\left[\frac{|z_1|^2}{\left(1+|z_1|^2\right)^2} + \frac{|z_2|^2}{\left(1+|z_2|^2\right)^2} - \frac{z_1\overline{z}_2 + \overline{z}_1 z_2}{\left(1+|z_1|^2\right)\left(1+|z_2|^2\right)}\right]$$

$$+ 4\frac{\left(|z_1|^2 - |z_2|^2\right)^2}{\left(1+|z_1|^2\right)^2\left(1+|z_2|^2\right)^2}$$

$$= 4\left[\frac{|z_1|^2\left(1+|z_2|^2\right)^2 + |z_2|^2\left(1+|z_1|^2\right)^2 + \left(|z_1|^2 - |z_2|^2\right)^2}{\left(1+|z_1|^2\right)^2\left(1+|z_2|^2\right)^2}\right]$$

$$- 4\left[\frac{z_1\overline{z}_2 + \overline{z}_1 z_2}{\left(1+|z_1|^2\right)\left(1+|z_2|^2\right)}\right]$$

$$= 4\left[\frac{\left(|z_1|^2 + |z_2|^2\right)\left(1+|z_1|^2\right)\left(1+|z_2|^2\right)}{\left(1+|z_1|^2\right)^2\left(1+|z_2|^2\right)^2}\right]$$

$$- 4\left[\frac{z_1\overline{z}_2 + \overline{z}_1 z_2}{\left(1+|z_1|^2\right)\left(1+|z_2|^2\right)}\right]$$

$$= 4\left[\frac{|z_1|^2 + |z_2|^2 - z_1\overline{z}_2 + \overline{z}_1 z_2}{\left(1+|z_1|^2\right)\left(1+|z_2|^2\right)}\right]$$

$$= 4\left[\frac{(z_1 - z_2)(\overline{z}_1 - \overline{z}_2)}{\left(1+|z_1|^2\right)\left(1+|z_2|^2\right)}\right]$$

$$= 4\left[\frac{|z_1 - z_2|^2}{\left(1+|z_1|^2\right)\left(1+|z_2|^2\right)}\right]$$

$$\therefore d(z_1, z_2) = \frac{2|z_1 - z_2|}{\sqrt{(1+|z_1|^2)(1+|z_2|^2)}}.$$

B.2 Stereographic Projection on $\mathbb{R}^n$

Now we will establish the mathematical properties of the stereographic projection in a general setup as promised in the last section.

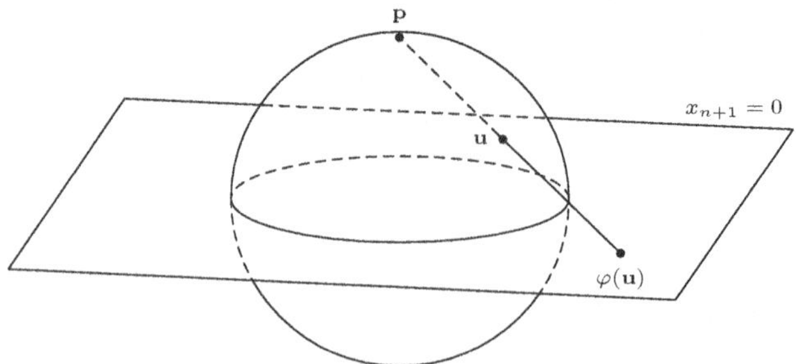

Fig. B.2 Stereographic projection for $\mathbb{R}^{n+1}$

Theorem B.1 *Let S^n denote the unit sphere in $\mathbb{R}^{n+1}$ given by*

$$S^n = \left\{ \mathbf{x} = (x_1, \ldots, x_n, x_{n+1}) \in \mathbb{R}^{n+1} : \|\mathbf{x}\| = 1 \right\}.$$

Let $\mathbf{p} = (\underbrace{0, 0, \ldots, 0}_{n \text{ times}}, 1) \in S^n$ be the north pole. Then $U = S^n \setminus \{\mathbf{p}\}$ is homeomorphic to $\mathbb{R}^n$.

Proof Observe that $\mathbb{R}^n$ is in fact the n dimensional hyperplane $x_{n+1} = 0$ embedded in $\mathbb{R}^{n+1}$. (This is in fact a homeomorphism as seen in Exercise 5.22(e).) Thus, our job converts to show that U is homeomorphic to $\mathbb{R}^n$ identified as

$$\mathbb{R}^n \equiv \left\{ (x_1, \ldots, x_n, x_{n+1}) \in \mathbb{R}^{n+1} : x_{n+1} = 0 \right\}.$$

Define the stereographic projection $\varphi\colon U \to \mathbb{R}^n$ by the map that sends the point $\mathbf{u} \in U$ to the point of intersection of the line joining $\mathbf{p}$ and $\mathbf{u}$ with the equator $x_{n+1} = 0$. (See Fig. B.2.) If $\mathbf{u}$ is given by its coordinates $\mathbf{u} = (u_1, \ldots, u_n, u_{n+1})$, the line joining $\mathbf{u}$ and $\mathbf{p} = (0, \ldots, 0, 1)$ is given by

$$\alpha(t) = \mathbf{p} + t(\mathbf{u} - \mathbf{p}) = (tu_1, \ldots, tu_n, 1 + tu_{n+1} - t), \quad t \in \mathbb{R}.$$

Therefore, the point of intersection of $\alpha(t)$ with the equator $x_{n+1} = 0$ is obtained by the value of t given by

$$1 + tu_{n+1} - t = 0 \implies t = \frac{1}{1 - u_{n+1}}.$$

Hence, for any point $\mathbf{u} = (u_1, \ldots, u_n, u_{n+1}) \in U$, we define

$$\begin{aligned}\varphi(\mathbf{u}) &= \alpha(t) = t\mathbf{u} + (1-t)\mathbf{p}, \quad \text{for } t = \frac{1}{1-u_{n+1}} \\ &= \left(\frac{1}{1-u_{n+1}}\right)\mathbf{u} - \left(\frac{u_{n+1}}{1-u_{n+1}}\right)\mathbf{p}. \end{aligned} \tag{B.1}$$

As for every point $\mathbf{u} \in U$, $u_{n+1} \neq 1$, the map $\varphi(\mathbf{u})$ is well-defined.

Now, let us find an expression for the reverse map

$$\psi \colon \mathbb{R}^n \equiv \left\{\mathbf{x} = (x_1, \ldots, x_n, 0) \in \mathbb{R}^{n+1}\right\} \to U.$$

Let $\mathbf{x}' = (x_1, \ldots, x_n) \in \mathbb{R}^n$ be arbitrary. Then again the line joining the points $\mathbf{x} = (\mathbf{x}', 0) \in \mathbb{R}^{n+1}$ and $\mathbf{p} = (0, \ldots, 0, 1)$ is given by

$$\gamma(t) = \mathbf{p} + t(\mathbf{x} - \mathbf{p}) = (t\mathbf{x}', 1-t), \quad t \in \mathbb{R}.$$

$\gamma(t)$ intersects S^n at the point obtained by the value of t given by

$$\begin{aligned} & \|\gamma(t)\|^2 = 1 \\ \iff\ & t^2\|\mathbf{x}'\|^2 + (1-t)^2 = 1 \\ \iff\ & (\|\mathbf{x}'\|^2 + 1)t^2 - 2t = 0 \\ \iff\ & t = 0, \frac{2}{\|\mathbf{x}'\|^2 + 1}. \end{aligned}$$

But $t = 0$ produces the trivial point of intersection $\gamma(0) = \mathbf{p}$. Therefore, the non-trivial point of intersection on U is given by

$$\gamma\left(\frac{2}{\|\mathbf{x}'\|^2+1}\right) = \left(\frac{2}{\|\mathbf{x}'\|^2+1}\mathbf{x}', 1 - \frac{2}{\|\mathbf{x}'\|^2+1}\right) = \frac{1}{\|\mathbf{x}'\|^2+1}\left(2\mathbf{x}', \|\mathbf{x}'\|^2 - 1\right).$$

We define for the point $\mathbf{x}' = (x_1, \ldots, x_n,) \in \mathbb{R}^n$ identified as the point $\mathbf{x} = (\mathbf{x}', 0) \in \mathbb{R}^{n+1}$,

$$\psi(\mathbf{x}) = \psi((\mathbf{x}', 0)) = \frac{1}{\|\mathbf{x}'\|^2+1}\left(2\mathbf{x}', \|\mathbf{x}'\|^2 - 1\right). \tag{B.2}$$

At this point, our job remains to show that both φ and ψ are continuous and inverse to each other. Now, both φ and ψ can easily be shown to be continuous using sequential criterion (apply Exercise 3.20 and Example 3.39.3). Hence, our proof will be done if we show φ and ψ are inverse of each other. We now proceed for that.

For points $\mathbf{x} = (\mathbf{x}', 0) \in \mathbb{R}^{n+1}$,

$$
\begin{aligned}
(\varphi \circ \psi)(\mathbf{x}) &= \varphi(\psi(\mathbf{x})) \\
&= \varphi\left(\frac{2}{\|\mathbf{x}'\|^2 + 1}\mathbf{x}', \frac{\|\mathbf{x}'\|^2 - 1}{\|\mathbf{x}'\|^2 + 1}\right) = \varphi(\mathbf{u}) = \varphi(\mathbf{u}', u_{n+1}),
\end{aligned} \tag{B.3}
$$

where $\mathbf{u} = \left(\frac{2}{\|\mathbf{x}'\|^2+1}\mathbf{x}', \frac{\|\mathbf{x}'\|^2-1}{\|\mathbf{x}'\|^2+1}\right)$, whose first n coordinates are clubbed in

$$
\mathbf{u}' = \frac{2}{\|\mathbf{x}'\|^2 + 1}\mathbf{x}',
$$

and the last coordinate

$$
u_{n+1} = \frac{\|\mathbf{x}'\|^2 - 1}{\|\mathbf{x}'\|^2 + 1}.
$$

Before calculating $\varphi(\mathbf{u})$, let us manipulate the following quantities with the last coordinate u_{n+1} of $\mathbf{u}$.

$$
\frac{1}{1 - u_{n+1}} = \frac{1}{1 - \frac{\|\mathbf{x}'\|^2-1}{\|\mathbf{x}'\|^2+1}} = \frac{\|\mathbf{x}'\|^2 + 1}{2},
$$

and

$$
\frac{u_{n+1}}{1 - u_{n+1}} = \frac{\frac{\|\mathbf{x}'\|^2-1}{\|\mathbf{x}'\|^2+1}}{1 - \frac{\|\mathbf{x}'\|^2-1}{\|\mathbf{x}'\|^2+1}} = \frac{\|\mathbf{x}'\|^2 - 1}{2}.
$$

Therefore, continuing from (B.3), using the definition of φ from (B.1), we obtain

$$
\begin{aligned}
(\varphi \circ \psi)(\mathbf{x}) &= \varphi(\mathbf{u}) \\
&= \frac{\|\mathbf{x}'\|^2 + 1}{2}\left(\frac{2}{\|\mathbf{x}'\|^2 + 1}\mathbf{x}', \frac{\|\mathbf{x}'\|^2 - 1}{\|\mathbf{x}'\|^2 + 1}\right) - \frac{\|\mathbf{x}'\|^2 - 1}{2}\mathbf{p} \\
&= \left(\mathbf{x}', \frac{\|\mathbf{x}'\|^2 - 1}{2} - \frac{\|\mathbf{x}'\|^2 - 1}{2}\right) \\
&= (\mathbf{x}', 0) = \mathbf{x}.
\end{aligned} \tag{B.4}
$$

Conversely, for a point $\mathbf{u} = (u_1, \ldots, u_n, u_{n+1}) = (\mathbf{u}', u_{n+1}) \in U$,

$$\begin{aligned}
(\psi \circ \varphi)(\mathbf{u}) &= \psi(\varphi(\mathbf{u})) \\
&= \psi\left(\left(\frac{1}{1-u_{n+1}}\right)\mathbf{u} - \left(\frac{u_{n+1}}{1-u_{n+1}}\right)\mathbf{p}\right) \\
&= \psi\left(\frac{1}{1-u_{n+1}}\mathbf{u}'\right) \\
&= \frac{1}{\frac{\|\mathbf{u}'\|^2}{(1-u_{n+1})^2}+1}\left(\frac{2\mathbf{u}'}{1-u_{n+1}}, \frac{\|\mathbf{u}'\|^2}{(1-u_{n+1})^2}-1\right).
\end{aligned} \tag{B.5}$$

Note that, as $\mathbf{u} \in U$, we have $\|\mathbf{u}\| = 1$, $\|\mathbf{u}\|^2 = \|\mathbf{u}'\|^2 + u_{n+1}^2$ and $u_{n+1} \neq 1$. With these facts, we manipulate the following quantities:

$$\begin{aligned}
\frac{1}{\frac{\|\mathbf{u}'\|^2}{(1-u_{n+1})^2}+1} &= \frac{(1-u_{n+1})^2}{\|\mathbf{u}'\|^2 + 1 - 2u_{n+1} + u_{n+1}^2} \\
&= \frac{(1-u_{n+1})^2}{2(1-u_{n+1})} && [as\ 1 = \|\mathbf{u}\|^2 = \|\mathbf{u}'\|^2 + u_{n+1}^2] \\
&= \frac{1-u_{n+1}}{2}, && [as\ u_{n+1} \neq 1]
\end{aligned}$$

$$\begin{aligned}
\frac{\|\mathbf{u}'\|^2}{(1-u_{n+1})^2} - 1 &= \frac{\|\mathbf{u}'\|^2 - 1 - u_{n+1}^2 + 2u_{n+1}}{(1-u_{n+1})^2} \\
&= \frac{-2u_{n+1}^2 + 2u_{n+1}}{(1-u_{n+1})^2} && [as\ \|\mathbf{u}'\|^2 - 1 = -u_{n+1}^2] \\
&= \frac{2u_{n+1}}{1-u_{n+1}}.. && [as\ u_{n+1} \neq 1]
\end{aligned}$$

Therefore, continuing from (B.5), we obtain

$$\begin{aligned}
(\psi \circ \varphi)(\mathbf{u}) &= \frac{1}{\frac{\|\mathbf{u}'\|^2}{(1-u_{n+1})^2}+1}\left(\frac{2\mathbf{u}'}{1-u_{n+1}}, \frac{\|\mathbf{u}'\|^2}{(1-u_{n+1})^2}-1\right) \\
&= \frac{1-u_{n+1}}{2}\left(\frac{2\mathbf{u}'}{1-u_{n+1}}, \frac{2u_{n+1}}{1-u_{n+1}}\right) \\
&= (\mathbf{u}', u_{n+1}) = \mathbf{u}.
\end{aligned} \tag{B.6}$$

Combining (B.4) and (B.6), we conclude that $\varphi^{-1} = \psi$. As both φ and ψ have already been proved to be continuous, U is homeomorphic to the equator plane $x_{n+1} = 0$ in $\mathbb{R}^{n+1}$. Hence by Exercise 5.22(e) and Exercise 5.23, U is homeomorphic to $\mathbb{R}^n$. ■

Appendix C
Weierstrass Approximation Theorem

Let $\mathcal{P}[a, b]$ denote the set of all polynomials on $[a, b]$ with real coefficients. Let $\mathcal{C}_{\text{sup}}[a, b]$ be as mentioned in Example 2.2.9. Then we have the following:

Theorem C.1 (Weierstrass Approximation Theorem) *$\mathcal{P}[a, b]$ is dense in $\mathcal{C}_{\text{sup}}[a, b]$.*

Proof WLOG we may assume $[a, b] = [0, 1]$. Let $f \in \mathcal{C}[0, 1]$ be arbitrary. We will show there exists a sequence $\{P_n(x)\}_n$ of polynomials from $\mathcal{P}[0, 1]$ converging to f in the uniform metric.

Assume the theorem be proved for all functions $g \in \mathcal{C}[0, 1]$ with $g(0) = 0 = g(1)$. Then for any $f \in \mathcal{C}[0, 1]$ chosen arbitrarily, we may write

$$g(x) = f(x) - f(0) - x[f(1) - f(0)], \quad x \in [0, 1].$$

Hence, $g \in \mathcal{C}[0, 1]$, $g(0) = 0 = g(1)$ and

$$f(x) = g(x) + f(0) + x[f(1) - f(0)].$$

Therefore, if a sequence $\{P_n\}$ from $\mathcal{P}[0, 1]$ converges to g, then the sequence

$$\begin{aligned}\big(P_n(x) + f(0) + x[f(1) - f(0)]\big) &\text{ converges to} \\ \big(g(x) + f(0) + x[f(1) - f(0)]\big) &= f(x),\end{aligned}$$

for each point $x \in [0, 1]$. However, for each n, $P_n(x) + f(0) + x[f(1) - f(0)]$ is again a polynomial. This shows that if the theorem is true for every such g, then it is also true for any $f \in \mathcal{C}[0, 1]$ as well. Thus, WLOG, let us assume $f(0) = 0 = f(1)$.

Also let us define $f(x)$ to be 0 outside $[0, 1]$. Hence, f is uniformly continuous over the whole real line.

S. Paul, *Metric Spaces*, University Texts in the Mathematical Sciences,
https://doi.org/10.1007/978-981-96-9259-0

For $n \in \mathbb{N}$, define

$$Q_n(x) = \begin{cases} c_n(1-x^2)^n, & \text{if } |x| \le 1 \\ 0, & \text{if } |x| > 1 \end{cases},$$

where c_n is so chosen that for every $n \in \mathbb{N}$,

$$\int_{-1}^{1} Q_n(x)\, dx = 1. \tag{C.1}$$

Observe that $Q_n(-x) = Q_n(x)$ and $(1-x^2)^n \ge 1 - nx^2$ for values of $x \in (0, 1)$. (The function $(1-x^2)^n + nx^2 - 1$ is 0 at $x = 0$ and has positive derivative in $(0, 1)$.) Then we arrive at the following estimate of c_n:

$$\begin{aligned} \int_{-1}^{1} (1-x^2)^n \, \mathrm{d}x &= 2\int_0^1 (1-x^2)^n \, \mathrm{d}x \\ &\ge 2\int_0^{1/\sqrt{n}} (1-x^2)^n \, \mathrm{d}x \\ &\ge 2\int_0^{1/\sqrt{n}} (1-nx^2) \, \mathrm{d}x \\ &= \frac{4}{3\sqrt{n}} > \frac{1}{\sqrt{n}}. \end{aligned}$$

But $1 = \int_{-1}^{1} Q_n(x)\, \mathrm{d}x = c_n \int_{-1}^{1} (1-x^2)^n \, \mathrm{d}x$ implies $c_n < \sqrt{n}$. Hence for any $\delta > 0$,

$$Q_n(x) \le \sqrt{n}(1-\delta^2)^n, \quad \text{for } \delta \le |x| \le 1. \tag{C.2}$$

Now set

$$P_n(x) = \int_{-1}^{1} f(x+t) Q_n(t)\, \mathrm{d}t, \quad 0 \le x \le 1. \tag{C.3}$$

Since $f(x) = 0$ for $x \in \mathbb{R} \setminus [0, 1]$, $P_n(x)$ reduces to

$$P_n(x) = \int_{-1}^{1} f(x+t) Q_n(t)\, \mathrm{d}t = \int_{-x}^{1-x} f(x+t) Q_n(t)\, \mathrm{d}t = \int_0^1 f(t) Q_n(t-x)\, \mathrm{d}t.$$

Observe that $f(t)Q_n(t-x)$ is a polynomial in x with coefficients in terms of t. Hence, an integration of $f(t)Q_n(t-x)$ w.r.t. t will still be a polynomial in x. Thus, for each n, $P_n(x)$ is a polynomial.

Now, as f is uniformly continuous over $\mathbb{R}$, for any chosen $\epsilon > 0$, there exists some $0 < \delta < 1$ such that the condition

$$|y - x| < 2\delta \implies |f(x) - f(y)| < \frac{\epsilon}{2} \tag{C.4}$$

holds for any pair of values x and y. Call $M = \sup\{|f(x)| : x \in \mathbb{R}\}$. Then we see that for each $x \in [0, 1]$, and for the obtained δ in (C.4),

$$\begin{aligned}
|P_n(x) - f(x)| &= \left| \int_{-1}^{1} [f(x+t) - f(x)] Q_n(t)\, \mathrm{d}t \right| && [\text{by (C.1)\& (C.3)}] \\
&\le \int_{-1}^{1} |f(x+t) - f(x)| Q_n(t)\, \mathrm{d}t && [as \forall t,\ Q_n(t) \ge 0] \\
&= \int_{-1}^{-\delta} \cdots + \int_{-\delta}^{\delta} \cdots + \int_{\delta}^{1} \cdots \\
&= I_1 + I_2 + I_3, \text{ say;}
\end{aligned}$$

where

$$\begin{aligned}
I_1 &= \int_{-1}^{-\delta} |f(x+t) - f(x)| Q_n(t)\, \mathrm{d}t \\
&\le 2M \int_{-1}^{-\delta} Q_n(t)\, \mathrm{d}t \\
&\le 2M\sqrt{n}(1-\delta^2)^n \int_{-1}^{-\delta} \mathrm{d}t && [\text{by (C.2)}] \\
&< 2M\sqrt{n}(1-\delta^2)^n; \\
I_2 &= \int_{-\delta}^{\delta} |f(x+t) - f(x)| Q_n(t)\, \mathrm{d}t \\
&< \frac{\epsilon}{2} \int_{-\delta}^{\delta} Q_n(t)\, \mathrm{d}t && [\text{by (C.4)}] \\
&< \frac{\epsilon}{2} \int_{-1}^{1} Q_n(t)\, \mathrm{d}t && [as\, Q_n(t) \ge 0] \\
&= \frac{\epsilon}{2};
\end{aligned}$$

and

$$
\begin{aligned}
I_3 &= \int_\delta^1 |f(x+t) - f(x)| Q_n(t)\, \mathrm{d}t \\
&\le 2M \int_\delta^1 Q_n(t)\, \mathrm{d}t \\
&\le 2M\sqrt{n}(1-\delta^2)^n \int_\delta^1 \mathrm{d}t && \text{[by (C.2)]} \\
&< 2M\sqrt{n}(1-\delta^2)^n.
\end{aligned}
$$

Hence,

$$
|P_n(x) - f(x)| < 4M\sqrt{n}(1-\delta^2)^n + \frac{\epsilon}{2}. \tag{C.5}
$$

Lastly, to show $4M\sqrt{n}(1-\delta^2)^n \to 0$ as $n \to \infty$, write $1 - \delta^2 = \frac{1}{1+a}$, $a > 0$. Then from the inequality

$$
na < 1 + na < (1+a)^n,
$$

we arrive at

$$
\frac{\sqrt{n}}{(1+a)^n} < \frac{1}{a\sqrt{n}}
$$

$$
\textit{[dividing the } 1^{\text{st}} \ \& \ 3^{\text{rd}} \textit{ quantities by } a\sqrt{n}(1+a)^n]
$$

$$
\implies \sqrt{n}(1-\delta^2)^n < \frac{1}{a\sqrt{n}}.
$$

But this shows that $\sqrt{n}(1-\delta^2)^n \to 0$ as $n \to \infty$. Therefore, for sufficiently large n, from (C.5), we may conclude that for each point $x \in [0, 1]$,

$$
|P_n(x) - f(x)| < \epsilon.
$$

Since $\epsilon > 0$ arbitrary, this proves the theorem. ■

Appendix D
Cantor Set

Theorem 3.10 says that the open sets in the real line are countable union of disjoint open intervals. Therefore, a closed set in $\mathbb{R}$ is attained by removing a countable collection of disjoint open intervals. This seemingly innocent process gives rise to one of the most intricate mathematical objects known as the *Cantor set* named after the German mathematician *Georg Cantor* who introduced it in 1883.

To construct the Cantor set, we begin with the closed unit interval [0, 1], denoted as F_0. We then remove the open middle-third interval $\left(\frac{1}{3}, \frac{2}{3}\right)$, leaving the set

$$F_1 = \left[0, \tfrac{1}{3}\right] \cup \left[\tfrac{2}{3}, 1\right].$$

Notice that F_1 remains a closed set. Next, from F_1, we remove the open middle thirds of each of its two subintervals, namely $\left(\frac{1}{9}, \frac{2}{9}\right)$ and $\left(\frac{7}{9}, \frac{8}{9}\right)$, yielding the closed set

$$F_2 = \left[0, \tfrac{1}{9}\right] \cup \left[\tfrac{2}{9}, \tfrac{1}{3}\right] \cup \left[\tfrac{2}{3}, \tfrac{7}{9}\right] \cup \left[\tfrac{8}{9}, 1\right].$$

This process continues indefinitely, where at each step, we remove the open middle thirds of all remaining closed intervals. This generates a nested sequence of closed sets $\{F_n\}$ satisfying $F_{n+1} \subset F_n$ for all n.

The limiting process thus calls in its final product as

$$\mathcal{C} = \bigcap_{n \geq 0} F_n,$$

which is known as the *Cantor set*.

The Cantor set $\mathcal{C}$ possesses several fascinating properties that make it a subject of deep interest in mathematics. We will establish the following ones in our present discourse.

S. Paul, *Metric Spaces*, University Texts in the Mathematical Sciences,
https://doi.org/10.1007/978-981-96-9259-0

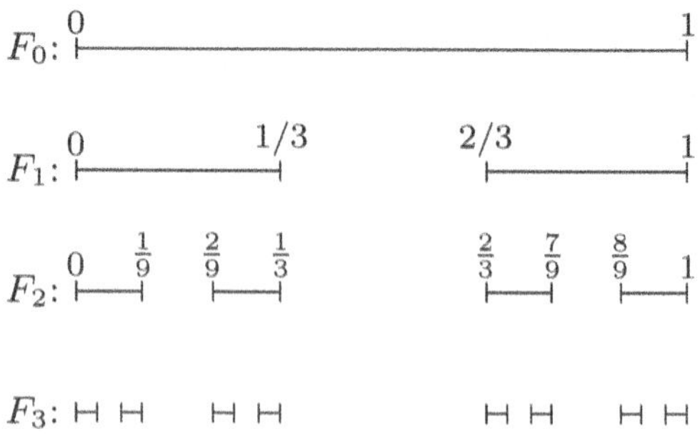

Fig. D.1 Construction of Cantor set

Theorem D.1

i) C has length (measure) 0, i.e., it has empty interior.
ii) C has uncountably many points.
iii) C is a perfect set.[1]
iv) C is compact.
v) C is totally disconnected.

To prove the second and third results in the theorem, we will later establish the ternary representation of the points in C in a separate section. However, the remaining facts follow more directly.

For the first result, we calculate the total length removed from the initial interval $F_0 = [0, 1]$ (refer to Fig. D.1). The portion removed at the first step has length $\frac{1}{3}$. In the second step, the total removed length is $2 \times \left(\frac{1}{3}\right)^2$. From F_2, the removed portion sums to $4 \times \left(\frac{1}{3}\right)^3$, and this pattern continues. More generally, F_{n+1} is formed by removing a total length of $2^n \times \left(\frac{1}{3}\right)^{n+1}$ from F_n. Summing over all steps, the total length removed in constructing C is

$$\sum_{n=0}^{\infty} \frac{2^n}{3^{n+1}} = \frac{1}{3} \times \frac{1}{1 - \frac{2}{3}} = 1.$$

This calculation shows that we have removed a total length equal to the entire interval $[0, 1]$, implying that the length (measure) of C is zero. Consequently, C has an empty interior, since any neighbourhood $B(x)$ of an interior point x would necessarily have positive length. (Examine the last two sentences in the light of the definition of a measure zero set given in Definition 1.124.)

The fourth and fifth results are even more straightforward. Since C is a closed and bounded subset of $\mathbb{R}$, it is compact by the Heine-Borel theorem. Additionally,

---

[1] A set $A \subset X$ is called a *perfect set* if $A' = A$, where A' is the set of all limit points of A.

because C has an empty interior, miscellaneous result 7.12 ensures that it is totally disconnected.

In the next section, we will explore the ternary representation of points in the Cantor set.

D.1 Ternary Representation of the Points from C

For this purpose, the person himself supplies our primary machinery: the nested interval theorem (Theorem 1.65). Let us record the statement here again for convenience:

Cantor's nested interval theorem: Let $\{I_n\}$ be a nested sequence of closed and bounded intervals, i.e., $I_n = [a_n, b_n]$, and $I_{n+1} \subset I_n$. Let $|I_n| = b_n - a_n$ denote the length of I_n. If $\inf I_n = 0$, then the intersection $\bigcap_{n=1}^{\infty} I_n$ contains precisely one point.

Let us first discuss how this theorem establishes the representation of any number. Imagine the following scenario. We want to obtain the exact length l (in meters) of a given object. Using a meter scale, we notice that l falls between 1m and 2m. Therefore, we conclude that l is a number between 1 and 2. After subdividing the scale into decimetres, suppose we observe that l is coming between 4dcm and 5dcm. Thus, l falls between $1 + \frac{4}{10}$ and $1 + \frac{5}{10}$. Going further, we record that l falls between 1cm and 2cm, i.e., l lies within $1 + \frac{4}{10} + \frac{1}{10^2}$ and $1 + \frac{4}{10} + \frac{2}{10^2}$. We proceed in this fashion, i.e., at each step, we subdivide the existing scale into ten equal parts, identify which subinterval l falls into, and repeat. (l may fall into two subintervals simultaneously, but we reserve that discussion for the next paragraph.) If $I_n = [a_n, b_n]$ denotes the nth subinterval that l falls into, then $I_{n+1} \subset I_n$, and $|I_{n+1}| = \frac{1}{10}|I_n|$. Thus $|I_n| \to 0$, and hence by the above theorem, there will be only one number sitting inside all the I_n's, which, in this case, is l. Since $a_n \uparrow l$ as $n \to \infty$ (a fact established in the proof of the nested interval theorem), we choose to represent the number l by the sequence $\{a_n\}$ of the left extremities. Thus, if l happens to be the number $\sqrt{2}$, then $a_n = \frac{[10^{n-1}\sqrt{2}]}{10^{n-1}}$ and $b_n = \frac{[10^{n-1}\sqrt{2}]+1}{10^{n-1}}$ ($[\cdot]$ is the greatest integer function). Consequently,

$$l = 1 + \frac{4}{10} + \frac{1}{10^2} + \frac{4}{10^3} + \frac{2}{10^4} + \frac{1}{10^5} + \frac{3}{10^6} + \dots$$

and is written as

$$1.414213562373\dots,$$

following the place value system. The last expression is called the *decimal representation* (or *decimal expansion*) of l.

Let us now attain the case when l falls into two subintervals simultaneously. Suppose $l = \frac{7}{5}$. Then in the decimetre scale, l falls into both the subintervals $[1 + \frac{3}{10}, 1 + \frac{4}{10}]$ and $[1 + \frac{4}{10}, 1 + \frac{5}{10}]$. How do we proceed then? As there is no obvious preference, we choose *any one* of them. Let us investigate both the cases individually:

If we choose $[1 + \frac{3}{10}, 1 + \frac{4}{10}]$, then in the successive steps, we *have to* choose the following subintervals:

$$\left[1 + \frac{3}{10} + \frac{9}{10^2}, 1 + \frac{3}{10} + \frac{10}{10^2}\right], \text{ i.e., } \left[1 + \frac{3}{10} + \frac{9}{10^2}, 1 + \frac{4}{10}\right];$$

$$\left[1 + \frac{3}{10} + \frac{9}{10^2} + \frac{9}{10^3}, 1 + \frac{3}{10} + \frac{9}{10^2} + \frac{10}{10^3}\right],$$

$$\text{i.e., } \left[1 + \frac{3}{10} + \frac{9}{10^2} + \frac{9}{10^3}, 1 + \frac{4}{10}\right],$$

and so on. This way, according to the arguments and convention set in the previous paragraph, l will be equal to the number

$$l = 1 + \frac{3}{10} + \frac{9}{10^2} + \frac{9}{10^3} + \dots, \text{ and will be denoted by } 1.3999\dots = 1.3\overline{9},$$

where $\overline{9}$ denotes the infinite recurrence of 9 after the decimal point. We call it the *non-terminating decimal representation* of l.

On the other hand, if we happen to choose the subinterval $[1 + \frac{4}{10}, 1 + \frac{5}{10}]$ in the decimetre scale, then we *have to* choose the following subintervals in the successive steps:

$$\left[1 + \frac{4}{10} + \frac{0}{10^2}, 1 + \frac{4}{10} + \frac{1}{10^2}\right], \text{ i.e., } \left[1 + \frac{4}{10}, 1 + \frac{4}{10} + \frac{1}{10^2}\right],$$

then

$$\left[1 + \frac{4}{10} + \frac{0}{10^2} + \frac{0}{10^3}, 1 + \frac{4}{10} + \frac{0}{10^2} + \frac{1}{10^3}\right],$$

$$\text{i.e., } \left[1 + \frac{4}{10}, 1 + \frac{4}{10} + \frac{0}{10^2} + \frac{1}{10^3}\right],$$

and so on. This way, l will be equal to the number

$$l = 1 + \frac{4}{10} = 1.4.$$

We call this the *terminating decimal representation* of l.

The above discussion stands on the latent assumption that $l > 0$. If $l < 0$, then $-l > 0$, and hence, will have a decimal representation, say $l_0.l_1l_2l_3\ldots$. Thus l will have the expression $-l_0.l_1l_2l_3\ldots$. However, in this case, the digits after the decimal point will *no longer* denote the sequence of left extremities of the successive subintervals enclosing l. Instead, we may obtain them by writing l as

$$l = [l] + \{l\},$$

where $[l]$ is the greatest integer less than or equal to l and $\{l\}$ is a number from the interval $[0, 1)$. (This decomposition of l is unique.) Then

$$\begin{aligned} l = -l_0.l_1l_2l_3\ldots &= -(l_0 + 1) + (1 - 0.l_1l_2l_3\ldots) \\ &= [l] + (0.999\cdots - 0.l_1l_2l_3\ldots) \\ &= [l] + 0.s_1s_2s_3\ldots, \end{aligned}$$

where for each k, $s_k = 9 - l_k$ is the complement digit of l_k in the decimal number system. For instance, if $l = -1.2$, then $[l] = -2$, and $\{l\} = 0.8$. Therefore,

$$-1.2 = (-2) + (1 - 0.2) = (-2) + (0.\overline{9} - 0.2) = (-2) + 0.7\overline{9},$$

or equivalently,

$$-1.2 = -1.1\overline{9} = (-2) + (1 - 0.1\overline{9}) = (-2) + (0.\overline{9} - 0.1\overline{9}) = (-2) + 0.8.$$

From the entire above exercise, the following observations can be recorded:

- A decimal representation exists for every real number.
- The digits in the decimal representation vary within the set $\{0, 1, 2, \ldots, 9\}$. If $a = a_0.a_1a_2a_3\ldots$ is a positive number, where $a_0 = [a]$ is the greatest integer smaller than or equal to a, then

$$a = a_0 + \sum_{k=1}^{\infty} \frac{a_k}{10^k}.$$

- Some numbers have both the terminating and the non-terminating representations. In that case, the non-terminating representation will end in an infinite recurrence of 9's. (It is a good exercise to investigate which numbers possess this property.)
- The number "zero" has only the unique representation "0".

From ancient times, the number 10 has appeared as the most popular choice for the "base" of the number system, arguably because we have 10 fingers. Nevertheless, mathematically speaking, any positive number greater than or equal to 2 can be the base of a number system. If we choose the base to be 2, we obtain the *binary number system*, whereas the choice of 3 gives us the *ternary number system*. The construction process remains the same, with the following couple of modifications: If b denotes the base, we subdivide the intervals into b equal parts in each successive step; and the digits in the expansion vary within the set $\{0, 1, 2, \ldots, b-1\}$. Since the Cantor set is constructed by trisecting the intervals, it will be wise to choose the ternary system to represent its elements. For convenience, let us rewrite the above observations for the ternary representations of the elements from the starting interval $F_0 = [0, 1]$:

- Every number in $F_0 = [0, 1]$ has a ternary representation (or a ternary expansion).
- The digits in the ternary representation of a number vary within the set $\{0, 1, 2\}$. If $a = 0.a_1a_2a_3\ldots$, then

$$a = \sum_{k=1}^{\infty} \frac{a_k}{3^k}.$$

- Some numbers have both the terminating and non-terminating representations. In that case, the non-terminating representation will end in an infinite recurrence of 2's.
- The number "zero" has only the unique representation "0".

Now let us investigate the ternary representations of the points of $\mathcal{C}$. Refer to Fig. D.1. We remove the open middle-third portion $(\frac{1}{3}, \frac{2}{3})$ of F_0 to obtain F_1. The followings are true for this process:

- $(\frac{1}{3}, \frac{2}{3})$ contains all points from $[0, 1]$ *necessarily* having a "1" in the first place after decimal point in their ternary representation.
- Ternary representation of the remaining points in $[0, 1]$:
  - Either *necessarily has* a 0 or 2 in the first place after decimal point.
  - Or *can have* a 0 or 2 in the first place after decimal. This applies specifically to the points

$$\frac{1}{3} = (0.1)_3 = (0.0\overline{2})_3, \quad \frac{2}{3} = (0.2)_3 = (0.1\overline{2})_3, \quad \text{and} \quad 1 = 0.\overline{2}_3.$$

Therefore, all points in F_1 can be represented with only 0 or 2 in the first place after decimal in their ternary representations. Similarly, after removing the two open middle thirds $(\frac{1}{9}, \frac{2}{9})$ and $(\frac{7}{9}, \frac{8}{9})$ from F_1, we observe that:

- $(\frac{1}{9}, \frac{2}{9})$ and $(\frac{7}{9}, \frac{8}{9})$ contain all points from $[0, 1]$ *necessarily* having a "1" in the second place after decimal point in their ternary representation.
- Ternary representation of the remaining points in $[0, 1]$:
  - Either *necessarily has* a 0 or 2 in the second place after decimal point

– Or *can have* a 0 or 2 in the second place after decimal. Specifically, for the points

$$\frac{1}{9} = (0.01)_3 = (0.00\overline{2})_3, \qquad \frac{2}{9} = (0.02)_3 = (0.01\overline{2})_3,$$
$$\frac{7}{9} = (0.21)_3 = (0.20\overline{2})_3, \qquad \frac{8}{9} = (0.22)_3 = (0.21\overline{2})_3$$

and

$$1 = 0.\overline{2}_3$$

Therefore, all points in F_2 can be represented with only 0 or 2 in the second place after decimal in their ternary representations. Proceeding inductively and recalling that $C = \bigcap_{n\geq 0} F_n$, we can conclude that every point in C can be represented with 0 or 2 in their ternary representations. In other words, if we call the set

$$D = \left\{ a \in [0, 1] : a = 0.a_1a_2a_3 \cdots = \sum_{n=1}^{\infty} \frac{a_n}{3^n}, \text{ where } a_n \text{ is either 0 or 2} \right\},$$

then $C \subset D$. On the other hand, consider $a = 0.a_1a_2a_3 \cdots \in D$ arbitrarily. Then, as $a_1 \neq 1$, $a \notin (\frac{1}{3}, \frac{2}{3})$, i.e., $a \in F_1$. Similarly, as $a_2 \neq 1$, $a \notin (\frac{1}{9}, \frac{2}{9}) \cup (\frac{7}{9}, \frac{8}{9})$, i.e., $a \in F_2$. Proceeding inductively likewise, we obtain that for each $n \in \mathbb{N}$, $a \in F_n$. Thus $a \in \bigcap_n F_n = C$. Consequently, $C = D$. Thus, we can say that:

Proposition D.2 *The Cantor set C consists of all points from $[0, 1]$ whose at least one ternary representation does not contain the digit 1, i.e.,*

$$C = \left\{ a \in [0, 1] : a = 0.a_1a_2a_3 \cdots = \sum_{n=1}^{\infty} \frac{a_n}{3^n}, \text{ where } a_n \text{ is either 0 or 2} \right\}.$$

Remark A good exercise is to show that for any two points a, b from C as described in the above proposition with $a = 0.a_1a_2\ldots$ and $b = 0.b_1b_2\ldots$, $a = b$ implies $a_n = b_n$ holds for each $n \in \mathbb{N}$. In other words, both the terminating and non-terminating representations of the same number from the Cantor set are not listed in the above description.

With the ternary representations of points of C established, we now prove the remaining facts in Theorem D.1 in the next section.

D.2 Proofs of Theorems D.1(*ii*) and D.1(*iii*)

There are multiple ways to prove Theorem D.1(*ii*). We will discuss a couple of them here.

Consider the closed and bounded interval $[0, 1]$ in its binary representation, where any number having two different representations ends in an infinite string of 1's. Let

$$a = 0.a_1a_2a_3\ldots, \text{ where } \forall n \in \mathbb{N},\ a_n \text{ is either 0 or 1}$$

be chosen arbitrarily from $[0, 1]$ in its (non-terminating) binary representation. For each n, set $b_n = 2a_n$. Then

$$b = 0.b_1b_2b_3\ldots,$$

considered as a number in its ternary representation, belongs to $\mathcal{C}$. Indicating the respective bases and writing $2a$ for b, we see that the map $f\colon [0, 1] \to \mathcal{C}$ defined as

$$f((a)_2) = (2a)_3$$

is one to one. Hence, $\mathcal{C}$ is uncountable.

The other way of establishing the same is via Cantor's diagonal argument. Assume that $\mathcal{C}$ is countable. Let $\mathcal{C} = \{a_1, a_2, \ldots\}$ be an enumeration of $\mathcal{C}$, where

$$\begin{aligned}
a_1 &= 0.a_{11}a_{12}a_{13}\ldots a_{1n}\ldots \\
a_2 &= 0.a_{21}a_{22}a_{23}\ldots a_{2n}\ldots \\
a_3 &= 0.a_{31}a_{32}a_{33}\ldots a_{3n}\ldots \\
&\ \ \vdots \qquad \vdots \\
a_m &= 0.a_{m1}a_{m2}a_{m3}\ldots a_{mn}\ldots \\
&\ \ \vdots \qquad \vdots
\end{aligned}$$

Construct the number $b = 0.b_1b_2\ldots b_n\ldots$ as follows:

$$b_n = \begin{cases} 2, \text{ if } a_{nn} = 0 \\ 0, \text{ if } a_{nn} = 2 \end{cases}.$$

Then $b \in \mathcal{C}$, but b is not included in the enumeration $\{a_1, a_2, \ldots\}$. Therefore, $\mathcal{C}$ must be non-enumerable.

To prove Theorem D.1(*iii*), consider $a \in \mathcal{C}$ arbitrarily. For any chosen $\epsilon > 0$, let $k \in \mathbb{N}$ be such that $\frac{1}{3^k} < \epsilon$. Consider $B(a; \epsilon) = (a - \epsilon, a + \epsilon)$. Now, the closed set

F_k in the construction of $\mathcal{C}$ has 2^k closed intervals each of length $\frac{1}{3^k}$. Let $I = [p, q]$ be the interval from F_k containing a. Then $I \subset B(a; \epsilon)$. Since both the extremities of all the subintervals remain in the Cantor set[2], we must have $p, q \in B(a; \epsilon) \cap \mathcal{C}$. Therefore, $[B(a; \epsilon) \setminus \{a\}] \cap \mathcal{C} \neq \varnothing$, showing that a is a limit point of $\mathcal{C}$. As $a \in \mathcal{C}$ is arbitrary, and $\mathcal{C}$ is closed, $\mathcal{C}$ must be perfect.

---

[2] The left extremities end with an infinite string of 2's, whereas the right extremities have a terminating representation ending in a 2.

Bibliography

1. Apostol, T. M.: Mathematical Analysis, 2nd edn. Narosa Publishing House, New Delhi-Madras-Bombay-Calcutta (2002)
2. Chakraborty, A.: Metric Spaces, first reprint edn. Levant Books, Kolkata (2017)
3. Copson, E. T.: Metric Spaces. Cambridge University Press, Cambridge (1968)
4. Hewitt, E.: The role of compactness in analysis. Am. Math. Month. **67**(6), 499–516 (1960)
5. Kemp, T.: Cauchy's construction of $\mathbb{R}$ (2020). http://www.math.ucsd.edu/~tkemp/140A/Construction.of.R.pdf
6. Kumaresan, S.: Topology of Metric Spaces, 2nd edn. Narosa Publishing House, New Delhi-Chennai-Kolkata-Mumbai (2006)
7. Mapa, S. K.: Introduction to Real Analysis, 7th edn. Sarat Book Distributors, Kolkata (2018)
8. Munkres, J. R.: Topology, 2nd edn, pp. 163–164. Prentice Hall of India Pvt Ltd, New Delhi (2008)
9. Rudin, W.: Principles of Mathematical Analysis, 3rd edn. McGraw-Hill Book Company, New York (1976)
10. Sengupta, J.: Metric Spaces, 4th edn. U N Dhur & Sons Private Limited, Kolkata (2014)
11. Shirali, S., Vasudeva, H.L.: Metric Spaces. Springer, London (2006)
12. Simmons, G.F.: Introduction to Topology and Modern Analysis, twentieth reprint indian edn. McGraw Hill Education (India) Private Limited, New Delhi (2013)
13. Tao, T.: Compactness and compactification. https://www.math.ucla.edu/~tao/preprints/compactness.pdf

S. Paul, *Metric Spaces*, University Texts in the Mathematical Sciences,
https://doi.org/10.1007/978-981-96-9259-0

Index

S. Paul, *Metric Spaces*, University Texts in the Mathematical Sciences,
https://doi.org/10.1007/978-981-96-9259-0

The manufacturer's authorised representative in the EU is Springer Nature Customer Service Centre GmbH, Europaplatz 3, 69115 Heidelberg, Germany. If you have any concerns regarding our products, please contact ProductSafety@springernature.com

Printed and bound by CPI Group (UK) Ltd, Croydon, CR0 4YY
13/07/2026
02165032-0001